Biodiversity

Springer-Verlag Berlin Heidelberg GmbH

Wilhelm Barthlott   Matthias Winiger   (Eds.)

Biodiversity

A Challenge for Development Research and Policy

With 91 Figures and 24 Tables

Springer

EDITORS:

Professor Dr. Wilhelm Barthlott
University of Bonn
Institute of Botany
Meckenheimer Allee 170
53115 Bonn, Germany
e-mail: barthlott@uni-bonn.de

Professor Dr. Matthias Winiger
University of Bonn
Institute of Geography
Meckenheimer Allee 166
53115 Bonn, Germany
e-mail: winiger@giub.uni-bonn.de

MANAGING EDITOR:

Dr. Nadja Biedinger
University of Bonn
Institute of Botany
Meckenheimer Allee 170
53115 Bonn, Germany

Cover illustrations:
W. Barthlott (5) and J. Theisen (1)

ISBN 978-3-642-08370-9 ISBN 978-3-662-06071-1 (eBook)
DOI 10.1007/978-3-662-06071-1

1st edition 1998, Second corrected printing 2001

Library of Congress Cataloging-in-Publication Data
Biodiversity: a challenge for development research and policy / Wilhelm Barthlott, Matthias Winiger (eds.) p.cm. Includes bibliographical references and index.

1. Biological diversity conservation. I. Barthlott, Wilhelm. II. Winiger, Matthias.
QH75.B5224 1998 333.95-dc21 97-51324 CIP

http.//www.springer.de

Originally published by Springer-Verlag Berlin Heidelberg New York in 2001.
Softcover reprint of the hardcover 1st edition 2001

Typesetting: Camera-ready by the editors
Cover design: design & production GmbH, Heidelberg
Printed on acid-free paper SPIN 10797536 30/3130/is — 5 4 3 2 1

Foreword

Biodiversity, sometimes simply understood as "diversity of species", is a specific quality of life on our planet, the dimensions and importance of which have just lately been fully realized. Today we know that "biological diversity is a global asset of incalculable value to present and future generations" (Kofi Annan). Biodiversity is spread unequally over the world: in fact, the main share of biological resources worldwide is harboured predominantly by the so-called developing countries in the tropics and subtropics.

Therefore, Biodiversity – A Challenge for Development Research and Policy was chosen as the title for an international conference which was held in Bonn in 1997 as one of the first major events organized by the then newly established North-South Centre for Development Research (ZEF) at Rheinische Friedrich-Wilhelms-Universität Bonn (Germany).

Since the ZEF, founded by the Senate of the University of Bonn in 1995, has played a central role in turning Bonn into a centre for international cooperation and North-South dialogue. The Centre is a product of the Bonn-Berlin agreement of July 1994 which was adopted to offset the effects caused by the Parliament and much of the Government moving to Berlin. It fits in well with the double strategy to strengthen Bonn's position as an international science arena and as an eminent place for development policy and the national and supranational agencies dealing with this issue.

The ZEF symposium on biodiversity made a valuable contribution to these goals and scientifically supported the implementation of the Convention on Biological Diversity (Rio 1992) and the Global Plan of Action for the Conservation and Sustainable Utilization of Plant Genetic Resources for Food and Agriculture (FAO Conference, Leipzig 1996). Therefore, I am very happy to present to the public a second edition of the proceedings of this symposium. I am extremely grateful to my colleagues, Professor Barthlott and Professor Winiger, for having initiated and once again superbly edited this publication.

Professor Dr. Klaus Borchard
Rector
Rheinische Friedrich-Wilhelms-Universität Bonn

Preface to the Second Printing

Authors and publishers can be pleased when a book's reprint becomes necessary only two years after its first publication. It shows the avid and still growing interest of science, politics and a broad public in the topic. Behind this interest stands the insight, or at least the assumption, that biodiversity or diversity in a larger sense must be considered as inevitable prerequisites for thriving, survival and further development of symbiosis and ecosystems. At the same time, it becomes more and more evident that the essential aspects of "biodiversity" have to be investigated more basically and more fundamentally. The importance of biodiversity has to be put in concrete terms in connection with society and development. Efforts have been made worldwide and on a large scale to achieve this aim. An increasing number of highly specialised and popular scientific publications underline this trend.

This book is based, in large part, on both an international conference and a symposium on Bolivian megadiversity. As many authors are involved, a completely new edition is not possible or worthwhile at the moment. Therefore, a slightly modified reprint is presented here including new and updated references.

Bonn, January 2001

Wilhelm Barthlott
Matthias Winiger

Preface to the First Printing

Life and its extraordinary diversity is the unique wealth which distinguishes the planet earth from all other planets in the universe. Mankind is a critical element of this wondrous spectrum, and thus biodiversity in all its aspects represents the very foundation of human existence. We have only become aware of the real dimensions of this wealth during the last twenty years, and today we know some 1.7 million different species probably representing far less than 10 % of the actual diversity. At the same time it is increasingly evident that – due to the rapid growth of the world's population which is nearing six billion people and is driving the extreme exploitation of natural resources – this diversity is undergoing a dramatic change. Fundamental genetic resources, which evolved over three billion years, probably are being lost at a dramatic pace. Extinction is forever.

The growing recognition and knowledge of the importance of biodiversity and genetic wealth has become part of the public awareness of the dual role of biodiversity: as an economic resource, but as well as an essential condition for the survival of inividuals and biotic communities. Much attention is being given to the important ethical and aesthetical implications of biodiversity. However, it is becoming increasingly clear that the loss of biodiversity risks also to have serious ecological consequences and considerable economic and social costs. As a result, biodiversity is now seen as a critical component of global environmental change. Of course, the main purpose cannot be to save all species at any price. Nevertheless, each species constitutes a part of the diversity of biotic communities and therefore is part of the diversity of habitats. The most diverse ecosystems are not found in industrialised countries. The greatest variety of genetic diversity is located predominantly in tropical rainforests and in certain subtropical areas. This enormous contrast between megadeveloped countries and megadiversity countries reinforces the need for us to devote greater attention and priority to: Biodiversity – a challenge for development research and policy.

The day has passed delightfully. Delight itself, however, is a weak term to express the feelings of a naturalist who, for the first time, has wandered himself in a Brazilian forest. The novelty of the plants, the beauty of the flowers, but above all the general luxuriance of the vegetation, filled me with admiration. – It was not by accident that Charles Darwin wrote this remark in his journal of researches on February 23, 1832 when he, for the first time, entered a tropical forest in Bahia, Brazil. Today, however, Darwin would not find many remains of a natural forest in the vicinity of Bahia, where, like in many parts of the world, a radical transformation of the original vegetation cover and ecosystems has begun. It is thus not surprising that in particular the developing countries, with their huge and threatened reservoir of diversity, initiated one of the most significant international agreements: the Convention on Biological Diversity (CBD), Rio de Janeiro 1992. This convention mandates the preservation, exploration and sustained utilisation of biodiversity. It regulates the use of biological resources through a fair and well-balanced mechanism of benefit sharing. Almost all nations of the world have signed this binding convention anchored in international law and are engaged in its implementation with varied success. For industrialised countries with a comparatively low species richness, the message is again clear: Biodiversity – a challenge for development research and policy.

Responding to this challenge, we present a global distribution of approximately 270 000 vascular plants in map form for the first time in detail. This geographical perspective, i.e. spatial dimension of potential phytodiversity, reveals a remarkable feature: in addition to extensive areas of extreme species paucity, we are able to delineate equally impressively centres of enormous species richness. Despite uncertainties posed by this global view, we believe that the essential aspects of diversity are reflected precisely. Furthermore, the map indicates those areas where exploration and preservation should receive highest priority support; i.e. regions with high diversity within the tropics and in certain areas of the subtropics. The six global diversity centres are listed according to their tentative species richness: 1. Chocó and Costa Rica Centre, 2. Tropical Eastern Andean Centre, 3. Atlantic Brazil Centre, 4. Eastern Himalayan-Yunnan Centre, 5. Northern Borneo Centre, and 6. New Guinea Centre. Tropical Africa, with somewhat lower species numbers, shows two maxima, and Madagascar also deserves mention. Several additional species numbers maxima characterise in particular Mediterranean-type regions throughout the world: the Mediterranean itself, but above all southern Africa and southwestern Australia, with their highly endemic floras. Obviously, high biodiversity is related to historical factors (paleoclimate and history of vegetation, paleogeography), position of the locality (degree of isolation, types of zonobiome), climatic conditions (water availability and higher temperature) and an accordingly high ***geodiversity*** (diversity of abiotic factors within the area). These factors explain why mountains especially have become focal points of global biodiversity research.

Our knowledge of the diversity of vascular plants and their distribution is surprisingly advanced compared to all other large groups of organisms (i.e. insects). The map presented here is based on an evaluation of data from approximately 1 400 standard floras and checklists. In an ecosystemary context we mapped the diversity of producers (plants). Since the diversity of consumers and decomposers depends on producers the map probably reflects the basic patterns of the entire terrestrial biodiversity in its entirety with relatively high precision.

GLOBAL BIODIVERSITY: SPECIES NUMBERS OF VASCULAR PLANTS

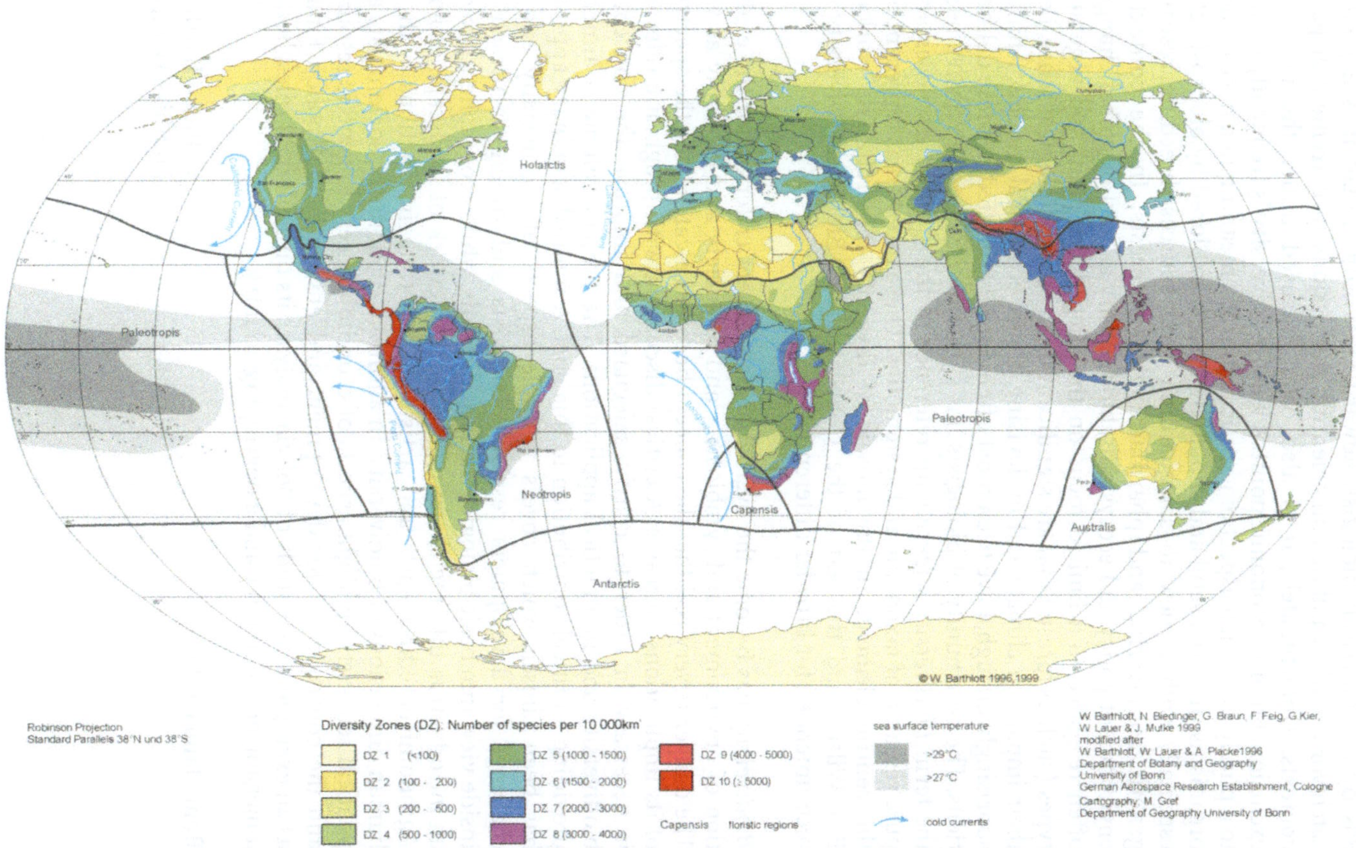

The term biodiversity was introduced in a comprehensive sense in the late 80s by the American biologist, Edward O. Wilson. Today, it is a political catchword used in public and scientific discussions in extremely disparate contexts. Unfortunately, it is also often misused under the mantle of exclusion. From an economic point of view, biodiversity is frequently reduced to a purely utilitarian aspect of genetic resources, meaning in a narrow sense organisms that are useful and should be further explored by bioprospection especially with a view to safeguarding food sources, breeding research, genetic engineering and pharmacological interests. In the context of environment policy and ecosystems, biodiversity as part of global change is often relegated to a limited focus on the alleged few plant and vegetation types (such as forests) that are significant in an ecophysiological sense, i.e. in their impact on climate and in devising global ecological models. Animals, suprisingly enough, have been most often entirely omitted in these deliberations. The community of biologists has sometimes claimed exclusive use of the term *biodiversity science* with reference to the – doubtless necessary – pure research of taxonomy and systematics. These three areas represent only the central elements in the mosaic of diversity research. The works of E.O. Wilson and the text of the Convention on Biological Diversity are recommended for broader differentiation.

Biodiversity has become the overriding assessment criterion for the evaluation of environment and symbiosis – and of society and economy as well. The danger is very real that the concept becomes a catchall as the spectrum of its implications increases. A critical discussion with pragmatic relevance is necessary in order to clarify, safeguard and improve the concept of biodiversity and its practical applications. The present volume is aimed at providing a critical contribution to this goal. A multi-disciplinary approach has been chosen which covers multiple aspects of biodiversity and its challenge for development research. Botanists and zoologists present an inside view into basic biological facts and trends, while geographers provide an introduction into the changing abiotic framework of landscape ecology. Agronomists, ethnologists and artists demonstrate interactions with the human dimensions of biodiversity. In addition, economists, sociologists and political scientists highlight aspects of economic access to genetic resources and the consequences for development and environmental policy. Finally, as an example, a case study of Bolivia analyses its difficult dual roles, both as a megadiversity country and a developing country.

Bonn, January 1998

Wilhelm Barthlott
Matthias Winiger

References

Barthlott W, Lauer W & Placke A (1996) Global distribution of species diversity in vascular plants: towards a world map of phytodiversity. Erdkunde 50(4): 317-327

Boyle TJB & Boyle CEB (eds) (1994) Biodiversity, temperate ecosystems and global change. Springer, Berlin Heidelberg New York

di Castri F & Younès T (eds) (1996) Biodiversity, science and development: towards a new partnership. CAB International in Association with the IUBS, Wallingford

Groombridge B & Jenkins MD (eds) (2000) Global biodiversity: earth's living resources in the 21st century. World Conservation Press, Cambridge

Heywood VH (ed) (1995) Global Biodiversity Assessment. UNEP, Cambridge

Messerli B & Ives JD (eds) (1997) Mountains of the world: a global priority. United Nations University Press, Tokyo

Mittermeier RA, Myers N, Gil PR & Mittermeier CG (1999) Hotspots: Earth's Biologically Richest and Most Endangered Terrestrial Ecoregions. CEMEX/Conservation International, Mexico City

Olson DM & Dinerstein E (1998) The Global 200: A representative Approach to Conserving the Earth's Most Biologically Valuable Ecoregions. Conservation Biology 12: 502-515

Porembski S & Barthlott W (eds) (2000) Inselbergs – Biotic Diversity of Isolated Rock Outcrops in Tropical and Temperate Regions. Springer, Berlin Heidelberg New York

Reaka-Kudla ML, Wilson DE & Wilson EO (eds) (1997) Biodiversity II: Understanding and protecting our biological resources. Joseph Henry Press, Washington, DC

United Nations (2000) The handbook of the convention on biological diversity. Earthscan, London

Wilson EO (1988) Biodiversity. National Academy Press, Washington, DC

Acknowledgements

Editors of books like this obviously owe a great debt to a large number of individuals and institutions. First of all, we would like to thank the Centre for International Cooperation in Advanced Education and Research (CICERO, Bonn), and the German Federal Ministry for the Environment, Nature Conservation and Nuclear Safety (BMU) for the financial support which enabled us to organise the symposium.

Second, we thank the authors and co-authors for their efforts in revising their contributions to the symposium and in making them available for this publication. The completion of this volume could not have been achieved without their patience and willingness to answer the multifold questions which arose.

The symposium has, to some extent, been developed out of two different ideas. First, the former Rector of the University of Bonn, Professor Max Huber, proposed to organize a symposium on biodiversity in connection with the foundation of the North-South Center for Development Research (ZEF). We are very grateful for his commitment and the ongoing support from his successor, Professor Klaus Borchard, and the University of Bonn.

At the same time one of our colleagues, Dr. Pierre Ibisch, currently Director of the Science Department Fundación Amigos de la Naturaleza in Santa Cruz, Bolivia, suggested convening a symposium on the "megadiversity" country Bolivia. We decided to combine both projects within one symposium on "Biodiversity - a challenge for development research and policy". This explains why Bolivia plays a significant role in this volume. We would like to thank His Excellency Orlando Donoso Aranda, Ambassador of Bolivia and especially Mrs. Eunice de Heins, First Secretary Culture for their untiring efforts and valuable advice.

The symposium could not have taken place without the competence and the involvement of the convenors and without the support of the University of Bonn and the Institute of Geography, with its excellent conference facilities. We acknowledge specifically the help of Professor Uwe Holtz, Dr. Martina Krechel, Mrs. Petra Duwe and in particular Dr. Hartmut Ihne, Director of CICERO, and his staff, above all Ms Susanne Klunkert for her excellent organisation.

We extend our thanks also to many colleagues of the botanical and geographical departments, especially to those of the Institute of Botany and the Botanical Garden, who have contributed to the realization of this project with their expertise in the field of biodiversity research: Dr. Stefan Porembski, Dr. Wolfram Lobin, Dr. Eberhard Fischer and Jens Mutke.

Finally, our very special thanks go to the Springer Verlag, and to the managing editor, Dr. Nadja Biedinger. Without Dr. Biedingers skills, expertise and commitment the compilation of this volume could not have been completed on time.

Wilhelm Barthlott
Matthias Winiger

Contents

Part 1. Introduction

Convenor: Wilhelm Barthlott

Part 4. Consequences for development and environmental policy

Convenor: Udo Vollmer

Part 5. Bolivia - A megadiversity country

Convenors: Alexandra Sánchez de Lozada, Michael von Websky, Horst Korn

1 The Biodiversity of Bolivia

3 Facing the Challenges of Biodiversity Conservation in Bolivia

4 Call for research and action

Part 1

Introduction

Biodiversity – Is there a second chance ?

Clas M. Naumann
Zoological Research Institute and Museum Alexander Koenig, Bonn, Germany

Abstract. The adaptation of living organisms to present abiotic and biotic global conditions is a process that has taken more than three billion years. It largely depends on genetic change by recombination which – in the higher forms of life – occurs during sexual reproduction and by natural selection, reducing the genetically produced large numbers of individuals. None of the extant species and their individuals can exist without hundreds of interrelations with other organisms, a fact that also applies to the existence of man and his ecology. Man is considered to be the only species that can actively change the breadth of its ecological niche through technical and cultural evolutionary processes. These processes depend on the availability of secondary energy which is made available mainly through fossil deposits. The latter are the result of biological activity (photosynthetic processes) over hundreds of millions of years. The rapid global-scale destruction of biodiversity being observed today, together with the overexploitation of fossil energy and the rapidly growing human population, are considered the most dangerous threats to man's continued existence. It is stressed that intensive research on global biodiversity together with continued, increasing effort to preserve as much as possible of the world's biological diversity will be the most future-oriented, urgently needed undertakings to ensure man's survival on earth.

Biodiversidad: ¿Hay una segunda oportunidad?

Resumen. La adaptación de organismos vivos a las actuales condiciones globales abióticas y bióticas constituye un proceso cuya duración abarca un período de más de tres billones de años y depende en gran medida de cambios genéticos por recombinación, los cuales - en las formas de vida superiores - tienen lugar durante la reproducción sexual, y por selección natural, reduciendo el elevado número de individuos producidos genéticamente. Ninguna de las especies existentes y sus individuos podría existir de no darse cientos de interrelaciones con otros organismos, un hecho que se puede aplicar también a la existencia del Hombre y su ecología. Se considera que el Hombre es la única especie capaz de cambiar activamente el alcance de su nicho ecológico mediante procesos evolutivos técnicos y culturales, los cuales dependen de la disponibilidad de energía secundaria puesta a disposición sobre todo a través de depósitos fósiles. Éstos últimos son el resultado de una actividad biológica (procesos fotosintéticos) que se ha venido desarrollando durante cientos de millones de años. La rápida destrucción de la biodiversidad a escala global que observamos en la actualidad, junto con la sobreexplotación de la energía fósil y el crecimiento acelerado de la población humana, se consideran las amenazas más peligrosas para la pervivencia de la existencia humana. Se destaca el hecho de que la intensa investigación de la biodiversidad global y el esfuerzo continuo y cada vez mayor por preservar lo más posible de la diversidad biológica del planeta habrán de ser las medidas más necesarias, urgentes y orientadas hacia el futuro si queremos asegurar la supervivencia del Hombre en la Tierra.

1 News about an old topic

Life on the planet Earth goes back a long way: Living organisms are associated with more than two thirds of the earth's 4.6 billion years of history. Microorganismic evolution made it possible for the earth's early CO_2-dominated atmosphere to slowly change into one with a high proportion of oxygen, thus creating the conditions necessary for the evolution of all higher forms of life, and especially for the photosynthetic activity of plants which, as producers of organic compounds and as suppliers of oxygen, create the conditions for nearly all forms of life today.

Despite drastic floral and faunal break-downs and recoveries, one can state that most of the major lineages of the animal kingdom, and the arthropods and vertebrates in particular, already existed by the time the fossil record started, i.e. by the end of the Precambrium era about 600 million years B.P. Major changes occurred with the colonization of terrestrial habitats, when higher forms of plant life developed and both the vertebrates and the arthropods, together with flowering plants started their glorious evolution of terrestrial forms of life which led to the extant forms of biodiversity. The process of evolution is one that necessitates interaction between species and can only take place through the comparatively slow process of varying and selecting genetic information from one generation to the next.

Even today we are far from having an accurate picture of the species numbers that contribute to the magnitude of biodiversity that existed, mainly undisturbed, let us say until the end of the last century. Since the day of Linnaeus (1758) – who recorded about 4,200 species – more than 300,000 species of higher plants and approximately 1,200,000 species of animals have been described. Thus, comparative biology has already created an enormous wealth of information on the biosphere. On the other hand it has become clear during the last two decades that our present knowledge is far from being complete. In some taxa (microorganisms, algae, nematodes, insects, to name just a few) specialists accept as a fact that between 5 to 20 times of the number of presently described species have yet to be discovered. Robert May, one of the leading and most influential biologists of the United Kingdom, stated that we tend to know more about the composition of the Milky Way and the universe than about the key elements of the ecosystems of the globe, the individual species.

Maximum species densities are to be found in the tropical rain forests (central and western Africa, Central and South America, eastern Asia) as far as terrestrial habitats are concerned, and in tropical and subtropical coral reefs, as well as in some parts of the deep sea, another only slightly explored part of the biosphere. Exceptionally high species numbers have also been recorded from the most southern part of Africa, with no less than approximately 2,300 species of flowering plants to be found on the Cape peninsula alone (by comparison, the Flora Europaea mentions only approximately 12,000 species for the whole of Europe).

2 The potential of biodiversity

Man, who represents something of a latecomer on the planet, has always known how to use a variety of forms of life in order to sustain himself. About 20 species of cereals account for more than 90% of the world's harvest (three of them produce more than 70 %). One wonders to what use the remaining thousands of species of angiosperms might be put once they are carefully studied and analysed with this aim in mind. But it is not only the potential use of plants and animals for food, fuel and clothing that makes biological species so important for the future of mankind. One also has to think about the hundreds of thousands of plant compounds occurring as secondary metabolites in plant and animal tissues. Only a few of these have been used for the treatment of illness, as medicine, antibiotics, phytohormones, or weed and other pest control agents. I want to mention just two examples that illustrate the immense potential of secondary plant and/or animal compounds: It was only a few years ago that two compounds hitherto unknown to science were discovered in common tropical plants found in Madagascar, *Cataranthus roseus* (Apocynaceae): Vinblastin and Vincristin. Using these two compounds to treat children's leukaemia, it has been possible to reduce the mortality rate in certain forms of this disease from 90 % to just 10 %. In certain species of the Afrotropical bush *Putterlickia* spec. (Celastraceae, another compound, Maytensene, was found to be influential in reducing the growth of certain forms of breast cancer.

One may not only wonder about what future use could be made of the thousands of unknown compounds in plants. What about the possible application of compounds yet to be discovered in microorganisms of which only 4,000 species have been described out of an estimated one million species?

Increasing portions of known and unknown diversity of life on earth are being destroyed by over-exploitation and destruction of entire ecosystems such as the tropical rain forests or the large marine basins which have been emptied without regard to their renewability and sustainability. The great tragedy of our generation is that on the one hand we have come to the realization that none of the biological and non-biological resources that we make use of is of an unlimited nature, while at the same time, under the intense participation of the modern mass media, we are witnessing the highest species extinction rates ever on earth. Within just a few generations, the surface of the globe will have changed from a once widely untouched paradise to a deserted planet that harbours just a few universal species. By destroying this tremendous wealth of living organisms around us, we have also severed the roots of life for future human generations.

3 A basic view of environmental concepts

Individuality is one of the basic concepts that makes the higher forms of life (plants and animals) so unique when compared to chemical compounds or minerals, that is to say, to the elements of the a-biosphere. This idea may seem strange to many laymen who would generally agree that there are differences between individual human beings in a classroom, a settlement or a nation. We would also agree that there are individual differences between our neighbour's and our own pet cat or horse. It is generally less well understood that the same type of individuality applies to practically all populations of higher plants and animals – a few exceptions consisting of cloned organisms occurring under certain ecological conditions notwithstanding. The main reason of this individuality is genetic and can be briefly explained as being caused by the different allelic expression of genes occurring in a single population. Assuming that an organism such as Drosophila harbours at least 50,000 genes, half of which occur in at least two, frequently many more allelic states and knowing that alleles are freshly arranged during the process of sexual reproduction (meiosis = gamete formation and fertilization = combining genetic information originating from two different organisms), it is clear that even over the course of hundreds of generations the probability of the creation of two genetically identical organisms originating from different lineages is practically zero. To conclude, each and every bisexually produced specimen of higher plants or animals does represent an individual. In other words, it is unique in space and time.

This also means that the individual may be considered as a temporary vessel containing part of the genetic information of the local population it is part of. Thus the existence of individuals will always remain limited in time. It starts with the fertilization of an egg cell, undergoes embryonic and postembryonic development, and will eventually reproduce and die later, possibly after a period of senescence.

In general, local populations of plants and animals consist of such genetically-produced individuals. The summary of their genetic information content varies from specimen to specimen. It represents the ecological wealth of a population and forms a population's main foundation for adapting from one generation to the next. Such adaptations are required because the local environment never remains constant: There are not only climatic differences from year to year, but also differences in the form and availability of food and prey, breeding and hibernating space, etc. In short, a population that loses its wealth of genetic information and a high percentage of its individuals will also be reduced in a way that it can react to environmental changes, both on a short-term and on a long-term basis. Thus, a population's genetic polymorphism is one of the key elements that enable it to survive in a given habitat and to interact with other species.

Species interact and interdepend in a multitude of ways within ecosystems. In short there are chains of energy and material flows ranging from the producers – the photosynthesizing plants – to various forms of consumers (herbivores and predators), many of which interact in many different ways with the former (be it as pollinators, seed dispersers etc.). Finally, the ecological cycle is closed by the destruents which are mainly, though not exclusively, mircoorganisms living on decaying matter. They thus recycle organic material and make it available to producers once again.

Not a single species on earth can exist without having such complex and manifold interrelations with other types of organisms. Plants, animals and microorganisms all depend on the action and metabolic products of other plants, animals and microorganisms, be it directly or indirectly. And this basic insight applies to the species of man in the same way it applies to the daffodil, the stinging nettle or a *Ficus* tree in a tropical rain forest, to the orange-tip butterfly and its caterpillar in the same way as the Komodo Varan on a remote Southeast Asian island or to the Blue Ara on the banks of a South American river. Man represents a unique species on our planet, but not because he does not depend on other species' ecological potency and action, but rather in the way that he interacts with nature and forms his own species-specific ecological niche.

4 The extinction of species past and present

When considering the present rate of species extinction as discussed by many modern ecologists, one is frequently confronted with the argument that species extinction is not a recent phenomenon. As a matter of fact species have been becoming extinct as long as evolution has occurred. The extinction of many species (e.g. at the Permian/Trias or at the Cretaceous/Tertiary interfaces) has been of a dramatic nature, and it appears that considerable parts of entire ecosystems have been wiped out. The extinction of some species has been due to catastrophic events, such as the last major wave of extinctions at the Cretaceous/Tertiary transition that led the extinction of the dinosaurs and ammonites. This wave was probably due to a meteorite impact that caused enormous atmospheric clouds of dust which absorbed much of the solar radiation and thus led to a global temperature decline. All such mass extinctions have subsequently been followed by a recovery of the earth's flora and fauna and new increases in species numbers.

However, when considering the present phenomenon in which entire ecosystems are becoming extinct in massive numbers, one has to keep in mind that the extinction of species in the past was only partial in nature: While parts of an ecosystem were being destroyed, other species that were better adapted to the newly developing climatic and ecological conditions were favoured by selection and led to new forms of adaptations. The large cold-blooded dinosaurs appear to have been replaced first by the smaller, homoithermic mammals which ultimately produced a number of fairly large species again. In any case, there has always remained a considerable potential of species whose genetic information content allowed them to take over and to adapt again. This process of genetic adaptation and speciation also required long periods of time because adaptation is a comparatively slow process.

In contrast, entire ecosystems are becoming extinct today, leaving very few or even no endemic species. And: The period of time that the ever-growing human population would allow for adaptation is practically zero compared to the long periods that would be needed to refill the gaps extinction has caused.

A few species will doubtless always remain, usually the so-called euryoecous species which can cope with a great variety of ecological conditions. Such species will even survive man assuming that earth will one day become inhabitable for human beings.

5 Man and nature – Or nature without man ?

As already explained, the basic background of the human species is of a biological nature. All the basic prerequisites described above apply to man just as they apply to other species: Human populations consist of individuals that differ in genetic information content and which have an individual life span and are bound into physiological as well as ecological processes, making man just one part of the biological environment. And still this brief description does not sufficiently cover the role that this single species plays in nature, because – in contrast to other species – man does not create his ecological niche in coordination and in response to his environment. Man appears to be the only species that can and does change his niche in a planned and controlled way. We are used to call this side of our nature 'technical evolution' or 'cultural evolution' in contrast to the biological evolution described above.

Cultural evolution is an energy-dependent process as we can easily see from the energy consumption rates that are linked to man´s early and present evolution. All the technological progress that we as a species have achieved has always been dependent on the availability of secondary energy, that is to say, on forms of energy that have been generated from fossil deposits or other sources that are only indirectly linked to the activity of solar energy (e.g. wood consumption). Consequently, during the evolution of man the availability of such secondary energy has become the main constraint limiting human population growth. Looking at the history of man one can easily ascertain some of the main constraints that man has overcome and which have subsequently led to new evolutionary thrust. Thus, with the management of fire man was able to substantially extend his nutritional basis; the invention of tools allowed early populations to exploit a larger spectrum of the natural environment and to improve their nutritional situation; in a similar way the change from a hunting/-collecting life style to that of the nomad and later on to that of the farmer improved man's ecological base and led to an ever-increasing population at the same time. I am convinced that by the end of the Middle Ages and after the large-scale destruction of continental and Mediterranean habitats the 'discovery' of the tropical and subtropical continents allowed European man to increase his range of exploitation of living and non-living resources and thus in a similar way allowed the continuation of European population growth. Later on, the invention of the steam engine led to the industrial revolution, that is to say, to large-scale production and exploitation of natural resources which has undoubtedly increased the survival chances of thousands of people. It has, at the same time, led to the never-ending spiral of exploitation of natural resources on the one hand and continued human population growth on the other. The invention of artificial fertilizers also broadened man's nutritional base and simultaneously allowed for more population growth.

Given that all the earth's natural resources (in theory this even applies to solar energy) are of a limited nature, it is clear that human population growth which is a consequence of constantly forced techno-cultural evolution in conjunction with an ever-increasing exploitation of natural resources is the greatest threat to man's long-term existence on earth. Specialists agree that the world's current economic and social situation, which makes a doubling of the world population inevitable, will ultimately lead to a global population of at least 12-13 billion people.

It is clear that such drastic changes in human population size will once again go hand in hand with an extreme intense and severe pressure on man's natural environment, that is to say, on the world's few remaining functioning ecosystems. At the same time however, these ecosystems represent the most important requirement if we are to cope with the drastic ecological, environmental and economic problems that will be facing us. Thus, any sort of future-oriented local and global policy should be directed to:

1. The conservation and preservation of as many of natural habitats as possible with the maximum possible content of global diversity, and
2. The immediate and systematic documentation, inventarization and analysis of biodiversity as the earth's most vital and precious resource.

Politicians will need to decide that the present degree of destruction and over-exploitation of our living resources is entirely incompatible with the aim of securing man's long-term existence on this planet. The miraculous new term 'sustainable use' points in the right direction, but still does not do adequate justice to the fact that we still have much to learn about the structure and functions of our natural environment. Any responsible policy will thus have to take into account the fact that research priorities will have to be redirected to large-scale and focused biodiversity research, starting with local and global programmes for the inventory and identification of our biological environment.

6 The valuation of the natural heritage in modern societies

Traditional societies all over the world developed well-adapted and carefully worked-out mechanisms for avoiding the over-exploitation of natural resources. In ancient times no Eskimo would have ever dared to harvest more whales, seals or fish than he would need for his family's immediate subsistence. In Nuristan (Afghanistan) one household was allowed to harvest one cedar tree a year which had to be selected under certain ecological and social criteria. Similarly, Tibetan nomads and the Yanomami Indians in the Amazonas lowland developed socially controlled mechanisms that would guarantee the environment's long-term existence. Thus, the mechanisms for achieving the goal of sustainable use had already been developed long before the modern term 'sustainable use' came into use and clouded political thinking as the new ecological answer to our needs. From this we can learn that in principle man is capable of developing control mechanisms that allow him to live in peace and harmony with his natural environment. The question is whether this also applies to modern industrial society or whether the latter has already become too distanced from a intellectual and psychological relationship with nature to find the necessary understanding and intellectual insight into this necessity.

It is odd to observe that our rich industrial societies always find it easy to provide considerable financial resources for the acquisition and preservation of our cultural heritage, be it in the field of arts, music or literature. This is especially evident when looking at the important decisions made over the past ten years in the city of Bonn regarding a number of important and highly valued cultural projects, such as historical and art exhibition halls, technological museums and handy craft research projects. At the same time, convincing political decision-makers to set aside even small amounts to launch major nature education and exhibition projects appears to be nearly impossible. One wonders whether attitudes like this which are widespread in this and other countries already reflect the above mentioned alienation from the natural resources of life.

One will also have to ask whether the short-term rhythm of political thinking and environmental consciousness which are found in the political echelons of modern democratic systems and allow only short periods which projects do not have to take election campaigns into consideration does in fact block the establishment of a policy that is truly orientated towards the long-term future of mankind. For example, the 1997 budget of the German Ministry for Education and Research lists about 10 top priority areas for research support, the largest being space research which swallows about 15 % of the entire budget, or about DM 1.5 billion. The issue of biodiversity research which is considered to represent the world's most urgent future-oriented issue does not even turn up in this list of primary research areas. It is clear that such fields will not be able to create thousands of jobs on a short-term scale unless the long-range potential of biodiversity (also in cooperation with modern biotechnology) is taken into consideration.

7 To conclude: A parable

Far back in the Middle Ages there was a wealthy community with a prospering economy, a luxurious social life and many facilities that looked after the well-being of the residents and their guests. Virtually nothing was lacking, but much of the city's wealth came from sources outside their own walls and sphere of influence.

One day a very old woman carrying twelve very old books under her arm came to the city magistrate and offered these books for sale. "Look," she said, "these books have been inherited from my ancestors and have been passed down from one generation to the other in my family for centuries. These books contain all the wisdom and the knowledge on earth and they are well suited to solve all the problems of the future. Now, that I am getting old and have no children I am going to sell them and offer them to the fathers of my city for just one pound of gold."

The lord mayor and the magistrate shook their heads and stroke their beards and said, "What a lousy old woman you are. You offer us just twelve dirty old books with broken spines and torn pages, and ask a tremendous amount of money for them. No, we will not buy these books."

The old lady went away and said, "What a pity for them because they might have made use of these books for the well-being and the future of this country. But since I cannot carry them around any longer I shall have to burn half of them." And so she burnt six of the books and kept the remaining six.

One year later she went to the magistrate again and offered him the remaining six books. But now she asked for two pounds of gold. "What a crazy old woman you are," the Lord Mayor of the city said to her. "Last year you offered us twelve books for one pound and now you come and ask for two pounds for just half of what you offered last year. No, we will not buy your books because we are living in a market economy age and are not ready to enter such games." So the old lady went away again and said to herself, "What a pity for now I shall have to burn another three of the books because I am getting old and cannot carry them around forever ." And once again she burnt half of the remaining books containing half of the world's knowledge and wisdom.

In the next year the city's situation had already changed. When she came to the magistrate and asked for four pounds of gold for the remaining three books, they said "We might well be in need of your books if they really contain all the remaining knowledge and wisdom about the earth. But, as you can see, we now have to cope with some economic irregularities, and prices are rising. There is also some social instability which costs us a lot of money. So we simply cannot afford to buy your books. Actually, we should have done so last year, but now it is too late. We can no longer afford to spend so much money." So, the old lady went away again and burnt another two of the remaining three books because she was getting old and could not carry them around any longer.

The next year, there was just one single volume left which contained only one twelfth of the world's knowledge and wisdom. The economy of the community had taken a change for the worse. Unemployment had sharply risen, there was political turmoil, there was even some talk that those in power might have to accept an opposition party. So the magistrate did not know how to solve their various problems. But they decided that if the old woman's last book actually did contain all that remained about the world's wisdom and unanswered questions then it would be the only option they had for overcoming their problems. And so they had to buy this one last book – for just eight pounds of gold. How much more might they have gained a few years earlier?

Knotted ropes, rings, lattices and lace: Retrofitting biodiversity into the cultural landscape

Newton Harrison, Helen Mayer Harrison
University of California, La Jolla, California, USA

The title of this lecture refers to the forms or patterns that we, as artists, have proposed or used to reintroduce biodiversity to places where nothing was left or to make connections between those special places where a biodiverse landscape still exists. These forms were developed out of considerable experience in recognizing existing or constructing new patterns. It was Dr. Wilhelm Barthlott who suggested that we might be extending the notion of saving hot spots by creating these new patterns and thus recontextualized our work in terms of ecological theory and invited us to speak.

THE LATTICE

For example, The Serpentine Lattice, a work on the death of the North American temperate coastal rain forest, addresses the clear cutting of about 55,000 square miles. This clear cutting not only destroyed most of the once enormous and singular forest containing some of the oldest and largest and most endangered trees in the world, but, concomitantly, seriously disturbed watercourses that were measured in the tens of thousands of linear miles. The text and images of the Serpentine Lattice, in addition to presenting a concept for restoration, make totally clear the brutality, the violence, the ignorance, the mindless, single purpose, self-interest involved in the creating of a new cultural landscape at this scale. Until recent times this was not so obvious, as it happened more slowly, bit by bit, over millennia.

The Serpentine Lattice began as a museum installation with interacting text and image. There is a two and a half meter by ten meter slide presentation with fifty-five sets of panoramic images, slowly dissolving from one to the other, revealing the destruction of the Pacific North West temperate rain forest. This conifer forest until relatively recently covered the Pacific coast from San Francisco Bay in California up through Oregon, Washington and British Columbia to Yakatat Bay in Alaska. The clear cutting modified soil chemistry and altered climate and made an approximately 2,000 mile long corridor, creating about 55,000 square miles of disturbed terrain. The installation, in addition, contains a two and a half meter by ten meter hand drawn map of the affected area, texts and drawings and photography.

We hold that every place is telling the story of its own becoming, which is another way of saying that it is continually creating its own history and we join that conversation of place. However, that conversation can only be joined effectively by invitation. In 1992, invited to Oregon by Reed College to develop an exhibition, as we usually do in exploration of place, we flew over the area in a small plane. The Serpentine Lattice developed as a response to that trip and three questions we posed as a result.

The questions were:

1. Can it be that everybody here permitted the clear cutting of 95% of the North American coastal temperate rain forest, seriously accelerating the rate of destruction over the past 25 years?
   The answer was obviously 'Yes'.
2. What are the principal ecologists trying to do about this?
   They were, in the main, working to save the few last areas of old growth.
3. And finally the unanswered question of who is looking after the 55,000 square miles that have been cut?

The first question was transformed into the following commentary:

> "Now there is enough new information about
> And enough old wisdom around
> For anybody who thinks about these things
> To know that the death of a great forest
> is a global tragedy."

After reflection on how we, as artists, might respond to destruction at this scale, we made the following arguments, presented here and distilled from a rather dense poetic text.

1. It is necessary to tune the response to the size of the event being addressed. If industry, which is after all a cultural invention, has removed this much forest, disturbed this many small drain basins, killed the life in this much topsoil and stressed the life in the rest, is any form of remedial action available? Can cultural invention find the way to make restoration at this same huge scale for an area in which forest has been removed, drain basins disturbed, and the life in the soil endangered or destroyed? If culture cannot, then the resulting enormous erosion, interspersed with chance succession ecologies and tree farms, does not look like a happy outcome. Certainly it presently seems as if market forces can not or will not provide a satisfactory resolution.
2. After an event of this scale, the destruction of almost the entire rain forest, a new kind of response is required. Thus an eco-security system is proposed, not unlike the social security system, wherein a percentage of the gross national product (which is made largely at the cost of the gross national eco-system) is

released to restore the environment. One percent of the gross national product, in 1992 terms, would amount to about 57 billion dollars. This could be spent annually, on the country as a whole, for restoration. This would create the first regenerative environmental feedback loop at continental scale. For, if one element of culture as a whole, in this case industry, is permitted to consume irreplaceable resources at this scale, what elements in culture can establish limits? What elements of culture can play a restorative role?

3. For instance, assuming an eco-security system, a cultural entity could buy or otherwise acquire, a swathe a mile or more wide along the ridge line from Southern Alaska to Northern California - from Yakatat Bay to San Francisco Bay. Given that much of this ridge land can never be returned to the original forest, nor even to tree farming, it has little value. Also, much of it is already in the public domain, so that the expense of acquiring this land would not be not exorbitant. Far less, say, than a four lane interstate highway.

> "We
> being grateful
> for the invitation to join this perilous conversation
> began seeing an act of restitution.
> You seeing a serpentine
> I seeing a lattice
> We beginning to image North-south continuities
> From Yakatat Bay
> To San Francisco.
> Continuities that would bespeak
> The eco-poetics of the whole."

4. Thereafter, one could begin the rehabilitation of the multitudes of drain basins in the area, restoring the streams, and permitting forest to return where possible, although due to the erosion of the topsoil, there may be severe limitations as to what can grow where. Thereafter, regenerating topsoil where possible, an altered, more ecological type of forest farming could interchange with areas that would be left to succession, alternating the operations of cultural activity with ecological succession and establishing the operations of culture as figures within a biodiverse field.

> "Then
> the gross national ecosystem
> could take its place
> privileged appropriately
> as the field within which
> the political systems
> social systems
> and
> business systems
> that comprise
> our eco-cultural entity
> can exist."

The title, The Serpentine Lattice, can be read literally, as a form or a pattern, a description of the lattice-like figure that emerges when the ridge line and the coastline are taken as the figuratively "parallel" boundaries to the land, and the watersheds between then become the "rungs".

> "Imagine the serpentine form
> of the crest of the coastal mountains.
> Imagine the serpentine form of the Pacific coast.
> Imagine some of the rivers
> really watershed ensembles
> extending from crest to coast
> connecting the serpentines
> as nearly leaf shaped rungs
> or cross members of the Lattice."

The title, The Serpentine Lattice can also be read metaphorically, as:

> "Where you said a lattice
> And I said a serpentine
> And you said, "Network the watersheds".
> And I said, "A game of Go."

All the works that are presented here are visions that set out to be self-fulfilling prophesies. Most are also practical and doable, and several of them are presently being enacted on the ground. The Serpentine Lattice has been exhibited many times and exists in a German translation as well as the original English. It was designed, as are all of our works to, at the least, influence the conversation of place and to expand the rather limited ideas and vocabulary of possibilities seen all around us in regard to restoration.

Most interesting, however, for ourselves as artists was the pattern that emerged or was created as we looked at the map. The pattern had certain properties, once it was recognized. It revealed the magnitude of the problem. It suggested a reclamation form at a scale never before put on the table, which had both ecological and cultural values. Thus, once the questions were posed and the scale determined, the pattern actually emerged from a perception of the salient topographical features - ridges, coastlines and the lines formed by rivers, grouped in little leaf- like drain basins.

MOSAIC

The next work to be discussed, the meadow on the roof of the Kunst-und Ausstellungshalle der Bundesrepublik Deutschland, entitled Future Garden, Part I, The Endangered Meadows of Europe, is the most complex, complete and condensed of our meadow works. It began with an invitation from the Kunst-und Ausstellungshalle to consider the open space on the roof. It took form after a discussion about meadows. We imagined the loss of the mosaic of meadow lands across Europe and set out to do a work that addressed the consequences of that loss. To do this, an extended narrative, taking the form of a series of stories about meadows, was constructed. As art, the installation operates as representational sculpture as well as a continuously changing, living, color field that covers about 4,000 square meters on the center of the roof. The roof space thus is transformed into a field upon which a complex drama is enacted. The drama begins with the decision to save a 400 year old meadow from the Eifel, transplanting it to the roof instead of letting it be torn up and replaced by a housing project. This decision sets the stage. Thereafter, sections of other endangered meadows, a wet meadow, a dry meadow and a stone meadow, are combined to make a diverse ensemble. Conversations between the artists and various botanists and ecologists, meadow masters as it were, fictionalized in a series of texts in both German and English are set in wooden housings in a fence structure, bespeaking the history and cultural function of meadows Photographic images, printed on tile, of meadows from other parts of Europe reflect the texts and confront the living meadow. Certain fencing structures hold story and image and also function as seats. They all together confront and co-join, presenting this drama which unfolds when walking or promenading the meadow perimeter. Given its proximity to the Bonn seat of government, the site was politically appropriate.

As artists, we have tended to be interested in discovering the unique within the commonplace. The ecological narrative is used to give voice to a unique element of the European cultural landscape. We understand that the meadow lands of Europe are a recent phenomena from an ecological point of view. They have developed over a number of centuries as a result of forest clearing, and are maintained in their present form by the grazing of livestock and/or the annual cutting of hay. However, they can also be perceived as a kind of agricultural model for a spontaneous, although unconscious, collaboration between humans and nature. The cutting and grazing has set the stage for a meadow ecosystem in Europe of considerable complexity and stability, one of the most successful collaborative and sustainable ventures between our species and the rest of the ecosystem. This collaboration and its history are of great interest to us.

Text XV

First one and then another then many
asked what we meant by this new metaphor.
The meadow is a teacher for the future and a model.

You said
The meadow represents an old wisdom.
It is humble and unassuming.
Yet it is a vast cultural landscape
a sensual and pleasing many purposed landscape
constructed over millennia
an aesthetic that is no longer being remembered
an information ground that is rapidly being lost.

And I said
Imagine a new forest
constructed intentionally over time
with the meadow as a model.
So that this new forest
like the meadow
will be a cultural construction
and will supply food for people and livestock
and wood and whatever
as well as cleansing waters and air.

And this newly constructed biodiverse life web
will be useful for many species
for the amphibian the reptile the bird
the insect
and the life in the soil.

And it is in this way
one of us said
that the meadow becomes a model for the future.

And the other said
In its present state
dense with information,
the meadows bespeaks the possibility
of constructing a multi species collaboration
with ourselves and all others involved.

And we said
It is in this sense
first in one way
then another
that the meadow is teacher and model and prophet.
....

In fact
I imagine monoculture Europe
reconstructing itself in total
over time
and the cultural landscape is transformed
so that human activity becomes a figure
within a biodiverse field
and the meadow continues
taking its place
within a future eco-space
for which
it had once been the model.

At the same time as biodiversity was disappearing from many European Meadows due to overcutting, the question of restoration at a vast scale was being confronted in areas of the east of Germany and in the Czech Republic and in Poland that were the sites of giant pit mines.

Now, just as the meadow has been a repeated theme in our work, so have the open pit mines of the Black Triangle. We were invited by the Cultural Foundation of the Free State of Saxony and the Cultural Program of the Siemens Corporation to travel to the borders of Poland, the Czech Republic and Germany to took at the brown coal open pit mines with the idea of seeing how artists might respond to that environment. This was known terrain for us, having been concerned with such ideas and worked with such spaces at the Academy in Prague and at Bauhaus Dessau. We were yet again struck by the size and the number of these vast excavations, situated both close to each other and close to the borders of the three countries, the bleak, nightmarish landscape with its deep pits of scoured earth, the clanging of coal cars on the moving belts, the mega-machines towering, eating the earth, has haunted us. But, above all, we were struck by the absence of any comprehensive vision for closure of these mines which thus appeared to be done mine by mine. We began looking at this excavation fragmented terrain searching for a grouping that wanted to or needed to or could become one place.

Such a grouping emerged for us south of Leipzig in the vast open pit mining operations that had happened and were still happening there.

THE LACE

The Shape of Turned Earth: A Brown Coal Park for Südraum Leipzig

The work was done at the joint invitation of the Kultur Stiftung of Siemens AG and the Kulturstiftung des Freistaats Sachsen. Although it has been exhibited several times, it is still more a sketch than a finished work, consisting of about eight images and texts.

"I began imaging a park that would take its shape and meaning from "turned earth".

You began saying that the shape of the park would be the shape of the brown coal that lay under the land and was, is or will be extracted.

I said that this park would function as memorial.

You said this park would function as reclamation and a new way of recreating the cultural landscape.

*I said that **The Shape of Brown Coal: A Park for the Area South of Leipzig**, would address ecological problems.*

You said there will be serious economic problems to be addressed.

I said there will be formidable political issues.

You said there will be issues of human need.

I said there will be issues of human greed.

You said and produce a regional identity.

Clearly it was the moment of an epiphany."

Confronted by the moonscape, crater scape, vast machinery, large smoke stacks, dirty sky landscape of the brown coal region of South Leipzig, the single most salient feature to us was the pattern of turned earth. By mapping this earth, an amazing iconic baroque shape of about 300 kilometers was produced. Internally there existed 30 square kilometers of infrastructure, 50 square kilometers of excavation that would become lakes, 35 square kilometers of farmland and 170 square kilometers of land belonging to the mining companies. Manifestly, extreme liberty had been taken with this land, even more brutal than the construction of the cultural landscape that was formerly the Pacific Northwest temperate rain forest Manifestly, extreme liberty must again be taken with the remains of the land, but liberty of a far more gentle kind.

One of the requests we had received when talking with the people and the planners of the area had been to restore the landscape so that it became whole again. Once the area south of Leipzig had been a lake district, a holiday area of some beauty. There had been many little towns and hamlets, at least 60 of them, that had been situated above this or that vein of coal.

"Then you said again
The life web is most diverse at the boundaries
Between ecotypes
And Witznitz and its neighborhood are both metaphor
And model
For a poetry of the whole.

I said
As the water level rises
All excavations will become lakes.
Therefore a lake region is inevitable.

You said
About 145 kilometers or more
Belong to the mining agency and are public lands
And can if people are willing become available.

Then you said
If the earth is neither poisoned nor too rich
Ecosystems will respond to disturbances
By moving rapidly across those surfaces
As at Mount Saint Helen's or Tagebau Bochwitz
And this is as inevitable as the rising of waters.

And I said
The shape of turned earth becomes an icon
In the cultural landscape
When boundary conditions are made
Clear available and useful.

Thereafter
It was not difficult to see how
The catastrophic event that transformed these lands
Could flip into its reverse
Where the randomness of a chaotic state
Becomes reorganized and transformed.

For
If and when the waters rise in the excavations
Even though variously polluted
And much of the public lands now held
By the mining agency
And presently not farmed
Are permitted to become succession forest and wetlands
With areas of meadow.
And
All waters and wetlands and forests
Are connected to each other.
And the propensity of nature
To rapidly cover disturbed earths
Is not interrupted
Then
a new cultural landscape can come into being
Where
Succession ensembles of birch and poplar and aspen
By every roadside
Can become the seed source
for other disturbed earths.
And
The available fragments of undisturbed earth
Can become the source
For the succession
Of original oak and hornbeam and beech triads
That formed the historic overstory.

And
As available public lands reforest themselves
Beginning their natural work of sequestering carbon
An image emerges of a transformation
That is literally a reversal of ground
Where
The cultural activities
Of existing farming
And existing towns
And industries and infrastructure
Collectively become the figures
In a biodiverse lace like field
Setting the stage for
A renewed eco-cultural landscape.
That enhances the quality of life
For all the communities that are here
or will come here."

Reviewing the final image which suggested a new cultural landscape at scale was possible here, the emerging salient feature was a pattern, lace like in nature, with culture acting as a figure within a biodiverse field.

THE KNOTTED ROPE

But what about hotspot regions? They appear to still exist in middle and eastern Europe although not at the size of the remnants of the great Amazon Basin or parts of North America and Australia. The one which we are most familiar with is a long lived cultural landscape, presently consisting of a flood plain, an oak forest, small villages and farms. The area was once the no-man's land between the Austro-Hungarian and Ottoman empires and so remained relatively untouched for hundred's of years. It still exists in a 280 square kilometer site along the Sava River, about 30 kilometers down river from Zagreb. We were brought there in 1988 by Dr. Hartmut Ern, a botanist who had been the Research Director of the Berlin Botanical Gardens. He had looked at our previous work and thought we might do something that could help to enable the making of this remarkable place into a nature reserve.

The place, surrounded by industrial farming, we perceived as an island that would soon be pressed by the farming methods that surrounded it, a place under attack, as it were, by Industrial Agriculture. To preserve it from simplification and death by isolation, we thought to connect it to all the other small areas of relatively natural growth from wetland to forest to field that existed up and down the river. The pattern suggested making the entire aue of the Sava River a nature reserve. The work was called Atempause für den Save Fluss. We proposed that it was not yet too late to save the region around the Sava River, which although disturbed, was still relatively unstressed. There were a few industries creating serious pollution, but if cleaned up, existing wetlands extended and new ones created, old oxbows restored as temporary reservoirs, off-stream storage used for flood control, with the possible inclusion of the fishponds along the banks, an image of a knotted rope emerged. In it, the flood plain emerged as the rope and all the ecologically provident areas, including sections of old forest, became the knots. Hence, the knotted ropes of the title.

The river itself interested us profoundly. We saw the river as an information processing system that was being asked to process its own degradation, which became a second guiding metaphor. We saw the river being forced to create a new history for itself - and the history was potentially quite negative.

The river is asked to process new information
when it hits the alluvial flood plain
and the information is mechanical.
A new shape has been constructed for the river
by the construction of levees and dams
so that the waters are permitted to rise and fall
but not to spread.
And the topology for a giant farming system is created thereby.
For the river it is the shape of catastrophe.

The river is asked to process new information
when it hits the alluvial floodplain
and the information is biological.
A change of state has been constructed for the river
by the disappearance of the life that once pervaded it
which depended upon the periodic spread
and withdrawal of waters
and although an act of compensation has been made
through the creation of a nature reserve -
for the river it is the state of catastrophe.

The river is asked to process new information
when it hits the alluvial floodplain
and the information is chemical
and the information is toxic
and
where the information is most toxic
by an unexpected congruence of circumstances
and an unexpected confluence of waters
there is an intersection with the nature reserve.

The pattern of nature reserve, forest, fish pond and meadow dotting the Sava aue, held together by land and water was attractive to many, particularly in the Croatian Government. Even the World Bank got interested in cleaning up some areas of the river. But then came the war.

THE RING

As artists, we have long been disturbed that many of the attempts made by ourselves and others, to enhance and preserve biodiversity and/or cultural diversity appear to run into a stone wall. The wall appears to be made up of systems of belief - of belief in an all powerful system of economics called the "free market" and of belief in the individual's lack of empowerment in the face of effects of this international corporate "free market". Overwhelmed by the scale and ubiquitous nature of some of its effects, good ideas are proposed, yet few of them are enacted.

Put simply, we have come to believe that, through their actions and literature, international corporate cultures announce that they perceive themselves as having intrinsic value. Conversely, the culture of place, the ecology of place or even the economic uniqueness of place is often treated as having, in the main, little or no intrinsic value. Thus, in our present context of a world dominated by an economically driven corporate culture, anything any group, any existing culture, any eco-system, any historically determined farming system, any and all long term relatively stable cultural or ecological entities, may become the object of exploitation. Indeed, with rare exception, any corporate entity is vulnerable to the predation of any other. There are exceptions, of course, and the economic value of being good corporate citizens is spreading. None the less, the exploitation, being for the most part economically driven, can devalue or lead to the termination of the entity exploited. Are we looking at a runaway feedback loop?

We believe that new and complex eco-cultural forms at scale can be brought into being that can help to reverse this trend, at least at the local level, and lead to re-empowerment. These forms need to be both stable and sustainable. We believe we may have found one such form, we have tentatively named it an eco-cultural *stability*.

Many comments at this recent biodiversity conference highlighted these problems in the context of the difficulties of maintaining bio-diversity in the face of growing marketplace activity and increasing population. They feel stymied by the problem that arises when scientists' informed opinions are ignored by political processes. This precisely underlines part of the issue. The distress expressed by our colleagues in the biological sciences resonates with the distress expressed by our colleagues in almost all other disciplines at the various universities we have recently been in contact with. Aside from the occasional monetary value of a new idea, or of works of art, anything but mass market culture and new technologies has been virtually marginalized.

Opposition to the primacy of the marketplace, in regard to the economic effects of the Euro, is of course, coming to pass. For instance, first there were the highly publicized workers' strikes in Germany and France. Now there are the results of the elections in Britain and France. As of this writing, many Frenchmen still see themselves as living in a cultural island, needing defense against the forces of the international, or more particularly the European common market, which they perceive as attacking their farming methods and gustatory preferences on one side, and their historic intellectual and linguistic styles and preferences on the other. The consumption of 95% of the North American Pacific Coast Rain Forest showed how benefit flowed from the region in massive disproportion to benefit returned. It is the same with forests of the Amazon Basin. These are other poignant and immediate examples. The ongoing reductions of the social contract in both the governmental and industrial safety nets in most of the western cultures in order to meet the standards of the common monetary system is still another.

In 1994, we were asked to address one stressed system, in practice and on the ground - the imminent probable destruction of a stable long term cultural landscape. The invitation came from the Cultural Council of South Holland. The request was to give definition to the great Green Heart of Holland and make evident its value, so that it might be preserved. The Green Heart, essentially the 900 to 1,200 square kilometer farming area and open space, dotted with small villages, around which the cities of Amsterdam, Rotterdam, Haarlem, Den Haag, Dordrecht, Utrecht, et. al., are located. This ring of cities is called the Randstad. The problem the area faced was economic. And the problem the area faced was ecological. And the problem the area faced was cultural. And the problem the area faced was development.

The issues were quite straightforward. Demographic studies indicated that 600,000 plus houses with relevant infrastructures would be needed in the area over the next fifteen years or so. We did the addition. The need for 600,000 houses and relevant support systems would cause an economic engine or force of perhaps 120 billion dollars to come into being. There was despair among many planners and politicians and large sections of the educated public, if, as was then expected, market forces unmediated determined how these moneys were to be spent. It seemed obvious that the developers would opt for the cheapest land which offered the greatest profit and this land would obviously be the Green Heart. This meant fragmenting the green center. The "green lungs" of the heavily populated cities would be further stressed with increased traffic and pollution. The historic Dutch dike/sloot/farm/small village life-style would be threatened as massive development moneys flowed in and then as much of the profit flowed out, out of the country into international global development coffers. Moreover, many feared a new city and new infrastructure with diverse suburbs would be built in the green center known as the Green Heart, to compete with the remaining cities of the Randstad, to the benefit of few and the disadvantage of many. Urban sprawl would become rampant.

This was the basis of the fear among many planners and politicians that Holland faced the risk of loss of its overall quality of life. Of course, there were others who saw great gain in developing the Green Heart and Los Angelizing the Randstad. This was roughly the state of the discourse when we engaged it.

Confronting this, in early 1995, we invented an array of propositions and forms that we call "cultural icons at scale" and expressed them in a work we called A Vision For the Green Heart of Holland. The installation consisted physically of two 4 m^2 maps of Holland, one printed in reverse, five 2 $meter^2$ drawn images, seven texts, a 7 meter in diameter aerial photograph printed on tile on the floor so that all could see their homes in relationship to the concepts presented, an eight minute video element and a slide presentation. The maps were all reworked. The work of art was produced as an exhibition in the town of Gouda, which opened to a discourse in the town hall by politicians. It was also translated into catalog form with text both in English and Dutch that discussed the issues and it appeared again as a large poster containing text and icons distributed to over 3,000 planners and politicians in Holland. An extensive public debate over the issues involved followed in the various media. (See the Dutch section of the original catalog as translated into English at the back) After a period of time most of the concepts embedded within "The Green Heart Vision" have worked their way into the public planning process.

"The Green Heart Vision" a set of word/image relationships, expresses the following elements.

1. The Green Heart is defined by the construction of a Biodiversity Ring as an external border approximately 140 kilometers long and about 2 kilometers wide.
2. The principal city boundaries at the perimeter of the Green Heart: Rotterdam, Delft, Den Haag, Haarlem, Amsterdam, Utrecht, et. al., collectively known as the Randstad, are separated, defined by long, ribbon like, public parks, approximately two kilometers wide and averaging 12 kilometers long.
3. All boundaries have clear existence and explicit uses. Additionally, all boundaries co-join, becoming a new eco-social, eco-political, eco-economic, eco-aesthetic form that we initially called A Cultural Icon. (A sun sign emerging from the map of lowland Holland).
4. Both the Green Heart and the Randstad are expressed imagistically and poetically as one place, a reciprocating figure-ground relationship, a yin-yang totality. In contradistinction to normal mapping operations, the biodiversity argument is wedded to structure here.

The icon made immediately apprehendable the fact that there was more than sufficient open land for development between the ring cities of the Randstad, without needing to infringe on the Green Heart. Neither was there a need to infringe on the vast proposed new park system in order to construct the 600,000 houses said to be necessary. (People had been sold the belief, manifestly untrue as the work demonstrates, that the Green Heart was the last possible available land for development.)

However, this new vision, could only be made possible on the ground if an external force, such as the government, were to legislate extending the present infrastructure of the ring cities to serve the new construction. And, conversely, if this external force were to legislate that only minimum growth would be permitted in the newly defined Green Heart, and that tax money lost would be replaced by some kind of government subsidy. (This, we have been informed, has since been agreed upon.)

Thus, the market forces would be given a clear sign that the lands available and the rules for use have been biased in favor of the perimeter; concomitantly, the Green Heart becomes unavailable for massive development. And, as should be clear, the amount of money spent would be approximately the same, although the infrastructure would be constructed differently. And, insofar as the money spent would be spread throughout the great circle of the perimeter, more small and medium sized businesses could make this money and, therefore, more money would be fed back into the local economic communities and less would flow directly into the global market. Thus the local communities are given the opportunity to build as slowly or as quickly as they wish in response to the rate of population growth. Finally, part of the work of the sustainability icons is to keep the parts from destroying the whole in a sort of auto-cannibalism as well as to keep outside forces from exploiting the whole to its own disadvantage.

There are five beliefs that guided the conceptual development of this work:

1. Ecosystems require temporal continuity and topographical contiguity in space and time to survive and develop the complexity to continue and evolve their potentialities. These needs can vary from a puddle to a 100,000 square kilometers.
2. Most cultural groups and entities that have developed identities over time require framing by rather knowable boundaries in space and/or mind and a feeling of constancy to continue and evolve.
3. Many planning systems, political systems and economic systems as they have developed in contemporary western culture fail to respect the needs of ecosystems for these continuities and fragment them.
4. Many planning systems, political systems and economic systems fail to sufficiently respect the need of cultures for framing or clarified boundary conditions, often making them continuous or letting them become continuous by permitting urban sprawl.
5. The observations made in 3 and 4 are not necessary conditions.

ON A STABILITY

We tentatively call this form we have invented a stability. This has been an idea nascent in our own works for many years. It appears to resonate, at least in our minds, with Prigogine's notion of dissipative structures far from stability. The lens that helped us envision this was the discourse on a new synthesis of these ideas presented by Fritjof Capra in his book, The Web of Life.

In discourse with the linguist and cognitive scientist, George Lakoff, we came up with the notion of a stability.

Reviewing Green Heart Vision we envision a stability as an emergent form in the cultural landscape. "It may require the continual (ongoing) embodiment of its systems pattern of organization." This concept is reflected in the Green Heart Vision, which is a mapping of the pattern within which the various Green Heart icons are representations of the initial elements that we suggest may become self-sustaining and self-regenerating feedback loops.

The following conditions apply:

1. A stability will only emerge when resources are available to meet a collectively perceived crisis in the environment. In fact, such a region formation may generate new resources or may make unrecognized resources visible.
2. A stability will only develop when the culture or cultures involved, however different,are sufficiently in tune or in agreement that well-being, economically, environmentally or culturally, is endangered and it is clear something needs to be done.
3. A stability will only develop when a coherent simply stated vision is put on the table of what it is, why it needs to be, and the consequences of it not happening. The vision itself must have certain properties

   A. It must emerge from and be supported by a culturally credible source.
   B. The values expressed in the vision must be perceived as advantageous on many levels and the perceived disadvantages manageable.
   C. It must be free of self interest and able to undergo public scrutiny and comment and be modified within the context of its structure.
   D. It cannot emerge exclusively from the internal operation of the culture itself.
   E. It must clearly address issues at the scale they exist and the proposed transformation must be visible, (i.e. cognizable) to or by all who so desire.
   F. The scale must be large enough to tolerate ambiguity, stress and dissonance without negating the pattern.
   G. It must have the creativity or new invention visible.
   H. It must be only minimally critical in nature.
   I. It requires a situation that is exhibition like, permitting space and time for a non-linear experience of the iconic array to take place, with a linear text reading as a second response.
   J. The presentation form, the iconic array, must be designed to bring rapid understanding of extremely complex issues. This permits a community of interest to form within a time frame as expressed in the "Urgency of the Moment".

On the Urgency of the Moment

Looking at the map of Holland.
Seeing it as the expression of a moment
in 1,200 years of contested history
about who will command the land
and why and how.
Seeing it as a metaphor
for yet another contest
as to who will shape the future
of this physical terrain
understood to be the Randstad
and the Green Heart.

Where in a ten year moment
less than one percent of the time
of its whole history as a civilization
the people on this ground
must construct a response
in physical terms
to intense population pressure
coupled to an expansion committed
economic engine in such a way
that these two self-reproducing forces
mutually energizing and interrelated
will consume
much of these lands available
in the Green Heart
which do not have specific ecological
or historical or other civic designation
and unless or until
a new direction is set in place
an alternative consensus agreed upon
by governmental and economic
and civic institutions
on limiting growth.

For
in the absence of
such an alternate consensus
clearly expressed on the ground
the outcome for the Green Heart
the Randstad
and the lowlands of Holland
appears to be
unfortunate in the extreme
mostly unnecessary
but mostly inevitable.

These very preliminary notions are based on hindsight, observations and analysis of the Green Heart Vision, which itself is based on ideas implicit in earlier works. The Green Heart Vision is explicitly referred to, as it appears in the main to be in the process of happening. As a result of this analysis, we have come to the tentative conclusion that the creation of a stability is a viable concept that in diverse contexts can be seen as wanting to happen. The potential properties of a stability are, in an eco-cultural context, the protection of large scale local environments and cultures, while at the same time directing, or in some way controlling the direction of market forces and growth. We suspect that, should an array of such stability regions come into being, they would automatically affect the market, moving it towards respecting cultural diversity, economic diversity and ecological continuities.

We use the dissipative structure observation and foundation of the work of Ilya Prigogine metaphorically, but not literally, i.e., a metaphor evokes identification - one is the other, although never completely the other. None-the-less, the Green Heart icon brought to mind the concept embedded in Prigogine's model of a whirlpool. As long as the energies of a certain type flowed through the system, acting as feedback loops, the system had constancy in space and time, even though the form was far from equilibrium. We first used his example as a metaphorical lens to see if there were physical and metaphysical congruences between his work and our own. After all, the Green Heart iconic array is a singular form that, should it be enacted on the ground, would be held in stability by the forces of law, economics, culture and belief, as well as by environmental elements. It is a form born of high social, economic, environmental and political stresses. It is a form that can fall back into environmental norms if the forces holding it in stability are removed or fade away, that is to say it has the appearance of a stable form far from equilibrium.

For us, the presentation of these ideas requires a compressed language structure tuned to a many layered iconic image. Therefore, the visual compression of the icon and the information compression of the form "story-metaphor-parable" afforded by the poetic process together act as vision carriers. While the information in each is richly connotative and layered, placed in each other's presence icon and text form an experience that is richer in meaning and affect than either separately. The whole is more than the sum of its parts and the experience offered by the whole operates in space and time as in an exhibition where all images can be scanned almost simultaneously and are grasped in a non-linear, all at once experience and thereafter sequentially and by grouping and regrouping.

We suggest that a "sustainability icon", if understood by many and enacted on the ground, makes manifest a process of continual co-creation, co-evolution as it were. For us the question is, "Can a visionary work of art, The Green Heart Icon, be also thought of as an emergent self-organizing pattern initially generated as a survival response by a well-defined social organism, Holland itself, to a perceived extreme threat? Further, can it set the terms for retrofitting biodiversity into a rather heavily stressed cultural landscape and thereafter can such biodiversity co-evolve, protect and be protected by the diverse cultural elements with which it shares the terrain?"

Finally, an element rarely considered, is the need to develop a new aesthetic based on relinquishing control and accepting and valuing the chance arrangements produced by the processes of succession. As this essay makes manifest, we see both cultural diversity and biodiversity as having intrinsic value. We see both forms of diversity as the yin/yang of a new cultural landscape well past the middle of its formation. We have come to believe that the quality of survival in the largest sense has become a matter of precisely how the co-evolution of these forms of diversity proceeds.

NOTES

1. We tentatively define a stability as a new eco-cultural pattern emerging in a region that could help to resolve or reverse the loss of cultural diversity and biological diversity. A stability depends upon creation of complex forms that establish new boundary conditions and economic and ecological opportunities. While the region is in the process of compensating for that which may have been lost, it also must contain sufficient feedback loops to protect it from future loss of coherence under political and economic pressures.
2. We define icon in its original and complex sense. For instance, a carefully constructed religious icon, one of St. Francis, showing him wearing the dress of the clergy and in relationship to animals, has an aesthetic of its own, an explicit narrative, referential to the bible. Moreover, a far more complex narrative is referred to within which the biblical narrative is nested. The whole icon carries within it codes of behavior which are suggestive as to how one might conduct a life. For the St. Francis icon, a written narrative beyond the title is not necessary. It is presumed sufficiently well known. A sustainability icon has many of these same qualities compressed within it; however, a brief text must be supplied as the narrative cannot be presumed to be known.

   The eight principal sustainability icons in the Green Heart Vision establish the limits to growth on the ground, but within these limits all elements can remain dynamic. Exactly how and how much can only be touched on in these brief comments. Sustainability in The Green Heart Vision is the outcome of enacting the stability on the ground. Its form is expressed in the eight icons. The Green Heart Sustainability Icons repeat the same specific shape but with vastly different content. The space is flexible, having only the requirement of continuity of surface area.
3. There is an ongoing discussion as to the minimum space needed for eco-systemic survival, which differs vastly for primary producers and predators, but since even the very simplest of food chains require both, it is the space for the survival of predators that must be the determinant.

Teaching sustainable management of renewable resources

Dennis L. Meadows, Thomas S. Fiddaman
Institute for Policy and Social Sciences Research, University of New Hampshire, Durham, USA

1 Introduction

Since 1983 we have experimented with a variety of gaming formats in our effort to produce a powerful game-based teaching technology that will give senior public officials insights and principles, or heuristics, that will help them deal more effectively with a set of basic problems observed in resource and environmental systems of many countries. Our goal is to address the so-called universal problems, those that manifest themselves to some extent in most regions of the world and can be addressed locally. Examples would be soil erosion, inflation, deforestation, AIDS, and urban decay. These may be contrasted with the so-called global problems which are also observed in most regions of the world but which require concerted, international action for their solution. Marine pollution, the arms race, acid rain, and ozone depletion are examples.

In this paper we will explain the goals and methods of our application of simulation gaming to these problems. To illustrate our approach, we will describe the function, mechanics of play, and theoretical foundations of a simple game that teaches principles for sustainable management of renewable resources.

2 Gaming

Games may be designed to fulfil a variety of goals: creating a friendly and co-operative attitude among the participants, enhancing the players' communication or problem-solving skills, providing a set of metaphors and a meta language related to a group of phenomena represented by the game, providing pleasure, assessing the effectiveness of alternative decision strategies, or instilling a set of principles about the causes and consequences of basic phenomena observed in a real system. These and other objectives are described usefully by Greenblat & Duke (1981). The styles or modes of games used to achieve any of these goals may differ drastically – person-machine interactions, pure role-playing simulations, board games, physical exercises, and others. However, teaching principles for management of a real system generally requires the game to use a fairly sophisticated model of the underlying system. This model may be embedded entirely in the rules of the game, or be partially represented by a computer simulation. There are several advantages to using a computer in a complex game. A computer can increase the realism and richness of information in the game, while at the same time freeing the players from performing tedious manual calculations. It is important that the computer be an unobtrusive part of the game, however, as the players' interpersonal relations are important to the processes of building co-operation in the group and of learning about the system.

3 Approach

Our approach has been to identify a set of cause-effect mechanisms that lie at the heart of a specific universal problem; these constitute a dynamic framework of the system represented by the game. Then we develop a game that conveys that set of causal interactions to an individual in a way that makes them relevant to many different regional settings and to a class of dynamically related problems. Rules structure the actions and relations of the players. The computer model represents a subset of the system and handles any accounting and scoring necessary. We create paraphernalia, such as game boards, to represent elements of the real system in order to allow visual thinking about the system and to facilitate communication and negotiation. We develop the underlying structure of the game system and the accompanying computer model using the methods of a systems analysis technique called system dynamics (SD). SD is a comprehensive approach to the representation, diagnosis, and change of behaviour patterns in complex, dynamic systems. The SD method is based on concepts of information feedback, and it employs computer simulation of feedback models representing real world issues. The SD technique has roots extending back many decades into fields as diverse as operations research, social psychology, economics, and computer science. But the present philosophy and tools of the SD method originated in the 1960's with Professor J. W. Forrester and his associates at MIT in Cambridge, Massachusetts. It is now used throughout the world (Forrester 1961). Normally SD models are closed; their behaviour derives entirely from a set of rules embedded in computer equations. But in order to make a game one must open the model to influence from the players in each cycle of the game. Our fisheries game illustrates this approach.

4 Fish Banks, Ltd.

Fish Banks, Ltd. is a microcomputer-assisted simulation that teaches principles for the sustainable management of renewable resources. It was developed to serve as the introduction to a five-day workshop that conveys the fundamental insights required for corporate managers, public officials, and private citizens to use their regions' natural resources intensively without deteriorating the long-term productivity of their resource endowment. The game also can be operated very effectively without the other workshop materials. Nevertheless, the game is influenced strongly by our goals for the workshop. That training session is intended principally for reproduction and wide-spread use by relatively unsophisticated instructors in Third World training centres. As a consequence, the game was specifically designed to fulfil eight requirements: - The lessons instilled by the game should be interesting and relevant to participants from a wide variety of different regions that may be at different levels of technical, political, and economic development.

- The game should be suitable for use by groups of any size between 8 and 40 players
- The game's computer program should be easily converted for use on any of the widely-available micro computers
- A complete set of materials for the game should cost less than $50 to produce, preferably much less
- One operator should be able simultaneously to run the computer and supervise the play of all participants in one game
- The materials that are destroyed, lost, or used up in the course of a game should be easily and cheaply replaced with photocopies
- It should be possible to start the play of the game without an extended introduction
- A computer should be used to make it unnecessary for participants to make tedious calculations

But the machine should be unobtrusive – definitely the computer should not intrude on the social interactions that are at the heart of the game. This last criterion was established for two reasons. First, an important goal of the game is to introduce the participants to each other; thus most of their time should be spent on interpersonal discussions, negotiations, and debates and not on person-machine dialogues and data entry. Second, most participants in our workshops have total ignorance, even apprehension, about computers. Therefore, the workshop materials must be designed so that participants can take part in early sessions of the workshop without being forced to learn how to operate any computer. One important goal of our resource management workshop is to give its participants some facility in the use of computers, and this is addressed later in the week. The opening sessions nevertheless must be designed to respond to the fact that computer-based systems and exercises are intimidating to many. Through two years of testing and refining the game we achieved all eight goals for Fish Banks, Ltd.

4.1 Set-up

Introducing, playing, and debriefing the exercise requires about three hours, with minimal set-up time. The room used should be large enough to allow all the players to move freely about. The principal materials required include the following items:

- Introduction and Debriefing Transparencies
- One copy of Role Description and Opening Scenario per player
- One Decision Sheet for each company
- Money
- Ships (100 units total in ships of three different colours: white = 1, red = 5, gold = 10)
- Game Board (2' x 2')
- Fish Banks, Ltd. program disk and any required system disks, tested in advance
- Overhead Projector
- Computer and printer with appropriate operator's manual

With the exception of the computer and overhead projector, which are readily available in a normal school or conference facility, all materials are included in the game kit. Expendable materials may be reproduced by photocopying from the game manual.

4.2 Briefing

Players are divided up into teams of two to five members, each of which will operate its own fishing company in a common ocean. Each player is provided with a five-page role description which enumerates the team's goals and resources and describes the state of the fisheries system. A few samples of the text of a role description follow: Congratulations! You have just been hired to manage one of the principal fishing companies in your country. Together with the others in your company - captain and crew members - you will operate your fishing fleet each year according to policies you design to accumulate the greatest assets. The rules and information required for your success are provided below. Your bank balance is increased by income from fish and ship sales, and decreased by expenditures for ship purchases and operation, Additionally, your account is subject to interest earnings and charges. Your total assets equal the sum of your bank balance plus the salvage value of your ships at the end of the game (250 per ship). You may change the size of your fleet by buying ships at auction, negotiating to buy or sell ships from another company, and ordering new ships from the shipyard. Two fishing areas are available to you: a large deep-sea fishery (#1), and a smaller coastal fishery (#2). Biologists have estimated that area #1 could potentially support between 2,000 and 4,000 fish, while area #2 could support between 1,000 and 2,000 fish. The fish population is increased by natural births; it is decreased by natural deaths and by harvesting. The fertility of the fish and their lifetime are both influenced by the density of fish. Your total fish catch is influenced by the number of ships you send to sea, the fishing area chosen, and ship effectiveness. The information in the role descriptions is supplemented by a brief introduction provided by the game operator. A set of overhead transparencies is provided for this function.

4.3 Play

The players proceed through six to eight one-year rounds of the game. In each round, there is an auction, in which the players have the opportunity to acquire new boats, and a trading session, in which players may buy or sell boats to or from other teams. The players then allocate their ships among the two fishing areas and the harbour. Their decisions are recorded on a sheet provided for that purpose and passed to the operator. The operator enters the decisions into the computer, which simulates the outcome of the year's activities. It calculates the fish harvest based on the allocation of ships and determines the regeneration of the fish population. The computer then generates for each team a report of their year's catch, income, and expenditures, and their current bank balance and fleet size. In a typical game the players will realize quickly that a large fleet can earn more money than a small one. They will then compete to expand their fleets far beyond the level that can be sustained by the two fishing areas. The following is a typical team strategy: purchase many boats until average fish productivity starts to decline. When fish productivity goes down, fish other areas. After a few more rounds, the fish population will be severely depleted, and the teams' catches will decline quite suddenly. This "crash" of the fisheries is the most powerful part of the game experience. After it has occurred, the game is halted for a discussion of its outcome.

4.4 Debriefing

To realize the full potential of Fish Banks as a learning tool, it is essential to discuss the game thoroughly at the end of play. The debriefing allows the players to vent their emotions from the game and understand its outcome. More importantly, it relates the elements of the game system and its behaviour to real fishery systems.

The collapse of this anchovy fishery, once the world's largest, has been costly. Peru lost two export commodities – the fish meal and the guano from sea birds that depend on anchovy – that once dominated its foreign-exchange earnings. When this fishery was at its peak in 1970, exports of its products earned Peru $340 million, roughly a third of its foreign exchange. The disappearance of this vitally needed source of hard currency contributed to the growth of Peru's external debt; in the mid- eighties over 40 percent of the nation's exports are required merely to service its outstanding loans. And the world has lost a major protein supplement, once used in rations of hogs and poultry. (Brown 1985)

The simple interactions shown in the causal loop diagram (Fig. 1) that is the basis of the game model provide a conceptual framework for understanding the behaviour of fisheries and other renewable resources, such as groundwater, timber, soil, and ozone.

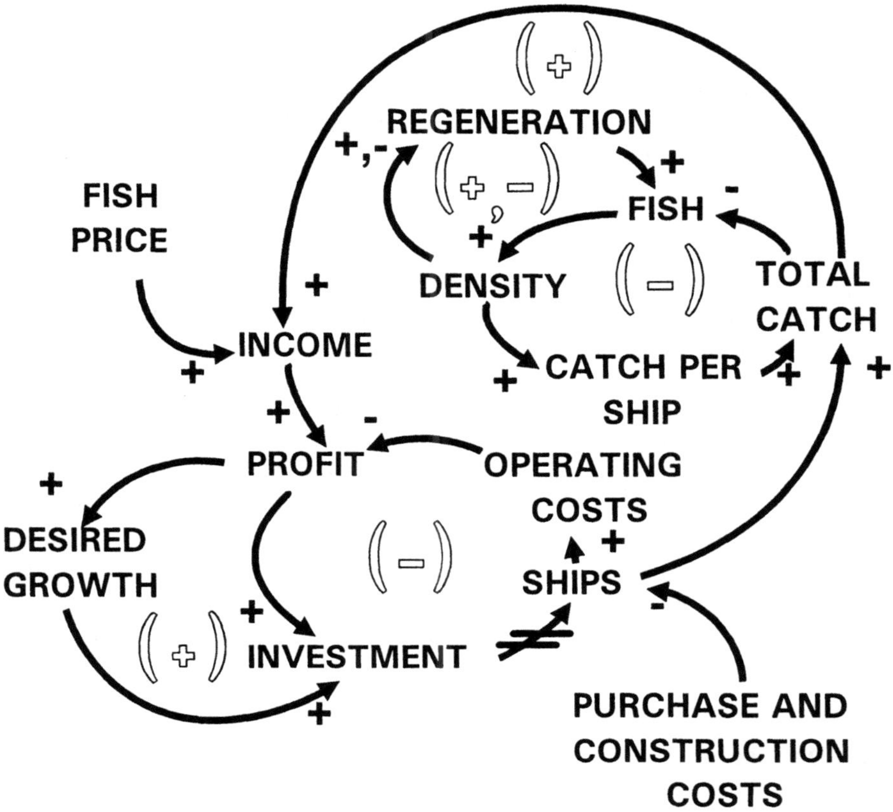

Fig. 1. The Fisheries System Structure

Finally, the game provides a metaphor for discussing solutions to the problem of resource depletion. Because they have directly experienced the collapse of a resource, entirely due to their own behaviour, the players feel a greater sense of responsibility for finding solutions to the real-world problems.

4.5 The Fisheries System

The basic theory of fish banks overexploitation used in the game is obviously not original to the authors of Fish Banks, Ltd. SD models of interactions between ship and fish populations were developed already in the early days of the field, in the late 1960's. These system dynamics models were based on even earlier work by resource economists and mathematical ecologists. The causal structure of the model is shown in Fig. 1. It is crucial to note, when examining the figure, that the structure portrayed here has an extremely wide applicability. If "Trees" were substituted for "Fish" and "Sawmills" were substituted for "Ships," the model would provide the foundation for explaining the widely-observed phenomenon of forest overcutting. Similarly, substituting "Groundwater Level" and "Irrigation Pumps" would give one a start on a good model that explains depletion of groundwater resources. Any structure that has this level of generality is called a generic structure. System dynamics provides a set of theoretical and graphical tools for identifying and analyzing such structures.

This approach has one important strength. Although generic structures can be developed in infinite numbers and varieties to represent causes of behaviour within elements of any system – economic, psychological, biological, political, industrial – the essential elements of any generic structure are very few in number. They consist of individual causal links, delays, and the feedback loops (closed chains of causal links) that govern the behaviour of the whole system. In this game there are five causal loops (Fig. 1).

This causal loop diagram shows the principal interactions governing the fisheries system. Arrows show influence or causality; a "+" indicates that the entities change in the same direction; a "-" indicates that the entities change in opposition. Delayed influences are indicated by "I I" on the arrow. Closed loops result in behaviour which may be self-reinforcing (a positive loop) or self-correcting (a negative loop).

The first (Fish – Density – Regeneration – Fish) governs the regeneration in the fish population. When the fish population is large, it acts as a negative feedback loop, to maintain the fish population within the carrying capacity of the fishery. When the fish population is smaller, it acts as a positive feedback loop, producing exponential growth of the fish population. Acting by itself, this loop would result in asymptotic growth until there was a stable fish population near the carrying capacity of the fishery.

The second loop (Total Catch – Income – Profit – Investment – Ships – Total Catch) is a positive loop, which produces exponential growth of the ship fleet. As the total catch of fish increases, income and profits increase, reinvestment of profits increases the ship fleet, and the total catch increases even more. This loop is accelerated by a third loop – Total Catch – Income – Profit – Desired Growth – Investment – Ships – Total Catch) which results from the players' competition. Because the profit from a season of fishing is nearly equal to the cost of a boat, the delay in constructing a new boat is short, and the players typically reinvest most of their profits, the ship fleet grows very rapidly. Though the initial boat fleet is less than half the sustainable limit, the fleet typically doubles three times in five rounds of the game.

A fourth loop (Total Catch – Fish – Density – Catch per Ship – Total Catch) acts to reduce the harvest as the fish population becomes depleted. As the total catch exceeds the regeneration of fish, the fish population declines. As the population declines, the density of fish in the bank decreases, and ships are less effective at catching fish. This reduces the total catch.

Finally, as the catch declines, the fifth loop (Ships – Operating Costs – Profit – Investment) becomes active. This is a negative loop which acts to restrain growth of the fish fleet when profits decrease. As the catch per ship declines, operating costs remain constant, and the income and profit from fishing are reduced.

Reinvestment in ships quickly drops to zero, and the fleet size stabilizes. However, by the time this occurs, the fish population is already severely depleted.

5 Conclusion

For the past twenty years I have taught formal courses on system dynamics for groups ranging from high school students to senior corporate and public officials. I have used the traditional forms of lectures, created one person - one machine simulations, and assigned pen and paper homework exercises to convey understanding about the causes of phenomena such as overfishing. None of these approaches has generated as many insights for my students as the exercise of working through this simple game.

6 Availability

A kit for operating Fish Banks, Ltd. may be obtained for $100, which includes the cost of shipping from:

Prof. Dennis Meadows, Director
Institute for Policy and Social Science Research
Hood House, University of New Hampshire
Durham, NH 03824, USA.

The kit includes a 113 page teacher's manual, players' materials, small wooden ships, game board, money, and program disks. In the package are also 48 masters for creating overhead projection slides; with these you can quickly learn to introduce and debrief the game. Please send cash or check and indicate whether you want a Macintosh or a 5 1/4" IBN-PC compatible disk.

References

Forrester SW (1961) Industrial Dynamics. MIT Press, Cambridge, MA
Greenblat, C, Duke R (1981) Principles and Practices of Gaming Simulation. Sage Publications, Beverly Hills, CA

Part 2

Biodiversity: Basis facts and trends

Marine biodiversity - Thoughts on the subjects and their investigators

Gerhard Haszprunar
National Zoological Collection, München, Germany

Abstract. A comparison of terrestrial and marine ecosystems in terms of their basic ecosystems and biodiversity characteristics (e.g., energy flow, trophic levels, species composition and knowledge) is presented to outline specific problems involved in the inference of marine biodiversity. Although the magnitude and percentage of unknown species is in same range as in terrestrial systems, the number of systematists for marine taxa is substantially smaller, mainly because there are (with the sole exception of conchology) no associated skilled amateurs. In addition, the systematic treatment of marine species generally requires more time and more expensive technical facilities on average than the taxonomic treatment of terrestrial taxa. Efforts must be made to increase the number of systematists for marine taxa at universities and museums and to improve their education, as well as to improve the policies of scientists themselves.

Biodiversidad marina - Reflexiones sobre los temas y sus investigadores

Resumen. Para perfilar los problemas específicos que plantea la biodiversidad marina, se presenta una comparación de ecosistemas básicos y de las características de la biodiversidad (p. ej. flujo energético, niveles tróficos, composición de las especies y conocimiento) encontradas en ecosistemas terrestres y marinos. Aunque la magnitud y el porcentaje de especies desconocidas vienen a ser los mismos que en el sistema terrestre, el número de taxónomos para los taxa marina es mucho menor debido, sobre todo, a la falta (con la única excepción de la conquiliología) de competentes aficionados asociados. Además, el tratamiento sistemático de las especies marinas suele requerir por lo general más tiempo y un material técnico más caro que el tratamiento taxonómico de los taxa terrestre. Hay que intentar aumentar el número de taxónomos para los taxa marina en universidades y museos y enriquecer su formación, así como mejorar las políticas de los propios científicos.

1 Introduction

Although oceans and estuaries comprise our planet's largest ecosystems, the level of knowledge on marine biodiversity is still much lower than comparable data on freshwater or terrestrial systems. Although it is a compilation of contributions from no less than 57 authors, it may be significant that the landmark edition by

Wilson & Peter (1988) "Biodiversity" does not include any significant example from the marine environment. The same is true of more recent volumes such as that edited by Gaston (1996). Today any literature search based on the keyword "biodiversity" reveals that the overwhelming majority of papers deal with tropical rain forests or freshwater systems, whereas articles on marine ecosystems and their faunal or floral biodiversity comprise less than 5% of all papers (e.g. Grassle 1991; Grassle *et al.* 1991; Angel 1993; Newmann & Cannon 1994; Maragos *et al.* 1995; Gosliner & Draheim 1996; Roy *et al.* 1996; see also the recent review "Understanding Marine Biodiversity" 1995).

The following thoughts on marine biodiversity and its investigators attempt to find an explanation for this phenomenon. After a brief survey of the similarities and differences between marine and terrestrial ecosystems, I will concentrate on the differences between the characteristics of terrestrial and marine fauna and the problems involved in investigating them. Ideas are then presented on how to increase the amount of research on marine biodiversity and how to increase our understanding of the marine environment in the course of doing so.

The reader should be aware that this comparison is being undertaken by a marine systematist, which also is an endangered species. This might consequently lead to some (in some cases, intentional) subjectivity in this paper's arguments.

2 Ecosystem characteristics of marine systems

Every textbook on marine ecology naturally outlines important general characteristics of marine ecosystems. However, my experience indicates that such textbooks are rarely read by terrestrial ecologists or limnologists. In light of this situation, the following paragraphs will outline some of the important differences between terrestrial and marine ecosystems (Table 1).

The characterization and naming of terrestrial ecosystems is generally based on vegetation type. Indeed, no terrestrial ecosystem has ever been named after its characteristic fauna. In contrast, most marine systems are named after their topographic (e.g. pelagial, marine caves, rocky shore, hydrothermal vents) or faunal (e.g. coral reef) characteristics, although there are certain exceptions such as the "Sargasso Sea" or "seagrass meadows." Compared to terrestrial ecosystems, marine ecosystems generally have a much lower level of geodiversity, i.e. the change and variability of abiotic factors (e.g. day-night, summer-winter), although there are exceptions such as the intertidal habitats.

The primary producers of the marine environment, mainly unicellular algae, are at least in part very small and often mobile. More significantly, the size range of the smallest (nanoplanktic: <1 μm) to the largest (brown algae: >100 m) primary producers extends over more than eight classes of magnitude (note that the respective surfaces square and the volumes cube these differences). In particular, primary holoplanktic producers are extremely sparse when compared with the situation in terrestrial ecosystems.

Table 1. Characteristics of ecosystems

	Marine Ecosystems	Terrestrial cosystems
Characterized by	mostly by faunal types (except e.g. "Sargasso-Sea")	always by vegetation type
Geodiversity	with exceptions (intertidal) generally high	generally low
Primary producers	small, often mobile size range >1:10.000.000 highly diluted	large, sessile size range 1:1.000 highly concentrated
Grazers, carnivores	complicated life-histories high reproductive output	complicated life-histories low reproductive output
Symbioses	mainly concern primary production	mainly concern consumption
Trophic interaction	very high between habitats	low between habitats
Biodiversity interaction	regularly over large distances: danger of extinction low	rarely over large distances: danger of extinction high

It should be noted that the marine environment alone shows ecosystems which are based on non-solar energy. "Hydrothermal vents" or "cold seeps" as well as anaerobic mud-beds are generally based on the primary production of various chemoautotrophic eu- or archaebacteria which use several anorganic energy and proton-sources (e.g. hydrogen sulfide, methane).

Recent investigations have shown that these systems are much more common than previously thought and although still poorly studied they comprise a significant part of diversity in primary production and consumption modes.

Grazers and carnivores very often show complicated life-histories in both terrestrial (holometabolic insects) and marine (various groups, see below) species. The latter are also very often characterized by a much higher reproductive output than terrestrial species. This is particularly true for large invertebrates such as oysters, large polychaetes or sea urchins, where an individual egg-release may exceed one million. Several fish species are also in this range. Terrestrial species have obviously been more heavily selected towards a reproductive k-strategy, probably by the fact that sperm cells in particular cannot directly survive in a non-aquatic environment.

Furthermore it is significant that symbioses of marine animals usually concern primary production (e.g. endosymbiotic bacteria or zooxanthellae), whereas symbioses of terrestrial species mostly concern consumption (e.g. bacterial symbionts in xylophagous insects or in ruminants) or the uptake of nutrients (e.g. mycorhizas).

In general there is high trophic interaction between the various marine ecosystems; the ecologist characterizes this bioenergetic condition as 'open'. Although pure consumer systems are extremely rare in the terrestrial environment, several types of pure consumer systems exist in the oceans of the world: the Watt, meso- and endopsammon or deep-water habitats (except those based on chemoautotrophic microbes, see above) are generally well-known examples.

Of course, this high degree of trophic interaction between marine ecosystems increases the possibility of negative external impacts and certain anthropogenic impacts in particular. For example, recent findings have revealed a particular fauna being associated with large whale cadavers. It is likely that whale cadavers serve, in several cases at least, as temporary islands for the survival and interaction of hydrothermal vent faunal elements. Thus, the dramatic anthropogenic decrease (or even extinction) of various holopelagic whale species will probably cause the extinction of a particular marine ecosystem with its characteristic fauna.

In marine ecosystems, biodiversity interaction occurs regularly over large distances. For example, the planktotrophic larvae of many gastropod species cross the Atlantic with the help of the Gulf stream in every generation. The same is true of other invertebrates of various phyla. Accordingly, very local endemism is generally rarer among marine species than among terrestrial species, which reduces the danger of the former becoming extinct.

The large amount of trophic and biodiversity interaction between various marine ecosystems and biogeographic areas also leads to more interaction and cooperation between scientists. Studying more or less "isolated" marine systems or species in a particular area is becoming less and less possible or feasible. This has dramatic implications for science policy (see below).

3 Characteristics of faunas

3.1 Species numbers

Current estimations of species numbers in terrestrial ecosystems concentrate on tropical rain forests and run in the range of 5 million to 50 million species, the majority being insects. Recent studies (e.g. Simon 1995; Basset *et al.* 1996) indicate that 5 million species is the more likely number. The situation is worse in regard to the marine environment: To my knowledge there are no serious estimates of true species numbers on the basis of any significant studies. The recent survey "Understanding Marine Biodiversity" by the U.S. National Research Council (1995) lists some examples and small pilot studies which are reproduced here (Table 2).

Among these examples I regard the cases of opisthobranch sea slugs and polyclad flatworms as particularly instructive: Both groups are at least partly composed of so-called "macrofaunal" (usually larger than 5 mm) species, most of which are particularly colourful and thus quite easy to detect. If the absolute and relative numbers of unknown species is as high as documented in small samples absolute number of marine species should be in the same range (5 million – 50 million) as estimated for the terrestrial environment.

Table 2. Examples of magnitude of undescribed marine biodiversity (modified from "Understanding Marine Biodivesity")

Site	Taxon	Undescribed species	Source
Gulf of Mexico (shelf site, 18 m)	Copepoda - Harpacticida	≈25 of ≈30	D. Thistle
New Guinea (one lagoon)	Gastropoda - Opisthobranchia	336 of 646	Gosliner & Draheim (1996)
Philippines (one island, multiple sites)	Gastropoda - Opisthobranchia	208 of 563	Gosliner & Draheim (1996)
Georges Bank (shallow shelf)	Annelida - Polychaeta	124 of 372	J. Blake
Hawai (6 liters of sand, one island)	Annelida - Polychaeta	112 of 158	Dutch (1988)
Great Barrier Reef (two islands)	Turbellaria - Polycladida	123 of 134	Newmann & Cannon (1994)

However, in light of the fact that less than 3,000 new species a year are described by all scientists worldwide, it is in my view rather unimportant whether the actual number of marine species is 5 million or 50 million (see also below).

3.2 Genetic diversity

Three aspects should be considered when comparing the genetic diversity of marine fauna with that of all terrestrial fauna: number and age of primary clades, substitution rates of DNA-nucleotides and frequency of speciation events.

According to the most recent paleontological and molecular findings, the origins of nearly all primary marine clades (phyla) date back to the Vendian to Early Cambrian periods (800-550 million years). Primary clades of terrestrial fauna date back to the Silurian to Devonian periods (450-400 million years) and are thus substantially younger. In addition, zoologists list 13 to 18 unique marine phyla (depending on the author), and marine taxa are the most basal groups of the remaining phyla. At the subphylum level only Stylommatophora (pulmonate land gastropods), the extant Arachnida (mainly spiders and mites), Myriapoda and Insecta, and the amniote Vertebrata (reptiles, birds, and mammals) are higher taxa with primary terrestrial radiation.

During the last ten years the "molecular clock hypothesis" (i.e. a more or less constant substitution rate of DNA-nucleotides in all clades) has been proven false. "Fast clock clades" as well as genetic "living fossils" frequently occur at all levels of organismic hierarchy and are clearly independent of the main habitat (marine, freshwater, terrestrial) of the representatives of these clades. On average (albeit with extremely high divergence), the substitution rates of marine clades do not significantly differ from those of terrestrial clades.

Table 3. Characteristics of faunal research

	Marine Ecosystems	Terrestrial Ecosystems
Species numbers	5-20.000.000 (estimation)	5-20.000.000 (estimation
Origins of primary clades	Vendian to Cambrian (800-550 Mio. years)	Silurian to Devonian (450-400 Mio. years)
Unique phyla	13-18 (depends on author)	none (3.5 subphyla)
Speciation	slow: in 1 - 5.000.000 years (average)	fast: in 10 - 50.000 years (average)
Species discrimination	mainly by internal characters, very different between taxa	mainly by external characters, often similar between taxa
Species desription	basically high technical equipment; much time per species necessary	basically low technical equipment; less time per species necessary
Amateur scientists	play no significant role (except conchology)	play a significant role (except very small animals)
Scientist: Biodiversity	extremely poor	magnitudes more, but still poor

Although data is highly divergent, speciation events among terrestrial species usually occur with much greater frequency – within much shorter time frames – than among marine clades. There are several causes for this phenomenon: (1) Biological causes include the comparatively reduced mobility of terrestrial animals (e.g. snails) or local restriction of their habitat (e.g. caves, islands) or of their resources (e.g. monophagous pterygote insects). (2) Isolation effects are much more common in a terrestrial environment than in marine ecosystems. In fact, local endemism as a phenomenon is extremely rare (but present) in marine species. (3) In addition, terrestrial habitats are generally less constant. Glaciation (e.g. in the European Alps) and desertification (e.g. in the Amazonian rain forest) in particular caused high numbers of speciation events.

4 Characteristics of marine biodiversity research

The marine and terrestrial environments differ significantly not only terms of their ecosystem and taxa characteristics, but also in the collection, preservation and investigation of their various faunas and species by scientists.

4.1 Collection of species

Whereas collecting marine species without the help of a boat or ship is rather limited, it is nevertheless possible to collect marine animals in their natural environment equipped with only a snorkel and flippers. "Low tide collecting" in areas with high tides is possible (albeit dangerous) even for non-swimmers. On the other hand, many marine habitats require more technical equipment for collecting animals. In any event – and even in "low-budget habitats" (intertidal, eulitoral) – the problems for the scientist start immediately after the specimens have been collected.

4.2 Preservation of species

Although nearly all terrestrial taxa have an external or internal skeleton, this is rarely true for marine species. Consequently, dry collections are restricted to conchology and a few other exceptions (e.g. stone corals, bryozoans, partly echinoderms). As a rule, marine species are preserved in formalin or alcohol. Apart from the problem of relaxing the animals, wet (alcohol) preservation involves (1) the high cost of glassware, (2) the problem of safe storage and regular control to avoid dessication and destruction, and (3) the loss of colour in the specimens. All this makes collecting marine animals problematic and unattractive for amateurs. It is therefore significant that marine prosobranch gastropods and shelled bivalves comprise more than 50% of all known marine species. In both cases, the soft bodies of less than 5% of the described species are known; all other species are known solely on the basis of their hard parts (shells).

4.3 Discrimination of species

Relatively inexpensive equipment – a stereo-microscope – is all that is needed to determine most terrestrial fauna. There are of course exceptions, such as the case of cryptic or sibling species. In contrast, much more sophisticated and expensive equipment is usually necessary for even basic determinations of marine species (again with the exception of conchology). This is because species of most taxa (e.g. nearly all "worms") are determined primarily by anatomical characteristics rather than by external morphology alone.

Simple dissection is seldom sufficient for studying the anatomy of marine species. In most cases, it is necessary to conduct anatomical reconstructions based on serial histological sections or interference-contrast studies of whole mounts. This accordingly requires highly technical equipment. In addition, it takes an average of ten to fifty times longer to determine a species by means of anatomical reconstruction than by external morphology.

Another factor in determination is the much higher basic diversity of marine taxa (see above). Whereas the basic handling and determination methods are quite similar for terrestrial taxa (90% insects), handling and determination methodology differs drastically among marine taxa (phyla). This poses a much greater threat to the continued tradition of expert taxonomic knowledge of marine animals should the number of scientists in this field continue to fall as it has in recent decades.

4.4 Significance of amateurs

In contrast to terrestrial taxa and with regard to entomology in particular, amateur scientists do not play a significant role in marine taxonomy (see above regarding the exception in conchology). The reasons for this have been already mentioned above: lack of expensive, highly technical equipment, wet collections with poor preservation of colours, and the large amount of time required for each determination and description of a marine species. It is significant that amateurs are also absent in terrestrial groups (e.g. microarthropods, oligochaetes) where these factors are also present.

As a consequence, the relation between the number of skilled scientists, be they amateurs or professionals, and the magnitude of work to be done is considerably worse in the field of marine animals than terrestrial species.

5 Consequences – what should be done and what should we do

These circumstances demand action at all levels. The following considerations attempt to assign responsibility not only to science policy but to scientists as well.

5.1 Science policy

Although systematic botany is well represented in European universities, systematic zoology as a discipline is largely restricted to the "reservations," as museums and research collections may be called. The latter institutions are (at least in Central Europe) generally poorly equipped for high-tech studies. Although there has been some progress in recent years, scanning and transmission electron microscopy, confocal laser scanning microscopy, enzyme electrophoresis, chromosome banding, DNA-hybridization and many molecular sequencing methods are still the exception in our museums. It is urgently necessary that the basic technical equipment of the professional systematic zoologist be upgraded: Research must become the partner of pure conservation.

The documentation of biodiversity, even when it makes use of computerized databases, is currently not regarded as "original research" (e.g. by the German Science Foundation) and consequently does not receive funding. In regard to the documentation of marine biodiversity with its even greater technical requirements and its lack of amateur input, this situation constitutes a crucial deficit and practically hinders modern biodiversity research.

As outlined above, the declining number of expert taxonomists poses the threat that traditional scientific knowledge could be lost forever. Universities in particular should train students in the various fields of taxonomic methodologies – but Germany currently does not have one single Graduiertenkolleg for biosystematics (we have just applied for the first one in Munich). Of course, any activity in this field should be conducted in close cooperation with a local museum or scientific collection. Because of the high diversity of marine systematic methodologies, such courses should be offered in a way that allows students from other universities to attend them as well. On the other hand, our governments must take steps to ensure that today's students will at least have a chance of finding a job after finishing their Master's degree or PhD.

The world's governments and national and international science foundations currently spend millions of dollars every year on activities that involve "biodiversity research." However, when it comes to the marine aspect of biodiversity, most projects are more interested in overall nutrient balances, biomasses, and energetics than species composition. It is vitally important that the ecologist be brought together with the systematist (and vice versa, see below).

5.2 Systematists

Although the above mentioned points are prerequisites for progress with marine biodiversity research, individual marine systematists (myself included!) and their respective institutions also have to make some changes in their behaviour and policies. Comparing the worldwide capacity for describing new marine species (less than 3,000 a year) with the estimated total number (5 million to 50 million species), makes it clear that even the attempt to simply catalogue the entire scope of marine biodiversity cannot succeed. We urgently need ideas on how we are to grasp 90% of a particular ecosystem although we are able to describe only 5% of all its estimated species. This is where the systematist has to be brought together with the marine ecologist and oceanographic scientist.

Concerning the internal policies on systematic zoology that institutions – and museums in particular – follow, there is still a dramatic discrepancy between the number of species in a particular group and the number of scientists working in the field of that particular group. This relation is often dramatically better for vertebrates than for marine animals, a field where there are virtually no skilled amateurs.

It also appears that we must increase the capacity of each marine systematist. Most of my colleagues do large amounts of work that could also be done to a large degree by less skilled persons. Wanted: The "parataxonomist" – an expert on local faunas or specific groups – who works with the professional so that the latter can concentrate on subjects that are simply too difficult for the others or require specific technical equipment. This help ranges all the way from sorting centres to undergraduate or graduate students. Regarding the latter, the systematist, who is not member of a university, must also assume some teaching activities in his particular field, even though this may not be one of his duties.

In general the possibilities and advantages offered by worldwide databases are still largely underestimated and inadequately used by phenotypic systematics, at least here in Europe. Contrary to molecular systematists, who save each sequenced piece of DNA or RNA in an international database, taxonomists do not do the same with descriptions of new species or, for instance, synonymy lists. Less than 5% of the specimens in European scientific collections are organized in a computerized database. Even the Zoological Record which is now available on CD-ROM covers only about 60% of all currrent species descriptions. Since marine taxonomy is largely the concern of professionals, there is the chance (and the need) to follow the example set by our molecular colleagues and use these facilities more efficiently than in the past.

6 Conclusions

Many aspects of marine biodiversity and its study differ significantly from conditions pertaining to terrestrial ecosystems. Marine biodiversity and systematic research entail different and usually greater requirements which demand new strategies of science policy (on the part of governments and countries), the various systematic institutions (universities, museums, scientific collections) and – last but not least – systematists themselves.

References

Angel MV (1993) Biodiversity of the pelagic ocean. Conservation Biol 7: 760-772
Basset Y, Samuelson GA, Allison A, Miller SE (1996) How many species of host-specific insects feed on a species of tropical tree? Biol J Linn Soc 59: 201-216
Gaston KJ (ed) (1996) Biodiversity: A Biology of Numbers and Differences. Blackwell Sciences, Cambridge MA
Gosliner TM, Draheim R (1996) Indo-Pacific opisthobranch biogeography: How do we know what we don t know? Amer Malac Bull 12: 37-43
Grassle JF (1991) Deep-sea benthic biodiversity. BioScience 41: 464-469
Grassle JF, Lasserre P, McIntyre AD, Ray GC (1991) Marine biodiversity and ecosystem function. Biol Intern Spec Issue 23
Maragos JE, Peterson MNA, Edredge LG, Bardach JE, Takeuchi HF (eds) (1995) Marine and Coastal Biodiversity in the Tropical Island Pacific Region. Vol. 1: Species Systematics and Information Managment priorities. Honolulu, Hawai
Newmann LJ, Cannon LRG (1994) Biodiversity of tropical polyclad flatworms from the Great Barrier Reef, Australia. Mem Queensl Mus 36: 159-163
Roy K, Jablonski D, Valentine JW (1996) Higher taxa in biodiversity studies: Patterns from eastern Pacific marine molluscs. Phil Trans R Soc Lond B 351: 1605-1631
Simon H-R (1995) Arteninventar des Tierreiches. Wieviele Tierarten kennen wir? Ber Naturwiss Ver Darmstadt NF 17: 103-121
"Understanding Marine Biodiversity" 1995 (Committee on Biological Diversity in Marine Systems). National Academy Press, Washington D.C.
Wilson EO, Peter FM (eds) (1988) Biodiversity. National Academy Press, Washington D.C.

Aspects of botanical biodiversity in southern African arid regions – An outline of concepts and results

Norbert Jürgens
Institute of Botany, University of Köln, Germany

Abstract. In times of rapid economical, social, political and environmental change, research on biodiversity has to encompass a wide range of goals and methods if biological sciences want to adequately describe reality and the factors that determine it. One approach to a more holistic view of biodiversity is presented here with examples from our research in the greater Namib Desert region. The article outlines the concepts and some findings of the various types of biodiversity research conducted by the Cologne research group with the goal of furthering cooperation with similar activities in other arid parts of the world.

Aspectos de la biodiversidad botánica en regiones áridas del Sur de Africa - un destaco de conceptos y resultados

Resumen. En una época en la que se producen rápidos cambios económicos, sociales, políticos y medioambientales, el estudio de la biodiversidad ha de abarcar una amplia gama de objetivos y métodos, si es que se desea que las ciencias biológicas describan adecuadamente la realidad y los factores que la determinan. Aquí se presenta un enfoque hacia una visión más global de la biodiversidad con ejemplos de nuestras investigaciones en la extensa región del Desierto de Namib. El artículo destaca los conceptos y algunos descubrimientos de varios tipos de estudio de la biodiversidad llevados a cabo por el equipo de investigadores de Colonia con el objetivo de promover la cooperación con actividades semejantes en otras partes áridas del mundo.

1 Introduction

Research on biodiversity should be based on the fact that biodiversity is and has always been controlled by ecological and evolutionary processes. Today it is even more imperative that biodiversity research include a broad spectrum of aspects rather than restrict itself to specialized topics:

- Over the course of millions of years, the general nature of the ecological factors that control biodiversity has experienced relatively little change, whereas human activities have led to rapid development during just the past few thousand years, and have become increasingly important during the past few decades. A comparison of these man-made processes with natural processes reveals fundamental differences.

Today we have the difficult situation where, for the very first time in history, the effects of very efficient, deliberate and systematic human activities and the "fuzzy" effects of natural processes overlap, with both contributing to the current situation.

Hence, good research on biodiversity should integrate both process levels into analysis in order to get the complete picture.

- Furthermore, in times of rapid economical, social, political and environmental change research on biodiversity should encompass a wide range of goals and methodological approaches if biological sciences are to contribute adequately to the pressing issues related to current environmental problems and nature conservation needs.

At present however, when plans for applied research are being made, such a holistic approach is often regarded as unfeasible. Consequently, it is important that basic research supply the necessary background.

2 Biodiversity research in the Namib Desert region

The following section describes a network of projects, all of which were established in the greater Namib Desert region of southern Africa and developed around the general themes "patterns," "processes," "temporal changes" and "management and conservation" of biodiversity.

The arid parts of southern Africa are of comparatively little economic interest and, as a consequence, are generally under-researched. Nevertheless, data on the biodiversity of the region is needed for dealing with obvious problems such as degradation processes arising from over-exploitation, desertification during drought years, rapid economic development, changing land use and its impact on existing ecosystems, as well as for dealing with the needs of modern nature conservation management.

Therefore, a number of research projects covering a wide range of aspects of biodiversity in the semiarid to arid parts of southwestern Africa are currently being conducted in close cooperation with various institutions in the Republic of South Africa (RSA) and the Republic of Namibia.

Furthermore, the challenges of global climate change necessitate monitoring the shifting desert boundaries and their biocoenoses. At the same time, basic research on the history of biodiversity can provide important information on the processes of degradation in the past, whether it be determined by climate or caused by human activity.

Finally, a detailed long-term monitoring programme which includes experimental approaches must be established in a variety of different ecosystems and land-use systems for studying the urgent question of how the degradation and restoration processes found in biodiversity function at ecosystem level.

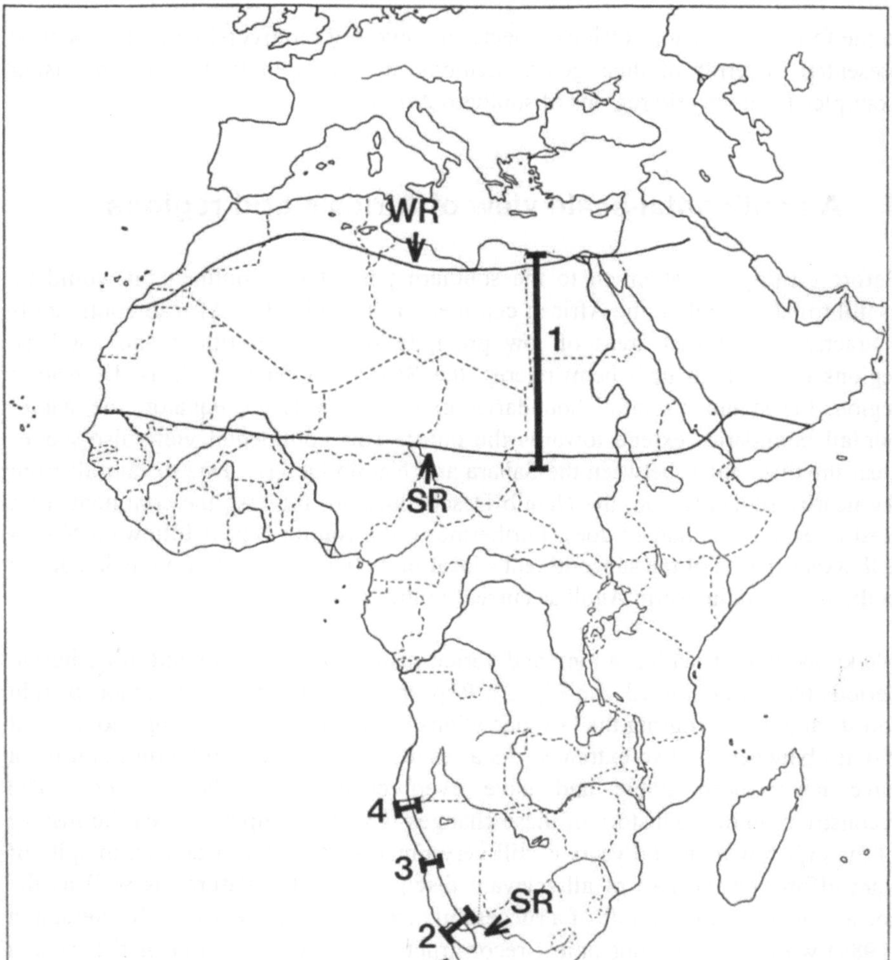

Fig. 1. Location of the subtropical hyperarid regions of the African continent (Sahara and Namib) between the winter and summer rainfall zones of both hemispheres. The lines (1-4) represent the monitoring transects for vegetation changes along the shifting desert boundaries (1 = Trans-Sahara monitoring transect, 2 = Southern Namib transect, 3 = Central Namib transect, 4 = Northern Namib transect)

Ilustración 1. Emplazamiento de las regiones subtropicales hiperáridas del continente africano (Sahara y Namib) entre las zonas de lluvias de invierno y verano de ambos hemisferios. Las líneas 1-4 representan las transecciones de control de los cambios de vegetación a lo largo de los límites cambiantes del desierto (1=transección de control Trans-Sahara, 2=transección del sur del Namib, 3=transección del centro del Namib, 4=transección del norte del Namib)

In the following section, various aspects and levels of biodiversity research will be presented in terms of their goals, methods and preliminary findings by using examples from the arid regions of southern Africa.

3 A continental-scale view of Africa's arid regions

Before turning our attention to the southern part of the continent, it would be useful to take a look at the African continent as a whole. The African continent is characterized by vast areas of low precipitation in its northern and southern regions (Fig. 1). In both hemispheres, the Sahara region as well as the Namib region, the summer rainfall boundaries extend towards the equator; the winter rainfall boundaries extend towards the poles. This continental view also makes clear the differences between the Sahara and Namib Deserts: Due to rainfall from the nearby Indian Ocean, the Namib Desert does not traverse the continent from west to east as the Sahara does. Furthermore, the Namib Desert follows a NNW-SSE axis because of the subcontinent's location between the cold Benguela current in the west and the warm Agulhas current in the east.

We know that Africa has a long and varied history of more arid and more humid periods (see section 6.2.3 and Fig. 2). Repeated expansion and retraction of rain forests and deserts during the last glaciations have been linked to migrations of all biomes between the two extremes. As a result, maps of African biomes and biota have never been stable and have even changed rapidly at times. The reconstruction of the history of these changes is still incomplete. Even the history of the past few thousand years is still very much the subject of debate, in spite of joint efforts on the part of all relevant disciplines in the natural as well as the social sciences. An example of a successful analysis was provided by K. Neumann (1989) with her archaeobotanical reconstruction of the vegetation of the eastern Sahara in a transect from Egypt to Sudan in 7,500 BP which showed that the area's major tropical vegetation belts – and human activity – have shifted more than 500 km southward since then. Archaeological findings (R. Kuper, pers. comm.) support the hypothesis that expanding human activity from the Mediterranean and from the Sahelian side did meet in the middle of the Sahara during the humid phase prior to 8,000 BP. The climate of the eastern Sahara deteriorated and developed its present hyperarid conditions during the past few thousand years. The question of whether future changes will produce more arid or more humid conditions is still open. However, considering the increased dynamics of global climate change, it is obviously of great importance that the changes in abiotic environmental factors and the vegetation of the shifting zones along the desert boundaries be analyzed.

Taking into account practical feasibility, regional political stability and the above-mentioned ecological patterns, the ideal location for continental transect projects for ecological analyses as well as for monitoring the shifting desert margins would be the western borders of Egypt and Sudan and in the area of the

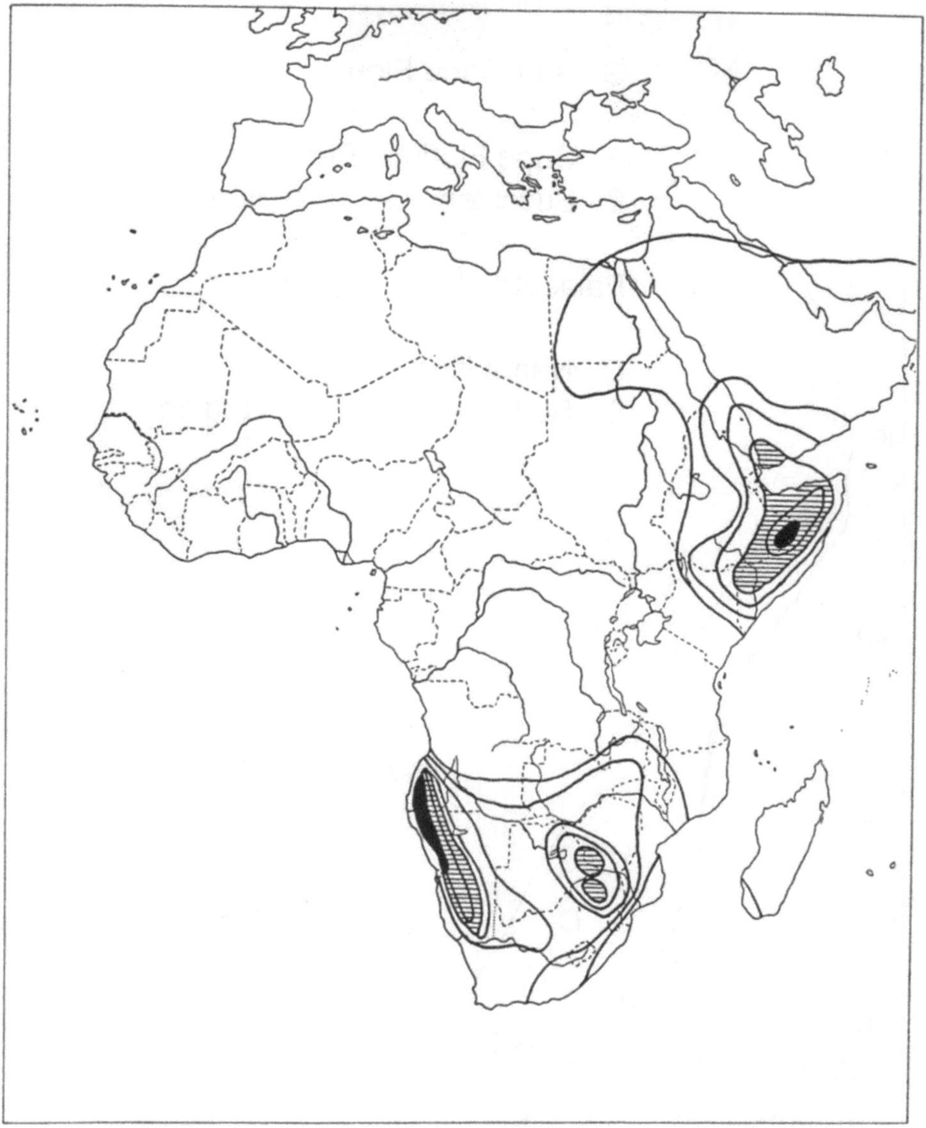

Fig. 2. Phytogeographical centres of similarity of eight species with a continental disjunction between NE- and SW-Africa (Jürgens 1997a)
Ilustración 2. Centros fitogeográficos de semejanza de ocho especies con una disyunción continental entre el noreste y el suroeste de África (Jürgens 1997a)

Southern Namib Desert that includes the Richtersveld and the Karasberge. Both regions include the entire range of typical gradients of desert boundaries encompassing both the winter rainfall climate and the summer rainfall climate (Fig. 1). An additional monitoring transect in the Central Namib enables an analysis of the role of the temperate and humid conditions in the coastal Namib,

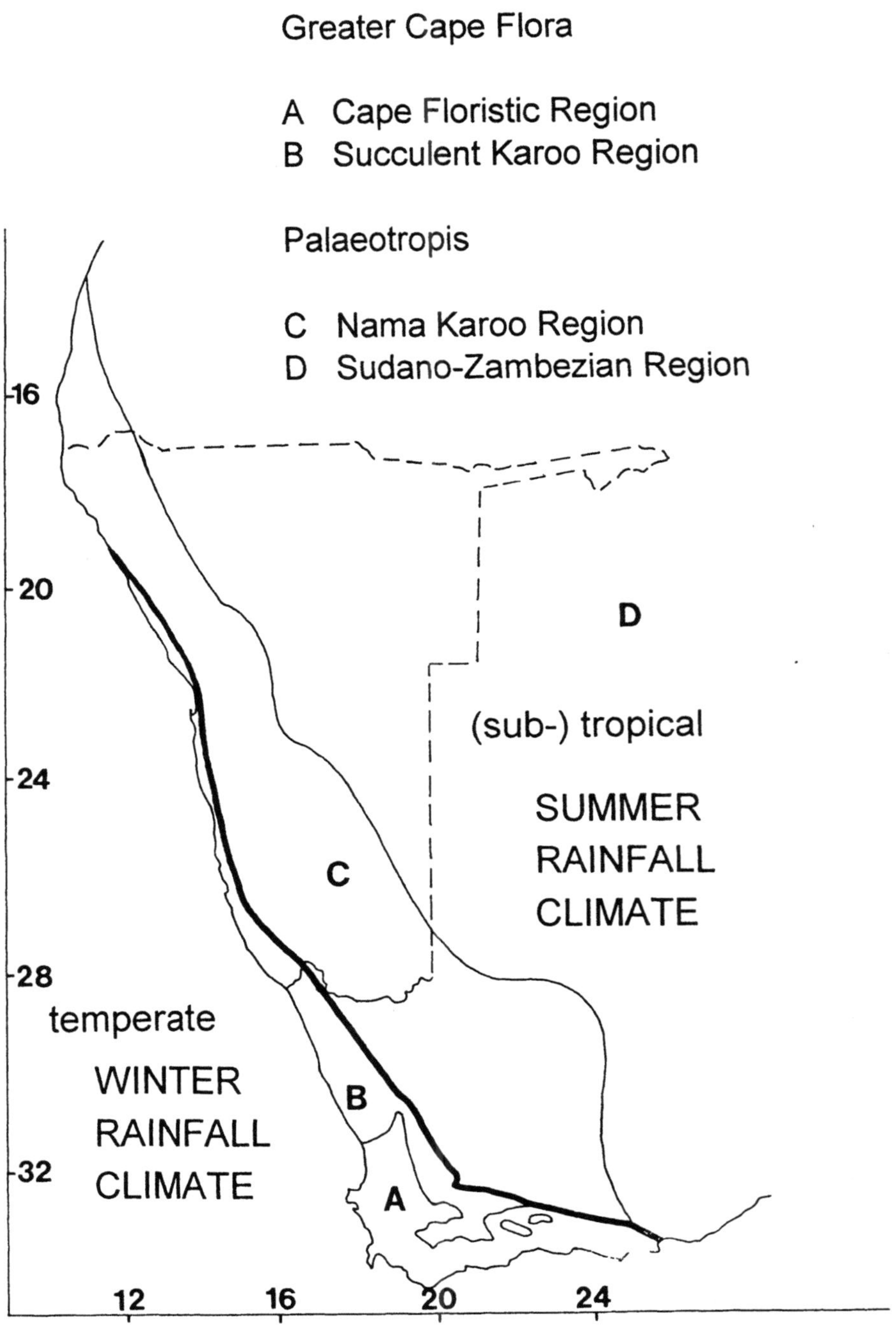

Fig. 3. Climatic and phytochorological subdivision of the arid regions of southern Africa. For details see text (section 1.2)

Ilustración 3. Subdivisión climática y fitocorológica de las regiones áridas del África austral. Para más detalles al respecto ver el texto (sección 1.2)

which also compensates for the diminished importance of winter rains in the more tropical climate of the northern parts of the Namib.

4 Introduction to the greater Namib Desert region

The shifting of the boundary between summer rainfall zone and winter rainfall zone in southern Africa during and after the last glacial period with its maximum around 18,000 BP was of extraordinary importance. Today the boundary (Fig. 3; bold-face line) between winter rainfall region and summer rainfall region runs from southwestern Namibia through northeastern Namaqualand to the Southern Cape region (Jürgens 1991, Jürgens *et al.* 1997). This line separates a temperate oceanic climate with high air humidity and cyclonic winter rains in the southwest from hotter subtropical inland climate with convective summer rainfall (thunderstorms) in the northeast.

This climatic boundary is also the main ecological boundary, regulating the patterns of flora and vegetation of the arid regions of southern Africa. In the southwestern part of this most important phytochorological discontinuity, leaf succulent chamaephytes, which belong mostly to the Aizoaceae family, are dominant in the flora of the Succulent Karoo Floristic Region (B on the map):

Low-growing chamaephytic plant formations are typical of the Succulent Karoo. The Succulent Karoo Floristic Region is an arid sister region of the (more humid) Cape Floristic Region, the traditional Cape Flora (A on the map).

Both units together form the Greater Cape Flora which has evolved (at species level) over the past 5 million years due to the development of the area's temperate winter rainfall climate.

Northeast of the subcontinental floristic discontinuity, tall stem-succulent shrubs, non-succulent shrubs, and grasses comprise the flora of the Nama Karoo Region (C on the map), an arid phytochorological region of the Palaeotropical Plant Kingdom.

These extraordinarily clear differences between the two arid floras – that of the temperate Succulent Karoo Region and of the tropical Nama Karoo Region – allow a very concise analysis of patterns of biodiversity along their boundary.

5 Spatial analysis

5.1 Current floristic biodiversity

Obviously, any approach is scale-dependent. The analysis of all of southern Africa requires a broad database of national herbaria. As a consequence the National Botanical Institute of South Africa and the University of Cape Town in particular have been working on this task (e.g. Cowling & Hilton-Taylor 1994, Cowling *et al.* 1997, Gibbs-Russell 1987, Hilton-Taylor 1987, 1996). With respect to the arid regions of southern Africa, the patterns of biodiversity have been analysed on subcontinental level (Jürgens 1991, 1997), regional level (Hilton-Taylor 1987, 1996; Robinson 1976, Williamson 1997) and local level. The very different degrees of floristic richness of the Succulent Karoo (Greater Cape Flora) and the Nama Karoo (Palaeotropis) is very obvious when one-square-km areas with roughly the same habitat diversity are compared (examples are shown in Table 1). The high biodiversity in the example provided by the Succulent Karoo calls for an explanation.

Table 1. Diversity on 1 km² (note the high level of diversity in the Succulent Karoo)

mm annual rainfall	383 mm	190 mm	68 mm (!)
Number of species	303	71	276 (!)
Genera	251	60	151
Families	45	31	51
Rainfall seasonality	summer rainfall	summer rainfall	winter rainfall
Locality	Ottukaru, Central Namibia	Gibeon,Namaland, Namibia	Numees, Richtersveld,RSA
Flora	Palaeotropis	Palaeotropis	GreaterCape
Floristic region	Sudano-Zambezian	Nama Karoo	Succulent Karoo
Source	Hoerner 1996	Jacobs 1996	Jürgens 1986

5.2 Analysis of the factors regulating spatial variability

Analyses of biodiversity can be linked to environmental data. For example, Fig. 4 (Jähnig 1993) shows the regular soil catena of the Namib Desert in a sequence from the coast (left hand side) to 90 km inland (right hand side). A gradient of soil properties, shown here by the composition of soluble ions, can be observed in this sequence. Salt crusts are predominant close to the coast and are replaced by gypsum crusts and calcretes further inland. This catena is controlled by a coast-to-inland climatic zonation, which is characterized by decreasing amounts of summer rainfall from inland toward the coast, and an increasing amount of fog precipitation close to the coast. Owing to these two sources of humidity, biodiversity in this area (Fig. 5, Hachfeld 1996) is at a minimum between the fog zone and the summer rainfall zone. The ordination of relevés by correspondance analysis (Fig. 6, Hachfeld 1996, Canoco, DCA) reveals that the distance to the coast which correlates with high incidence of fog and low temperatures is the most important determinant (first axis of similarity). Further inland, where summer rainfall levels are higher, additional gradients are caused by different soil types, which produces a second axis of similarity. In contrast, a predominantly soil-controlled vegetation landscape is presented with a correspondance analysis of vegetation relevés from Namaland (Gimborn 1997), that are arranged according to their growth form composition and show the soil properties of each relevé (Fig. 7).

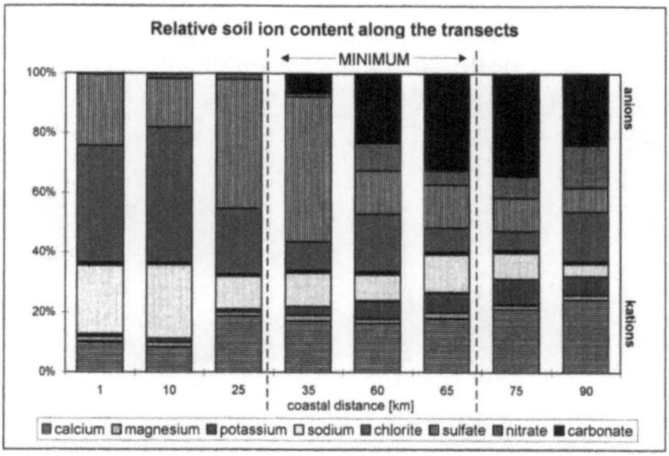

Fig. 4. Topsoil composition of soluble ions along a transect from the coast of the Atlantic Ocean through the Central Namib Desert between Swakopmund and Hentjiesbay: Notice the gradient in anion concentration that ranges from Cl^- to $SO4^{--}$ to $CO3^-$ from coast to inland. This gradient in chemical soil ion composition is linked to climatic zones. The fogs along the coast, the summer rainfall inland and the hyperarid conditions in between these two regions are all of great importance (Jähnig 1993)

Ilustración 4. Composición de la capa superior del suelo de iones solubles a lo largo de una transección desde la costa del Océano Atlántico a través del Desierto de Namib Central entre Swakopmund y Hentjiesbay: Obsérvese el gradiente en concentración de aniones que abarca desde Cl^- hasta $SO4^{--}$ hasta $CO3^-$ desde la costa hasta el interior. Este gradiente indicador de una composición química de iones terrestres está vinculado a zonas climáticas. La niebla a lo largo de la costa, el interior con lluvias de verano y las condiciones hiperáridas entre estas dos regiones son todos aspectos de gran importancia (Jähnig 1993)

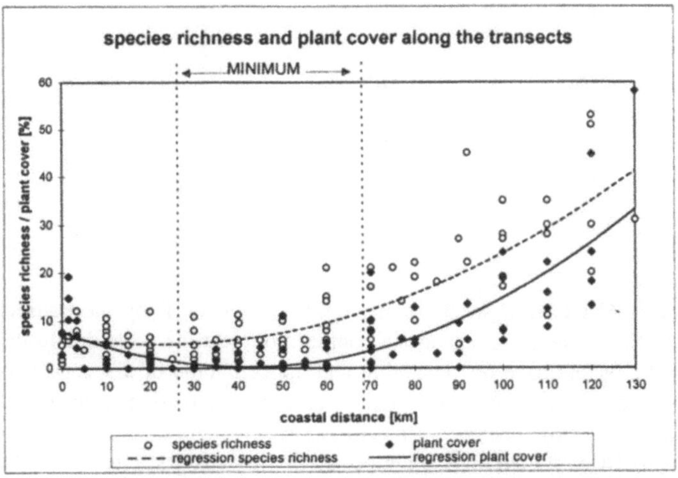

Fig. 5. Biodiversity along the transect shown in Fig. 4: A minimum of species diversity between 30 and 60 km from the coast (Hachfeld 1996)

Ilustración 5. Biodiversidad a lo largo del transecto mostrado en la ilustración 4: Un mínimo de diversidad de especies entre 30 y 60 kilómetros de la costa (Hachfeld 1996)

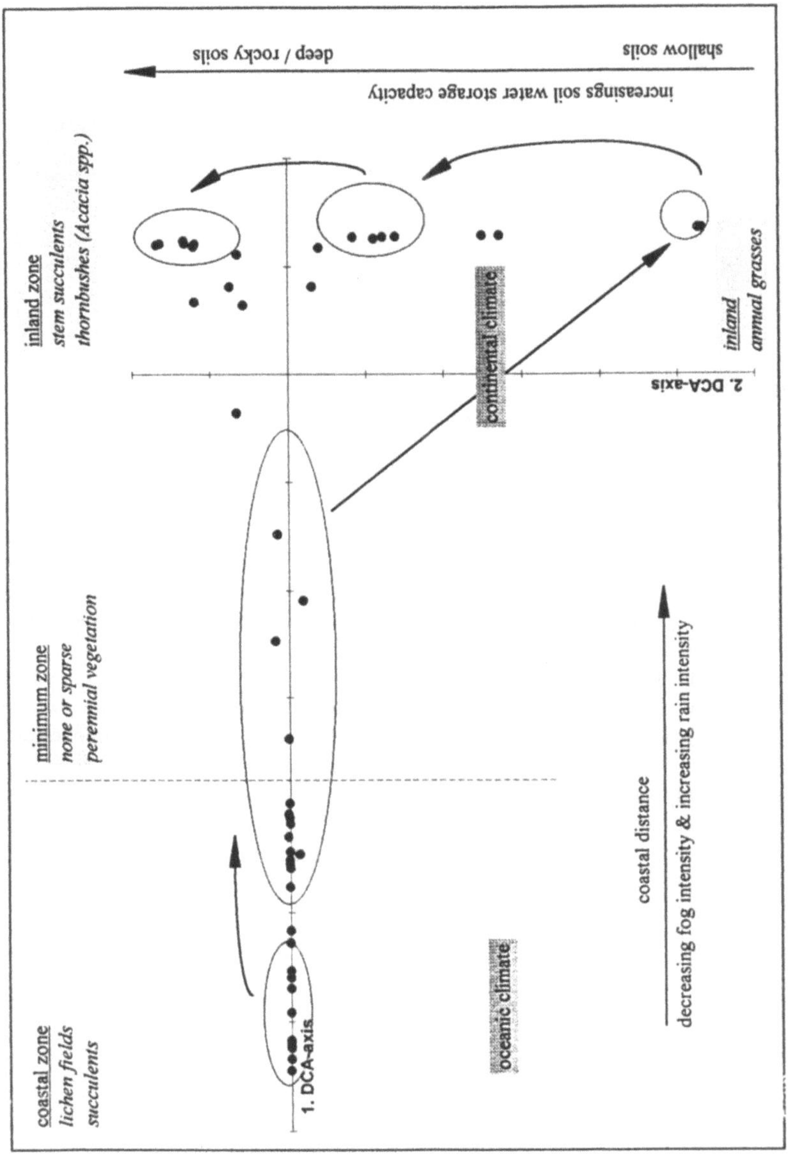

Fig. 6. DCA ordination (CANOCO) of species composition of relevés along the transect shown in Figs. 4 and 5: Notice the linear change in composition near the coast due to the impact that zonal climatic factors have on vegetation and the formation of a second axis further inland, comprised of different vegetation units due to the increasing influence of different soil types (Hachfeld 1996)

Ilustración 6. Orden de DCA (CANOCO) de especies de los relevés a lo largo del transecto mostrado en las ilustraciones 4 y 5: Nótese el cambio lineal de composición cerca de la costa debido al impacto que factores climáticos por zonas tienen sobre la vegetación y la formación de un segundo eje más en el interior compuesto por diferentes unidades de vegetación debido a la influencia creciente de diferentes tipos de suelo (Hachfeld 1996)

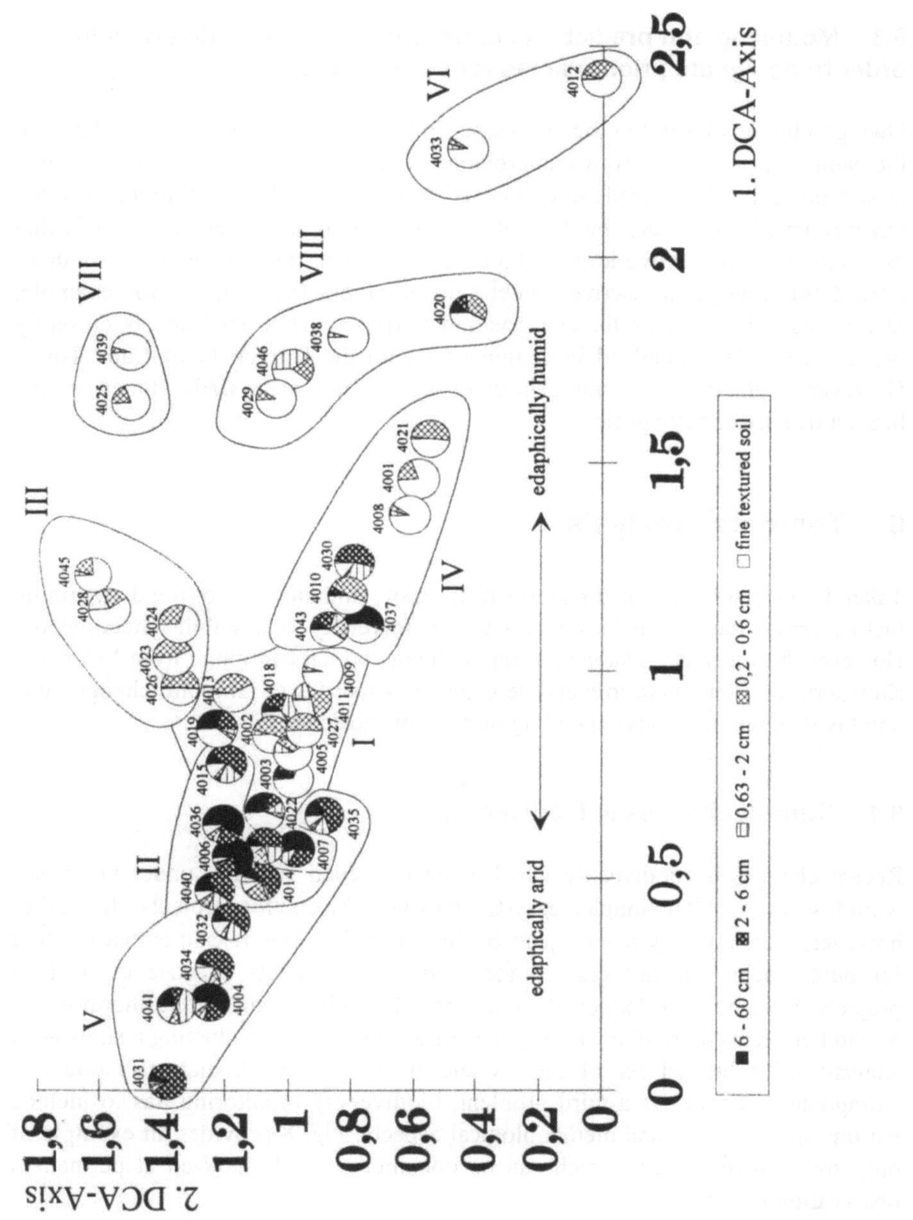

Fig. 7. DCA ordination (CANOCO) of the life-form composition of vegetation in a transect in Namaland (Namibia, compare Table 1). Individual circles represent fractions of soil particle size at relevés. In this landscape, soil factors clearly determine the composition of vegetation with respect to life forms as well as species (Gimborn 1997)

Ilustración 7. Orden de DCA (CANOCO) de la composición de formas de vida de la vegetación en un transecto en Namaland (Namibia, comparar con el cuadro 1). Los círculos individuales representan fracciones de tamaños de partículas del suelo de los relevés. En este paisaje, los factores de suelo determinan claramente la composición de la vegetación con respecto tanto a las formas de vida como a las especies (Gimborn 1997)

5.3 Modelling and predicting spatial variability of biodiversity in order to designate priority areas for conservation

Owing to the regularities of the landscape, soil properties and climatic gradients in the Namib region, the sum of numerous local and regional analyses as presented in sections 5.1 and 5.2 leads to a general knowledge of the correlations between environmental factors and biodiversity. These enable us to develop models that can predict the respective level of biodiversity. Given the vast areas of southern Africa, this type of predictive model can be of practical use in, for example, identifying priority areas for conservation purposes. This approach is currently being used in Namaqualand in conjunction with the University of Cape Town. However, alpha-diversity and beta-diversity are not necessarily linked to the highest density of rare species.

6 Temporal analysis

Taken by themselves, spatial patterns of diversity and analyses of the determining factors create the wrong impression that biodiversity is a stabile phenomenon. However, biodiversity is subject to rapid change which can range from reversible short-term oscillations to irreversible changes. Analyses of temporal change must employ different methods depending on the time scale used.

6.1 Current changes in biodiversity

Recent changes in biodiversity which have happened virtually under our noses would seem to offer simple, easy-to-analyze subjects for projects. In reality however, there are very few projects dealing with this. One reason is that funding for pure research is not granted for monitoring projects because even short projects have to run longer than a normal funding period. Furthermore, a researcher seems more likely to reap scientific rewards by conducting a short-term analysis of a limited set of factors and then using mathematical models to extrapolate his data. As a third problem, biodiversity monitoring has to include various spatial scales and methodological aspects. Fig. 8 provides an example of only some of the details which can be documented and analyzed at permanent observation relevés.

6.1.1 Regional monitoring and analysis of shifting desert margins at landscape level

Regional-scale monitoring is very important for the analysis of how global climate change affects arid and semiarid zonal habitats along and outside desert margins. In light of the evidence we have of rich Holocene vegetation in many now hyperarid and plantless desert regions, the predictive models of global climate change make a continental monitoring of the shifting desert margins of the "dry continent" Africa very important. For this reason, a number of partial projects involving the vegetation ecology and biodiversity of African arid regions have been linked in a network. For this reason, vegetation is currently being monitored along the shifting desert margins of one transect in the Sahara and three transects in the Namib Desert as already mentioned in Section 3 and shown in Fig. 1. With regard to method, this monitoring programme will be based on a remote sensing and numerous permanent observation relevés on the ground that are linked to automated weather stations. This approach will provide far better information than the annual ups and downs reported by remote sensing observations of the ephemeral effects of rare rainfall events because the use of monitoring relevés allows us to analyze the development of the important stabilizing elements in ecosystems (nurse plants, seed banks, soil properties). In addition, the arid plant species provide indicator information on the climatic conditions during their life time. This information is urgently required by developing global models of past climate conditions (PAGES-INQUA Palaeomonsoon Workshop in Siwa, Egypt, 11-22 January 1997).

6.1.2 Local monitoring, modelling and prediction of population dynamics

Even better data on the processes of vegetation rehabilitation or degradation near shifting desert margins can be derived from intensive monitoring of permanent observation relevés that are linked to automated weather stations. The documentation of live histories of all plants conducted at the individual plant level allows detailed analysis, modelling and prediction of population dynamics and their responses to changing climatic factors and soil properties (Fig. 8). It also produces more detailed indicator information on the rehabilitation and degradation phases of all these species. Furthermore, the role of (missing, existing or developing) seed banks, nurse plants (safe sites) and microhabitat properties can be studied in detail and supplemented by experimental approaches such as artificial clearing. This makes it possible to obtain further information about competition, coexistence, aggregation rules, succession sequences and the properties of plant functional types. The documentation of permanent observation plots in the southern Namib Desert over a 17-year period (Jürgens *et al.* 1997) reveals that a period of 17 years can simply be a period of increase (or rehabilitation) that starts with few species, plant functional types and individuals. The number of all these elements then increases more or less steadily over the 17-year period that followed two years of extreme drought in 1979 and 1980.

We do not know when a dry phase will supersede this humid phase of rehabilitation. And we do not yet know which rains are the important for high biodiversity –

- the sum for a whole year,
- the winter rainfall portion,
- or the summer rainfall portion?

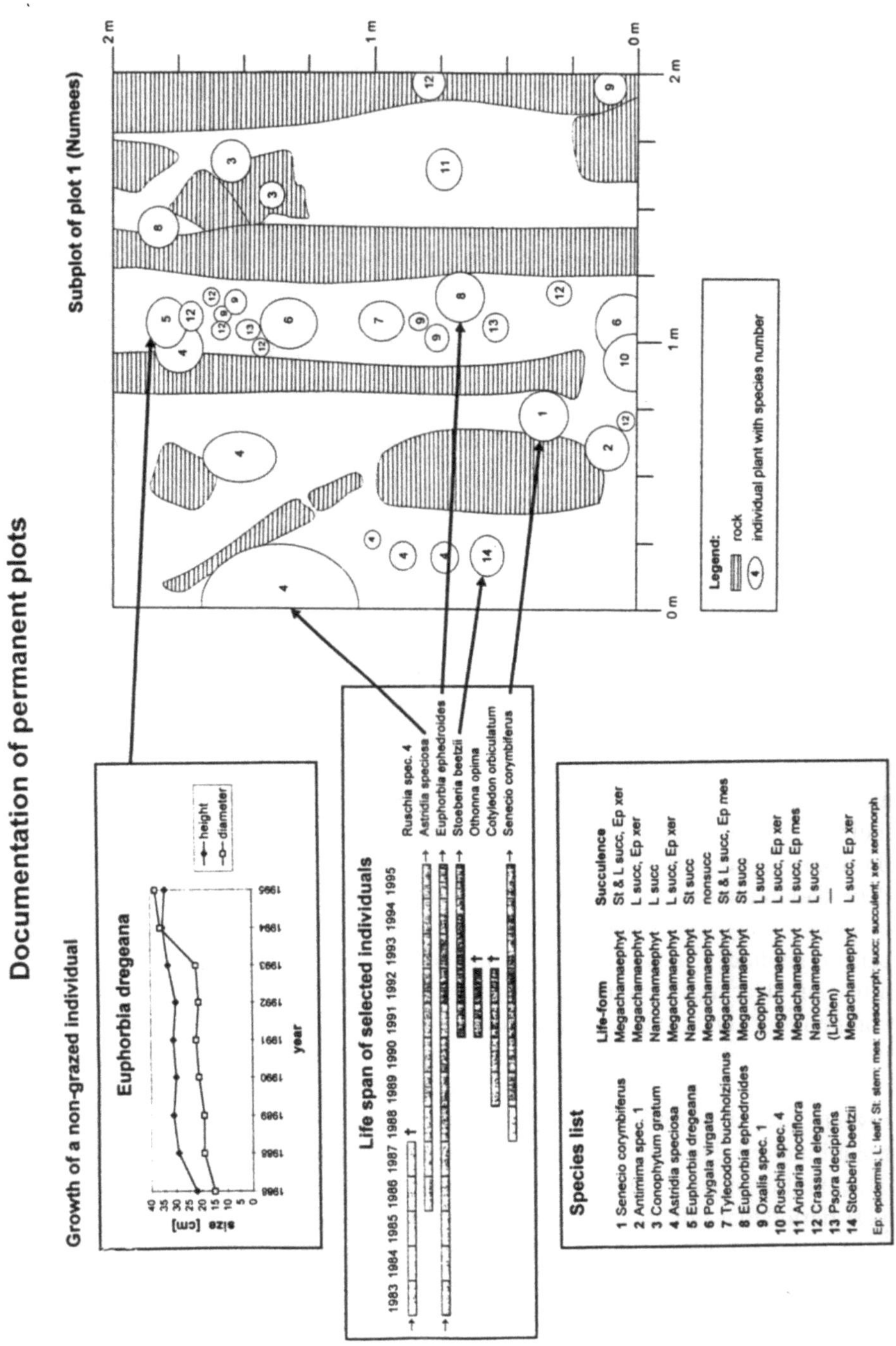

Fig. 8. Schematic sketch illustrating the scale of documentation at permanent observation relevés

Ilustración 8. Esbozo esquemático que ilustra el grado de documentación en relevés de observación permanente

We do not even know whether oscillations of this type are cycles which return repeatedly to more or less the same communities or whether we observe a history that is determined by more or less rapid processes which are changing the entire definition of the ecosystems and their organisms.

6.1.3 Analysis of the determining factors in biota-determined degradation, desertification and/or stability

As with spatial patterns of biodiversity, temporal change in biodiversity is also controlled by environmental factors. These factors comprise not only the variability of climate and soil properties, but the activity of biota which range from ants and termites to wild game and stock as well. A successful analysis requires the comparison of experiment-like situations where the biota exert different degrees of pressure (or abiotic controlling factors) as well as a sufficiently long time frame so that it is possible to differentiate between the influence of the various factors involved (by using multivariate statistics as a minimum). Due to the nature of the most important determinants, stock and wild game usually necessitate different research approaches.

- Stock grazing ecology uses stock management measures (fences, exclosures, stocking rates, rotation, different stock, nomadic herd monitoring) to control grazing pressure, which explains why detailed experimental scenarios are developed. Feedback regarding the results of natural resource management is often of great importance for stock keepers and can induce them to support research.
- Wild game grazing ecology is generally less intensive and is limited in most cases to analyses at exclosures. This also reflects the fact that there is less interest in nature conservation management.

6.2 Analysis of the historical background

Although all the above-mentioned aspects of patterns and processes examples for systematically planned studies exist, such as in cases like the phases preceeding the proclamation of national parks, the history of an area's current biodiversity is seldom the subject of research. Different time frames should be applied when conducting this type of analysis.

6.2.1 Analysis of the area's historical land use and its effects on the vegetation (over decades)

The history of resource utilization during the past few decades is still very much a part of local land users' memories. This is the scale of individual memory, therefore historical documents allow the reconstruction of changes that took place during this time. Hoffman & Cowling (1990) and Hoffman (1991) have supplied very interesting examples for successful analyses conducted on this level: By comparing current vegetation with conditions documented by old photographs, they were able to reject Acock's (1953) hypothesis that arid Karoo vegetation spread eastward in the course of man-made desertification.

However, as already stated in section 6.1.2, natural oscillations that follow periods of extreme drought can last more than 15 years. Whether they be man-made or natural, we should also remember that we observe not only oscillations, which are reversible. It is much more likely that our documentation describes unique history:

6.2.2 *Analysis of more recent climatic history and related changes in vegetation (millennia)*

Cattle farming was introduced to northeastern Africa some 8,000 to 10,000 years ago (Wendorf & Schild 1984) and spread to eastern Africa 6,000 years BP and southern Africa 2,000 BP. Wild game was eradicated by European hunters about 100 years ago. And exotic weeds have been introduced into the Namib Desert region. Considering these examples, would not it be possible that right now new combinations of species are being created under new combinations of selection pressures?

I believe that these processes are of great importance for every serious attempt to achieve a sustainable utilization of natural vegetation resources. Since these aspects have received very little attention to date, we should include not only more long-term studies in any research that is related to the management and utilization of natural vegetation. We should also direct more of our efforts toward reconstructing the past using adequate methods. The demographic structure of populations of the long-living species *Welwitschia mirabilis* and of *Aloe dichotoma* in the Namib Desert support the hypothesis that desert conditions have been spreading eastward over the past few hundred years (Jürgens 1992, 1997). Archaeobotanical research on this subject is now being conducted in Kaokoland in the northern Namib Desert in Namibia.

6.2.3 *Analysis of the area's floristic history (past 100,000 years) and the development of its flora*

Changes in biodiversity that took place in earlier history may be of less importance for the processes taking place today. On the other hand, their role in helping us understand migrations and extinction could be all the greater. Here again, new methods must be used. One example provided by Jürgens (1997) shows that the disjunct distributional patterns in the Namib region can be interpreted as having been caused by the survival of palaeotropical taxa in warm refuge areas during the last glaciation. These warmer lowland pockets were perhaps the only areas where the tropical arid taxa of the Nama Karoo were able to survive during the last glaciation some 18,000 years ago. Survival in relatively small refuge areas would also explain some patterns of the Succulent Karoo, e.g. the concentration of numerous peculiar taxa in some centres of endemism such as the Knersvlakte and parts of the Little Karoo. Convergent evolution has produced very similar life and growth form series in ecologically similar soil catenas of the

Knersvlakte and the Little Karoo, that are represented by different genera of the Aizoaceae in both regions, respectively (Jürgens 1996). Isolation in refuge areas might have been expressed more strongly in the past. It is even possible to make predictions about current ecology, on the basis of phytogeographical patterns of the past: Today, along the Atlantic Ocean coastal strip of the Central Namib Desert, a small but significant number of species from the temperate Succulent Karoo Floristic Region of the Greater Cape Flora are surrounded by the extremely dry and hot inland parts of the Namib Desert, the even more tropical northern desert fringes towards Angola and the Great Dune Field of the Namib Desert, south of the Kuiseb River. This fragment of temperate southern flora is able to survive along the coastline today because of the cool Benguela current which is responsible for the area's mild temperatures and high air humidity (and thereby imitates the climate of the Succulent Karoo Region in the actual winter rainfall region south of the Great Dune Field). A part of this flora probably reached the Central Namib by migrating through the dune field during periods when the climate was moister. This was probably the case *Brownanthus kuntzii* in the Central Namib, a relative of *B. arenosus* that occurs only south of the Dune Field. Another part of the temperate flora of the Central Namib Desert, which also includes the Amaranthaceae *Arthraerua leubnitziae*, is probably derived from palaeotropical flora which has been in contact with the temperate, air-moist desert for a long time. Taxa such as the *Brownanthus kuntzii* and *Arthraerua leubnitziae* are largely restricted to the temperate fog zone near the coast. Therefore it is almost impossible that the climate changed to such an extent during the past few million years that large amounts of summer rain extended to the coastal region of the Central Namib Desert. Until now, no other discipline has provided similar straightforward evidence that the arid climate of the Central Namib Desert existed over a long period, without any interruption.

Acknowledgements. The author is indepted to the German Research Foundation (DFG) and the DFG Sonderforschungsbereich 389 "Arid Climate, Adaptation and Cultural Innovation in Africa".

References

Acocks JPH (1953) Veld Types of South Africa. Mem Bot Surv S Afr 40: 1-128

Cowling RM, Hilton-Taylor C (1994) Patterns of plant diversity and endemism in southern Africa: an overview. Strelitzia 1: 31-52

Cowling RM, Richardson DM, Schulze RE, Hoffman MT, Midgley JJ, Hilton-Taylor C (1997, in press) Species diversity at the regional scale. In: Cowling RM, Richardson D (eds) Vegetation of Southern Africa. Cambridge University Press

Gibbs Russell GE (1987) Preliminary floristic analysis of the major biomes in Southern Africa. Bothalia 17 (2): 213-227

Hilton-Taylor C (1987) Phytogeography and origins of the Karoo Flora. In: Cowling RM, Roux PW (eds) The karoo biome: a preliminary synthesis. Part 2 - vegetation and history. S Afr Nat Sci Progr Report 142: 70-95

Hilton-Taylor C (1996) Patterns and characteristics of the flora of the Succulent Karoo Biome, southern Africa. In: van der Maesen LJE, van der Burgt XM, van Medenbach de Rooy JM (eds) The biodiversity of African plants, Kluwer, Dordrecht, pp 58-72

Hachfeld B (1996) Vegetationsökologische Transektanalyse in der nördlichen Zentralen Namib, Diplomarbeit, Universität Hamburg

Hoerner A (1996) Floristische und strukturelle Analyse eines Vegetationsausschnittes auf einer Farm mit Holistic Range Management in Namibia (Otukarru, Okahandja), Diplomarbeit, Universität Köln

Hoffman MT (1991) Is the Karoo spreading? Veld & Flora 77 (1): 4-7

Hoffman MT, Cowling RM (1990) Vegetation change in the semi-arid eastern Karoo over the last 200 years: an expanding Karoo - fact or fiction? S Afr J Sc 86: 286-294

Gimborn A (1996) Vegetationsökologische Analyse eines kleinräumigen Transektes von 5 km Länge im Namaland (Namibia), Examensarbeit, Universität Köln

Jacobs A (1996) Vegetationskundliche und ökophysiologische Untersuchungen an einem Landschaftsausschnitt der Nama-Karoo Region (Namibia), Examensarbeit, Universität Köln

Jähnig U (1993) Charakterisierung arider Böden in der Namib unter besonderer Berücksichtigung der Vegetation, Diplomarbeit, Universität Hamburg

Jürgens N (1986) Untersuchungen zur Ökologie sukkulenter Pflanzen des südlichen Afrika. Mitt Inst Allg Bot Hamburg 21: 139-365

Jürgens N (1991) A new approach to the Namib region. I. Phytogeographic subdivision. Vegetatio 97: 21-38

Jürgens N (1992) Namib - die afrikanische Wüste der lebenden Wasserspeicher. Uni HH Forschung 27: 68-76

Jürgens N (1997) Floristic biodiversity and history of African arid regions. Biodiversity and Conservation 6: 495-514

Jürgens N, Burke A, Seely MK, Jacobson KM (1997, in press) Desert (The Namib Desert). In: Cowling RM, Richardson D (eds) Vegetation of Southern Africa. Cambridge University Press

Neumann K (1989) On the holocene vegetational history of the Eastern Sahara In: Kuper, R. (ed) Forschungen zur Umweltgeschichte der Ostsahara, Heinrich-Barth-Institut, Köln

Robinson ER (1976) Phytosociology of the Namib Desert Park, South West Africa. MSc Thesis. Pietermaritzburg: University of Natal

Rutherford MC, Westfall RH (1986) Biomes of Southern Africa an objective categorization. Memoirs of the Botanical Survey of South Africa, 54: 1-98

Wendorf F, Schild R (1984) The Emergence of Food Production in the Egyptian Sahara. In: Clark JD, Brandt SA (eds) From Hunters to Farmers. Berkley 1984: 93–101

Williamson G (1997) Preliminary account of the Floristic Zones of the Sperrgebiet (Protected Diamond Area) in southwest Namibia. Dinteria 25: 1-68

The Surumoni project: The botanical approach toward gaining an interdisciplinary understanding of the functions of the rain forest canopy

Wilfried Morawetz
Institut für Spezielle Botanik, Leipzig, Germany

Abstract. Today, thanks to modern observation systems, it is now possible to investigate rain forest canopies without major disturbances to the ecosystem. For the Surumoni project (Upper Orinoco, Venezuela), an observation crane on rails was set up for the first time ever within a primary rain forest. This crane covers an observation area of ca. 1.4 hectares. The scientists themselves steer the gondola by remote control; the gondola can move quickly and silently in all directions. The botanical part of the international project (which is headed by the Austrian Academy of Sciences) is attempting to investigate mainly interactions within the plant community and between plants and animals in order to better understand how rain forests function. This consequently necessitates an interdisciplinary approach. The investigation focuses primarily on ant-plant interaction; genetic variability and gene flow; crown shyness; phenological systems; dispersal and germination; butterflies and beetles: positive and negative effects on plants; lianas: distribution and growth control; life forms, growth types and structural diversity in the canopy; what is found only in the canopy; fungi in competition with plants and animals.

El proyecto Surumoni: El enfoque botánico hacia un entendimiento interdisciplinario de las funciones de las copas de los árboles de la selva tropical

Resumen. Con sistemas modernos de observación resulta hoy en día posible investigar las copas de los árboles de la selva tropical sin grandes problemas. En el proyecto Surumoni (Venezuela, parte alta del Orinoco) se montó por primera vez una grúa de observación sobre railes en el interior de la selva tropical primaria, grúa que cubre una área de observación de aproximadamente 1,4 ha. La cabina es dirigida a través de control remoto por los mismos científicos y permite un movimiento rápido y silencioso en tres dimensiones. La parte botánica del proyecto internacional (dirigida por la Academia de Ciencias Austriaca) tiene como primer objetivo investigar las interacciones en la comunidad de las plantas y entre animales y plantas con vistas a comprender mejor la función de la selva tropical. Por consiguiente, es necesario un enfoque interdisciplinario. Los temas principales de la investigación son: interacciones entre plantas y hormigas; variabilidad genética y flujo de genes; timidez de las copas; sistemas fenológicos; dispersión y germinación; mariposas y escarabajos: efectos positivos y negativos sobre las plantas; lianas: distribución y control del crecimiento; formas de vida, tipos de crecimiento y diversidad estructural en las copas de los árboles; hongos en competencia con plantas y animales.

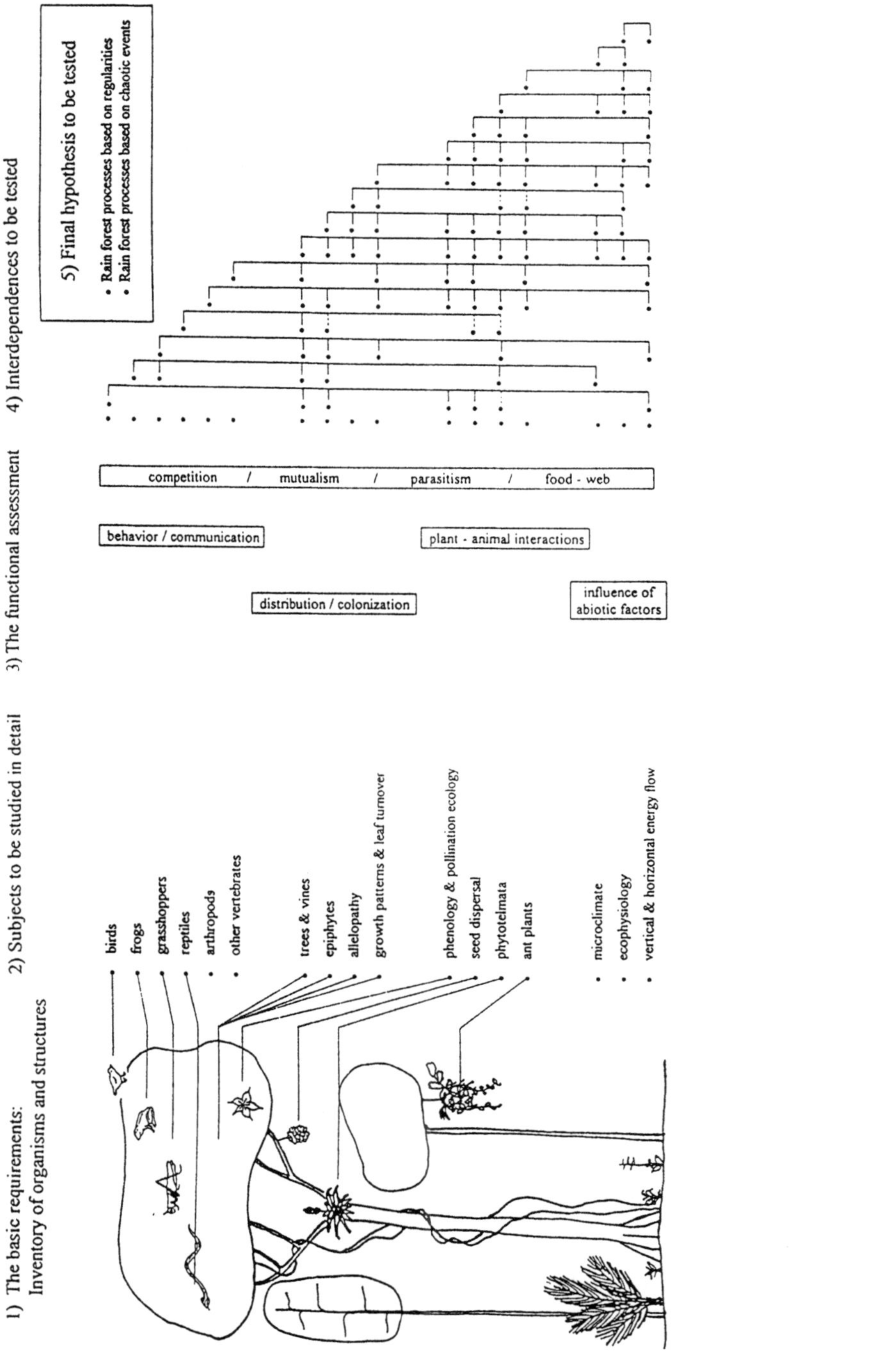

Fig. 1. The original plan for interdisciplinary investigation in the Surumoni project as developed by Morawetz and collaborators at the Austrian Academy of Sciences

Ilustración 1. El concepto original de investigación interdisciplinaria en el proyecto Surumoni desarrollado por Morawetz y sus colaboradores en la Academia de Ciencias Austríaca

1 Background

Tropical botanists have known for many years that the complex interactions between different organisms within the rain forest can be understood only through investigation of the canopy region. Techniques such as towers (e.g. in Manaus, Barro Colorado Island a.o.), walkways (Colombia, Malaysia), climbing (Berry a.o.) and rafts (Hallé *et al.* 1978) have without a doubt been a fundamental help in analysing the canopy region. One of the most important improvements in this area was the installation of a crane with an observation gondola in a natural park in Panama City. It has clearly demonstrated that this system is the best for measuring and observing tree crowns without overly influencing the surroundings.

In 1993 the author began to plan a crane system to be installed in a primary rain forest. The system was designed to cover an observation area of more than one hectare (Fig. 2). With the help of several technicians and engineers (the Liebherr company and engineer A. Waha), an optimal system was designed and constructed. It consists of a slim crane running on 120 m of rails with two different types of observation gondolas (Fig. 3) that are steered by remote control. The Austrian Academy of Sciences supported our goal of conducting tropical forest research. Therefore, once a highly interesting location in southern Venezuela had been found and the Venezuelan government and the Academy had signed a corresponding agreement, the Academy's Research Centre for Biosystematics and Ecology installed an observation system. In the meantime, a small group of tropical biologists (W. Morawetz and H. Winkler of Vienna, H. Römer of Graz, W. Barthlott of Bonn, D. Anhuf of Mannheim, J. Hafellner of Graz, and W. Hödl of Vienna) and their collaborators developed a theoretical plan for joining the various interests involved in the project into a mesh of interdisciplinary studies (Fig. 1). The author moved to the University of Leipzig which financed the final installation phase and the first months of crane activities and supported the start of the botanical-ecological project on the Surumoni. Winkler from the Austrian Academy of Sciences has served as main coordinator for the Surumoni Project to date.

2 Our main questions

2.1 Ant-plant interaction

Given their close connection with plants, ants are probably one of the most important animal groups in the rain forest. We estimate that the project region has several hundred ant species, most of them linked either with specific species, species groups or only plant structures. Their role as decomposers and turn-over agents of organic matter is evident. At present, typically adapted ant plants are mainly known from secondary sites, similar to those found in the Surumoni plot (*Cecropia*, *Cordia*, see also Morawetz *et al.* 1992). What we do not know

Fig. 2. A bird's eye view of the crane at the Surumoni site. Note the rails at the bottom
Ilustración 2. La grúa en el sitio de Surumoni a vista de pájaro. Nótense los rieles en el suelo

however is their role in the canopy, where they inhabit hollow twigs (adaptations, preadaptations?), build nests and ant gardens. Tree seed dispersal by ants obviously occurs but has been neglected in the rain forest in the past (pers. comm. F. Eichhorn). Space competition on tree trunks occurs between different ant species. Pheromones can be expected to have an influence on flower development and microphenological phenomena. Temporary predator defence by plant-ant activities seems to be more frequent and effective than permanent defence.

2.2 Genetic variability and gene flow

One of the most crucial questions is the genetic variability of rain forest trees. Thus far, studies have shown that genetic variability is probably greater than expected (Hemrick & Loveless 1986). These studies did not however take into account the ecological status of the trees investigated (secondary versus primary species). Presuming that major genetic variability also indicates a broad ecological amplitude of species, two scenarios can be imagined. Either you find enormous genetic variability in trees where seedlings have a good chance of sprouting under quite diverse conditions and of establishing themselves successfully. This would mean a continual adaptation of the genetic pool of tree species germinating in very different ecological niches that are constantly

Fig. 3. The smaller observation gondola at work (Winkler and Sax inside). This gondola is capable of penetrating even dense tree crowns without damaging them
Ilustracion 3. La pequeña cabina de observación en funcionamiento (dentro Winkler y Sax), capaz de penetrar también en espesas cimas de árboles sin dañarlas

produced anew as a result of competition and synecological phenomena. The other extreme would mean genetic combinations that vary only slightly and therefore allow trees to germinate only under very specific conditions, something like a jigsaw puzzle.

To understand the influence and interaction of genetic conditions on and with forest ecology, it is necessary to investigate gene flow in genetic terms as well as flower ecology phenomena (Bawa & Hadley 1990). Clear relations between pollination, gene flow, dispersal, fruit size and quantity, predators and final establishment of seedlings are to be expected (Gómez-Pompa *et al.* 1991). Furthermore, the fertility of trees must be seen in connection with the effects of climatic events, micro- and macrophenology. Although we know little about it, the continual predation of dispersal units needs an adequate biological buffer. We can also expect to find very closed and isolated systems of individuals that have few links to the other species and which act as founder populations.

2.3 Coevolutionary phenomena and mutualistic systems

Especially within the canopy we expect to find a broad range of interaction at various evolutionary levels, beginning with simple causal mutualisms, and

Fig. 4. Crown shyness from the observation gondola, ca. 6 m above the crowns at the Rio Surumoni area

Ilustración 4. La timidez de las copas desde la cabina de observación, aprox. 6 metros sobre las cimas de los árboles en el área del río Surumoni

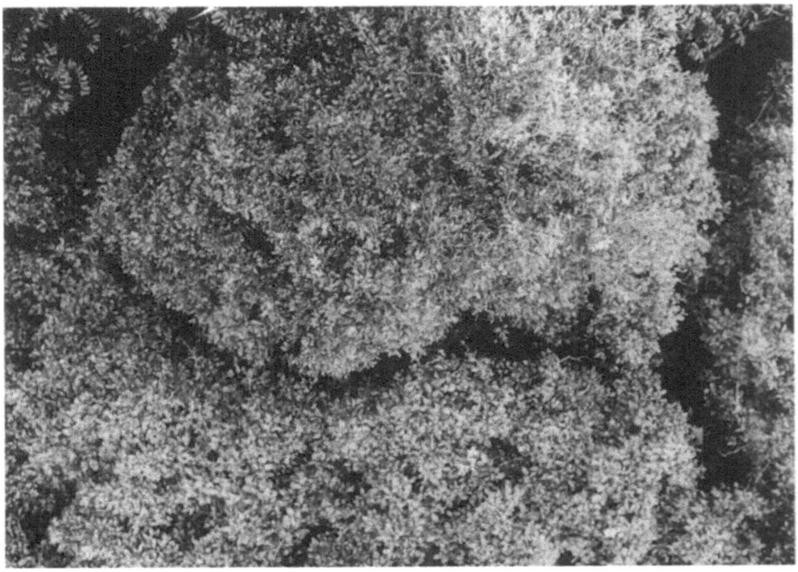

Fig. 5. Crown shyness region from a short distance. Note that most branches of the outer canopy stop growing or die back before reaching the neighbouring tree

Ilustración 5. Región con timidez de las copas desde una distancia corta. Nótese que varias ramas dejan de crecer o se acortan sin haber alcanzado el árbol vecino

extending to highly fixed and refined systems (Krügel 1993). Ants and ant plant, as well as phytotelmata (Fish 1983) are well known subjects for such investigations. Both fields will be considered in our investigations, although based on our initial observations, we expect a rather large number of events which signalise a more or less regular interaction between insects and plants. Once we have described these systems, our main task should be to test how much both partners depend on one another and how easily interactors can change partners.

2.4 Crown shyness

Anyone who has passed closely over a rain forest canopy knows that tree crowns scarcely touch each other and always keep a certain distance from their neighbours. This phenomenon is known as "crown shyness" and appears to us to be one of the most important factors for maintaining an equilibrium between different species in a rain forest and for regulating competition within the canopy region (Fig. 4 and 5). However, little is known about the mechanisms of canopy growth control, although there are two hypotheses which could be tested. One is more mechanistic (contact and wind movement stop growth), the other more allelopathic (gases or substances hinder bud growth within a certain distance to other trees). Crown shyness is an absolute necessity for both individuals and the forest as a whole to survive over long periods. For this reason, we will first map areas of crown shyness in order to see which species really maintain a substantial distance and which do not. We will test the plants' gas emissions in sito and in vitro and try to show their effect on different plant individuals. Finally we will set up an experimental crown shyness situation to test if one of the individuals involved is "stronger" than the other and to perhaps show how "communication" takes place.

2.5 Phenological systems

Gentry (1973) was the first to show that phenology within the large Bignoniaceae family varies distinctly. It became obvious that phenology is one of the most important characteristics of plant behaviour and plant biology (Bawa & Hadley 1990, Rhatcke & Lacey 1985). Although his major types (e.g. "big bang", "multiple bang", etc.) can be similarly traced in different families as well, it is obvious that a number of yet undescribed phenological types exist. Our experience indicates that some of these will be special cases of the general types set forth by Gentry (1973) and some will be completely new behavioural types. A list of the phenological spectrum of the whole plot will be drawn up, firstly, to obtain a realistic idea of phenological variation within the forest community and, secondly, to pick out those cases which have not been studied in detail to date. Different phenological types require different pollination systems, attract different types of predators, and form complex and well-defined evolutionary nuclei in the canopy.

2.6 Dispersal and germination

It is not necessary to point out that dispersal and germination constitute one of the most crucial events in a plant's ontogeny. Especially in rain forests, where the frequency of individuals is extremely low, we still do not understand how so many species avoid extinction. Direct observation of flowering and fruiting reveals a much more differentiated picture of the ripening and predation of fruits, e.g. the gradual ripening of fruit on a tree seems to be an important plant behaviour for escaping predation (Morawetz & Listabarth in prep.). We may expect a similar time pattern in the case of germination, so that both ripening and germination events occur with long time frames. It is now easy to investigate the number and biomass of all ripe fruit, and the size of single dispersal units. This will give an initial insight how plants establish themselves. Ecological tolerance and genetic variability (see above) shall serve as the basis for comparison.

2.7 Lianas: distribution and growth control

We know little about lianas, their growth, biology and behaviour in the crowns. Schenk (1893) was the first to outline the most important anatomical basis for liana growth and survival. Putz & Mooney (1991) have more recently published an excellent overview and bibliographic study on this subject and we for ourselves soon saw that technical difficulties in particular have hindered biologists in the past from being able to conduct thorough studies. We can now follow liana life and growth by using an observation gondola capable of moving in all directions. We expect to be faced with a number of new questions during our first investigations. The following topics however have priority at this time: (1) Why do lianas seldom overgrow their host and what would be negative for both if they were to do so? (2) What growing strategies do lianas use to reach the next tree, to climb up into the canopy and to establish seedlings? (3) Why are some trees completely overgrown by lianas, and others not? (4) Which interactions, mutualisms and coevolutionary phenomena can we suspect exist between lianas and birds (dispersal), insects (predation), ants (are there liana ant plants)? The first step will be to map lianas and their branching, identify their seedlings and small plants and observe their variation and differentiation throughout their life span.

2.8 Butterflies and beetles: positive and negative effects on plants

The most astonishing experience we had in the canopy was the fact that hardly any insects could be traced during day although we knew from our night observations that they existed in enormous abundance. Similarly, it was rarely possible to link damaged leaves to any animal. Therefore we plan to first get an idea of the life cycles of small animals such as insects, simply to know how they behave, where they live during daytime and nighttime, and how they influence forest growth. Based on our impressions, it would not be too surprising if negative and positive effects were to balance each other.

2.9 Life forms, growth types and structural diversity in the canopy

No environment is richer in life forms – plant or animal – than the tropical rain forest (Koepcke 1971, 1973, 1974, Jacobs 1981). Although a good survey on growth form differentiation among tropical trees already exists (Hallé *et al.* 1978, Oldeman 1990), our observations in the canopy showed that with our new technique a more detailed analysis will be possible, especially with regard to branching patterns and leaf exposure. Furthermore, epiphytic growth (see e.g. Ibisch 1996), hemiparasites and palm establishment will be placed in relation to crown dynamics and growth ontogeny during our project. Structural diversity and establishment in spatial niches will be of special interest.

2.10 What is found only in the canopy

It is quite clear that a wealth of phenomena exists within the canopy that cannot be found, traced or analysed anywhere else. Although there are well known examples of insects, mammals and birds that stay in the canopy nearly all of their lives, we have no well-documented figures (see Roubik 1993). Therefore one of the screening programmes in this project will concentrate on developing a list of all those plant and animal groups which can be found only in the tree crowns of tropical rain forests. This task should be highly interdisciplinary and should be developed together with specialists.

2.11 Fungi in competition with plants and animals

The rain forest is rich in fungi and possibly even dominated by them. As a matter of fact, fungi are one of the strongest regulatives and probably one of the most important keystone groups. Fungi play an essential role as mycorrhiza and we may expect to make findings similar to those made in San Carlos where nearly all trees were associated with mycorrhiza. They are also an important factor in the decomposition of dead organic substance. We find them both as epiphytes and parasites on twigs and leaves. Their influence on and interaction with plants and animals is large, as substantiated by our first observations in the area. We suspect that fungi *inter alia* influence the movements and colonisation of ants, are responsible in some way for leaf turnover and regulate some tree-growth patterns after having infected bark injuries.

References

Bawa KS, Hadley M (eds) (1990) Reproductive ecology of tropical forest plants. Man and the biosphere series. Vol. 7, Paris

Fish D (1983) Phytotelmata: Flora and fauna. In: Frank JH, Lounibos LP (eds) Phytotelmata: Terrestrial plants as hosts for aquatic insect communities. Plexus, Medford, New Jersey, pp 1-27

Gentry AH (1973) Generic delimitations of Central American Bignoniaceae. Brittonia 25: 226-242

Gómez-Pompa A, Whitmore TC, Hadley M (eds) (1991): Rain forest regeneration and management. Man and the biosphere series. Vol. 6, Paris

Hallé F, Oldeman RAA, Tomlinson PB (1978) Tropical trees and forests. An architectural analysis. Springer, Berlin Heidelberg New York

Hamrick JL, Loveless MD (1986) Isozyme variation in tropical trees: procedures and preliminary results. Biotropica 18: 201-207

Ibisch PL (1996) Neotropische Epiphytendiversität - das Beispiel Bolivien. Martina Galunder-Verlag, Wiehl

Koepcke H-W (1971, 1973) Die Lebensformen. Band 1, 1. Teil: Grundbegriffe. 2. Teil: Die Selbstbehauptung. Goecke & Evers, Krefeld

Koepcke H-W (1973, 1974) Die Lebensformen. Band 2, 3. Teil: Die Arterhaltung. 4. Teil: Die universelle Bedeutung der Lebensformen. Goecke & Evers, Krefeld

Jacobs M (1988) The Tropical Rain Forest. Springer, Berlin Heidelberg New York London Paris Tokyo

Krügel P (1993) Biologie und Ökologie der Bromelienfauna von *Guzmania weberbaueri* im amazonischen Peru, ergänzt durch eine umfassende Bibliographie der Bromelien-Phytotelmata. In: Morawetz W (ed) Biosyst Ecol Ser 2, Austr Acad Sci, Wien

Morawetz W (1982) Morphologisch-ökologische Differenzierung, Biologie, Systematik und Evolution der neotropischen Gattung *Jacaranda* (Bignoniaceae). In Kommission bei Springer, Wien New York

Morawetz W, Henzl M, Wallnöfer B (1992) Tree killing by herbicid producing ants for the establishment of pure *Tococa occidentalis* populations in the Peruvian Amazon. Biodiv and Conserv 1:19-33

Oldeman RAA (1990) Forests: Elements of silvology. Springer, Berlin

Putz FE, Mooney HA (eds) (1991) The biology of vines. Cambridge University Press, Cambridge New York Port Chester Melbourne Sydney.

Rathcke B, Lacey EB (1985) Phenological patterns of terrestrial plants. A Rev Ecol Syst 16: 179-217

Roubik DW (1993) Tropical pollinators in the canopy and understory: Field data and theory for stratum "preferences". J Ins Behav 6: 659-673

Schenk H (1893) Beiträge zur Biologie und Anatomie der Lianen, im besonderen der in Brasilien einheimischen Arten. In: Schimper (ed) Botanische Mitteilungen aus den Tropen 5: 1-271

Part 3

Biodiversity in change

Zonal features of phytodiversity under natural conditions and under human impact – a comparative survey

Michael Richter
Institute of Geography, University of Erlangen, Germany

Abstract. Principal features of alpha and beta diversity for nine ecozones and two altitudinal gradients are presented (Fig. 17 and 16), although the types of diversity in them may vary considerably depending on community structure, topographic variety and different evolutionary processes. Under extratropical climatic conditions (including Mediterranean conditions), natural forests general exhibit fewer species and a less pronounced community diversity (Fig. 4 and 5) than traditionally cultivated landscapes (Fig. 9). The Mediterranean region of old in particular must be regarded as a very important genetic pool for weeds. As the result of man's influence, plants originating there, and annuals in particular, have enriched neighbouring floristic regions. Under natural conditions, alpha and beta diversity are, in comparison, normally greater in ecozones with treeless tundras and steppes than in the forests of cold and temperate regions (Fig. 3 and 7).

In contrast to the situation in extratropical zones, it must be considered a general fact that any human impact on tropical regions and most subtropical regions with summer rain will reduce diversity. This is due to the loss of three characteristic life forms which constitute heterogeneity in tropical and subtropical forests: trees, climbers and epiphytes (Fig. 14 and 15). As a result, secondary successions lead to increasing species richness after abandonment – in contrast to Central Europe and the Mediterranean where it leads to a decrease (Fig. 6).

Consequently, when developing concepts for sustainability and environmental protection one has to consider these two contrasting zonal features of diversity: For the extratropical type of diversity, the maintenance of a dense pattern of different traditional land-use systems guarantees the highest level of alpha and beta diversity. In the tropics, greater diversity is ensured by combining nature conservation with optimized agriculture that imitates the structures and nutrition cycles of the region's predominant natural ecosystems.

Características de fitodiversidad según zonas bajo condiciones naturales y bajo impacto humano: Un estudio comparativo

Resumen. Las principales características de la diversidad alfa y beta en nueve ecozonas y dos gradientes altitudinales están presentes (figuras 17 y 16), aunque los tipos de diversidad pueden variar considerablemente dependiendo de la estructura de comunidades, la variedad topográfica y los diferentes procesos evolutivos. Bajo condiciones climáticas extratropicales (incluyendo las condiciones mediterráneas), los bosques naturales muestran por lo general menos especies y una diversidad de comunidades menos pronunciada (fig. 4 y 5) que los paisajes muy cultivados (fig. 9). La región mediterránea en particular ha de ser considerada una fuente genética muy importante de maleza. Como resultado de la influencia humana, las plantas ahí originadas, y en particulara las plantas anuales, han enriquecido las regiones vegetales colindantes. Bajo condiciones naturales, la diversidad alfa y beta suele ser también mayor en ecozonas con tundras despobladas de árboles y estepas que en los bosques de regiones frías y templadas (fig. 3 y 7).

Frente a la situación en las zonas extratropicales, hay que considerar el hecho general de que cualquier impacto humano en las regiones tropicales y en la mayoría de las regiones subtropicales con lluvias de verano reducirá la biodiversidad. Esto se debe a la pérdida de tres formas de vida características que constituyen la heterogeneidad tropical en los bosques tropicales y subtropicales, es decir: árboles, trepadoras y epífitas (fig. 14 y 15). Como resultado, sucesiones secundarias llevan a un aumento de la riqueza de especies después del abandono, frente a lo que ocurre en Centroeuropa y el Mediterráneo, donde llevan a un descenso (fig. 6).

Por lo tanto, a la hora de desarrollar conceptos para la sostenibilidad y protección medioambiental, hay que tener en cuenta estas dos características contrastivas de diversidad según zonas: Para el tipo de diversidad extratropical, el mantenimiento de un patrón denso de sistemas diferentes de uso tradicional del suelo garantiza el más alto nivel de diversidad alfa y beta. En los trópicos, se garantiza una mayor diversidad combinando la conservación de la Naturaleza con una agricultura optimizada que imite las estructuras y ciclos de nutrición de los ecosistemas naturales predominantes en la región.

1 Some principals regarding biodiversity

When considering the sense and nonsense to be found in biodiversity studies today, one might come across some rather ironical comments regarding the term 'biodiversity'. For example, Green (1979) described biodiversity as an "answer to a question which is yet unknown." Which means that in any good science, a description is the start of knowledge about facts and processes. So, what we do know is that considerable differences in diversity exist between ecosystems. We also know that diversity differs between natural ecosystems and those that have been influenced by humans. But we are not in the position to present convincing scientific explanations for why a high level of diversity should be better or worse than a low level.

Of course, high diversity is interesting from an economic standpoint to chemical or pharmaceutical enterprises because of the opportunities it offers for the exploitation of the drugs they contain. Some people have used this argument as a political instrument for protecting rain forests. As a result, the term biodiversity has become widely known as part of a moral stance. But these people are unable to provide a convincing argument on the basis of scientific evidence. Even equilibrium models which include the idea that a large number of different species means greater ecological stability are often not correct. Instead of this, we have to consider that the processes of patch dynamics create changing diversity patterns within the same ecosystem.

A presentation of aspects of zonal differentiation in phytodiversity must be preceded by some basic definitions of the term diversity (according to Whittaker 1972). It is necessary to distinguish between

- alpha diversity which designates species richness or the so-called "within habitat diversity" expressed by the number of species
- beta diversity means community richness or the so-called "between habitat diversity" and
- gamma diversity which is the total of alpha and beta diversity, and expresses a landscape's total diversity, or "ecodiversity" (according to Barthlott *et al.* 1996)

With regard to the zonal gradients of biodiversity, this paper considers only four assumptions which were the subject of a recent discussion about the so-called Rapoport's Rule which maintains that latitudinal ranges of animal and plant species increase with latitude (narrow definition of the term by Rohde 1996):

1. Low-latitude species typically have less environmental tolerance than high-latitude species. Furthermore, a larger number of "accidentals" (i.e. species that are poorly suited for their habitat) occur in tropical communities. The constant input of these accidentals artificially inflates species numbers in the tropics (Stevens 1989)
2. Greater "effective evolutionary time" – which is defined as a factor that is comprised of greater evolutionary speed at tropical temperatures and longer contemporary evolutionary time under relatively constant conditions – is the primary cause of greater species numbers in the tropics (Rohde 1992)
3. Present seasonal variability between sites separated by latitude or elevation drives the Rapoport phenomenon because high-latitude species are better able to tolerate seasonal temperature fluctuations than low-latitude species (Seasonal Variability Hypothesis, according to Stevens 1996)
4. There is evidence that the effect of the ice ages was particularly great only in the mid-to-high latitudes which would suggest that tolerance to temperature fluctuations during those periods was a primary factor in the extinction of less tolerant species in the extratropics (Rohde 1996)

Consequently, Rapoport's Rule provides an initial explanation for the geographical pattern that marks the two extremes of diversity: "a chemostat-like system tending to become monospecific, and a Noah's ark or museum situation with an infinite number of species each represented by just one specimen" (Margalef 1994). The same concept applies to the principal pattern of increasing species richness from the poles towards the equator. However, zonal features of diversity do not at all follow a continual gradient: Arid or perhumid areas as well as regional evolutionary retardations and accelerations cause a tendency toward undulating change patterns for any type of diversity in a given longitudinal band.

Apart from these brief general assumptions, this paper deals only with alpha and beta diversity features of natural vegetation as compared with the features found in vegetation affected by human intervention. Concerning the latter case of vegetation, human landscapes have to be divided into areas of cultivation with

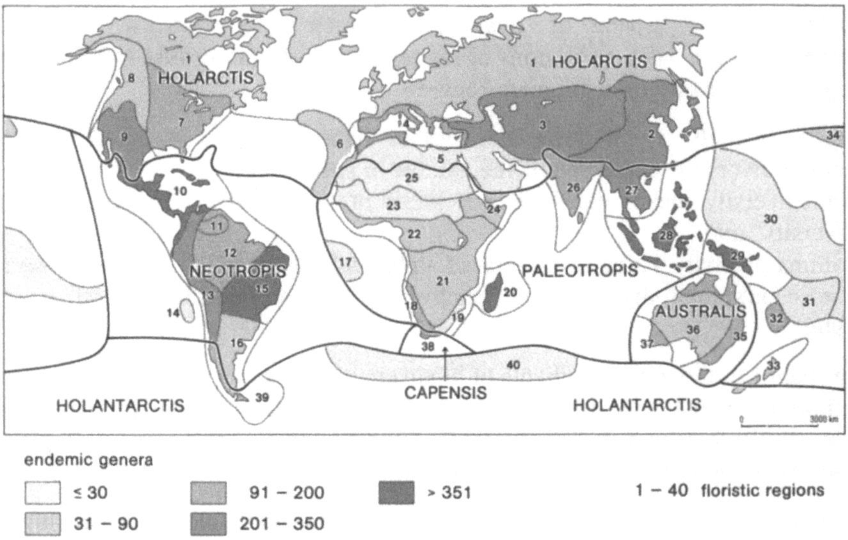

Fig. 1. Number of endemic genera in 40 different floristic regions (based on references made in Takhtajan 1985 and chorionomic categories compiled from information provided by various authors)

Ilustración 1. Número de géneros endémicos en 40 regiones florales diferentes (basado en referencias hechas en Takhtajan en 1985 y categorías corionómicas recopiladas a partir de información proporcionada por varios autores)

minor or major technical and chemical impact, and into rural or urban settlements that are characterized by various degrees of hemerobia (Richter 1997). A few principles concerning the trends of secondary succession in the middle latitudes and tropics also have to be presented.

2 Phytodiversity patterns in different ecozones

With regard to zonal biomes, the following nine types of ecosystems are based on the system developed by Schultz (1995). This system does not take special conditions of azonal or extrazonal complexes in any of the nine ecozones into consideration. Of course, the different types of diversity may vary considerably, depending on community structure and topographic variety.

In addition, problems arise from the considerable differences in taxonomic diversity that evolutionary processes have caused between the world's various floristic regions (Whittaker 1975). This becomes obvious when interpreting Fig. 1 which shows a marked deficit of endemic genera in the tropical zone of the Guineo-Congolian (22) and Zambezian regions (21) compared to the high numbers reported for the Caribbean (10), Brazilian (15) and Malaysian regions (28). The same is true of all other zones except the polar and boreal areas with their relatively scanty array of flora. However, regional patterns of genera

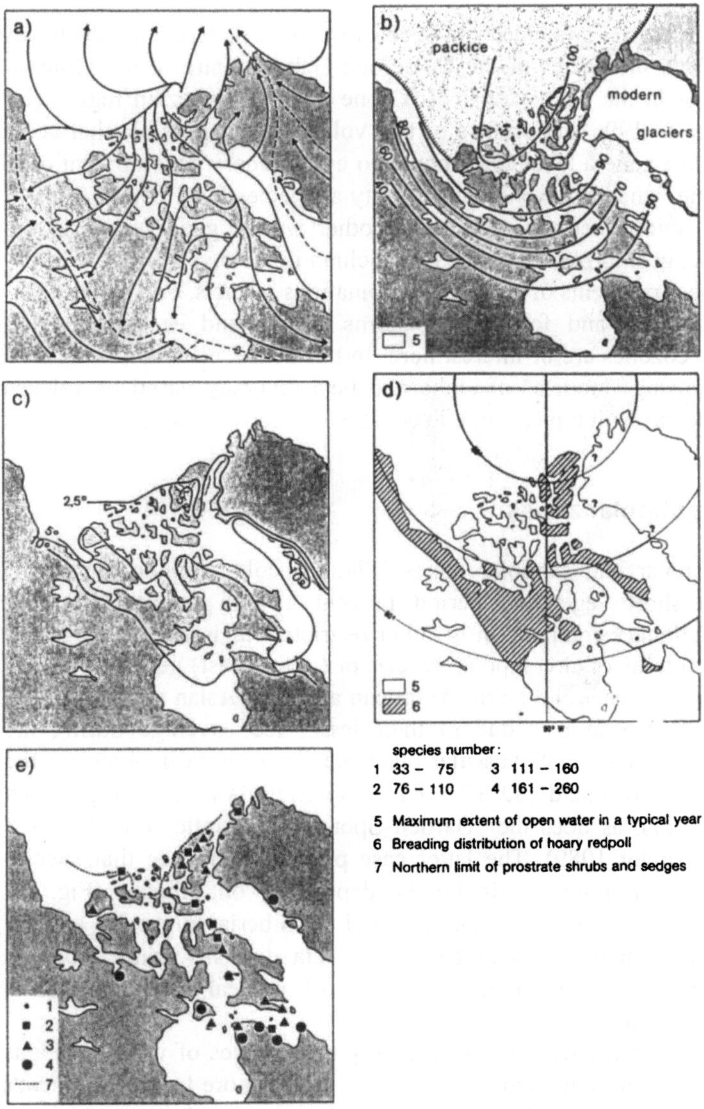

Fig. 2. Surface stream flux in July (a), frequency of occurrence of July arctic air masses and distribution of inland ice as an indicator of increased humidity (b), isotherms of the mean temperature for the month of July (c), breeding distribution of *Calcarius lapponicus* (d) and the northern limits of sedges and prostrate shrubs with species richness at different collection localities (e) on the Queen Elizabeth Islands, N.W.T. (a and b based on Edlund & Taylor 1989; c and e on Rannie 1986; d on Ouellet 1990)

Ilustración 2. Corriente por deshielo en la superficie en julio (a); frecuencia con que se producen las masas de aire ártico en julio y distribución del hielo del interior como indicador de un aumento de la humedad (b); isotermos de la temperatura media en el mes de julio (c); distribución de retoños de *Calcarius lapponicus* (d) y los límites septentrionales de arbustos de juncos y rastreros con riqueza en especies en diferentes lugares de recogida (e) de las Islas Queen Elizabeth, N.W.T. (a y b se basan en Edlund & Taylor 1989; c y e en Rannie 1986; d se basa en Ouellet 1990)

richness do not necessarily coincide with species-richness patterns, as illustrated by the large numbers of species in the Cameroon-Gabon centre in the Guineo-Congolian region or in the East African Rift zone of the Zambezian region (cf. map of Barthlott *et al.* 1996 and preface of this volume). It is possible that taxonomic diversity at the genera level corresponds to evolutionary development over long periods of time, whereas taxonomic diversity at species level corresponds to the parameters of abiotic heterogeneity, or in other words, geodiversity in the sense of orographic, geomorphological and mesoclimatic variety.

Disregarding these problems of regional combinations of facts, only "normal" – predominant – life-form and formation patterns of lowland ecosystems that represent specific ecozones are of interest here. In this sense, each zone is shown by illustrations showing "models" of alpha and beta diversity based on relevés (own unpublished data) with typical situations at the respective location.

2.1 Polar and subpolar zones

Starting with the polar and subpolar zones (Fig. 2), cold climatic conditions combined with a short vegetation period prevent a high species diversity. Furthermore, evolutionary competition is rather restricted, as is evidenced by the wide distribution patterns of circumpolar species or (at the least) genera. Floristic diversity is slightly greater in the North American and East Asian sections of the zone than in the European part due to their lesser ice coverage during the Pleistocene glaciation. With its 93 vascular plant species, even Peary Land on the northern-most tip of Greenland (80°33' - 83°41'N) exhibits a surprisingly high variety (Holmen 1957), as does the "Garden Spot of the Arctic" on Ellesmere Island (Edlund & Taylor 1989). The latter case provides evidence that species richness in a polar environment is highly dependent on climate (Fig. 2). According to Petrovsky (1988), Wrangel Island in Siberia might be another surprising location for high levels of alpha and beta diversity, since isolation combined with relatively mild climatic situations offers ideal conditions for an increased rate of speciation.

In the Antarctic on the other hand, where only two species of vascular plants exist, the region's extremely limited heterogeneity is due more to the continent's isolation than to its climatic conditions. And even though the subantarctic islands have a much more adequate climate for rich plant growth, their taxonomic diversity is also very limited and consists of a total of only 70 species, with Macquarie Island accounting for up to 33 of them.

Although anthropogenic disturbances in the arctic zone remain visible for much longer periods than in other ecozones, in the arctic zone at least the different stages of secondary succession are comprised of a greater number of species than at sites without human impact (Harper & Kershaw 1996). This fact is due to an additional amount of "exotic" invaders of synanthropic origin (e.g. *Taraxacum officinale* and *Poa annua* on the Kerguelen Islands).

Despite the fact that alpha diversity is limited in most cases, Fig. 3 presents a relatively high level of beta diversity. The reason for this varying community

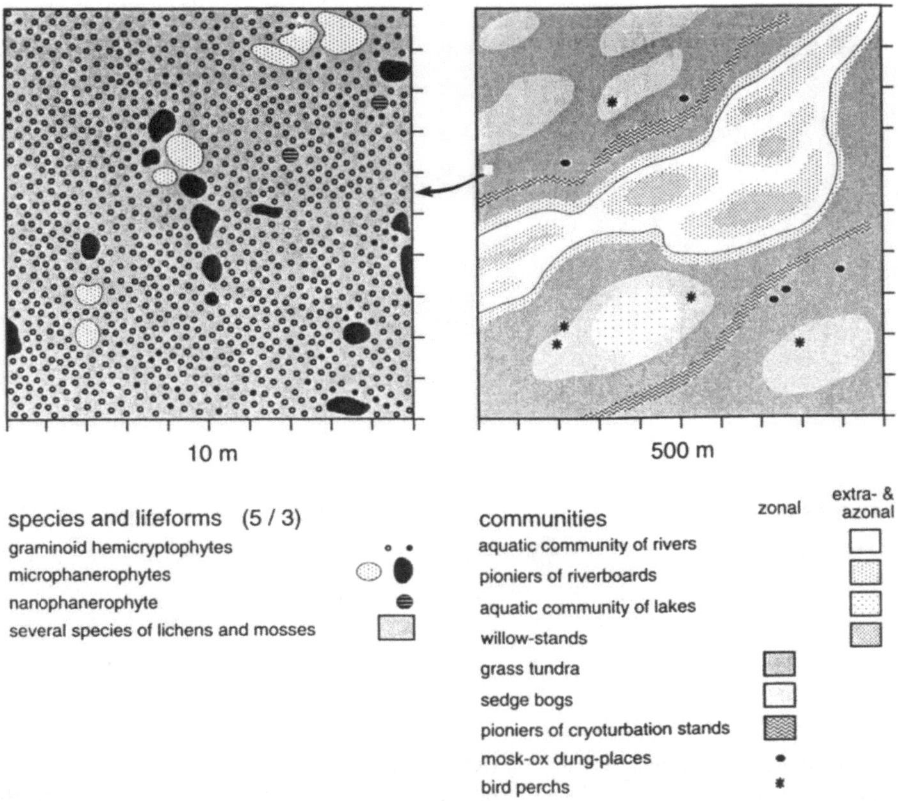

Fig. 3. Illustration of the plant distribution (left: 100 m^2) and community patterns (right: 2.5 km^2) in a tundra biome (models: Nome, Alaska and Kevo, Lapland)
Ilustración 3. Ilustración de la distribución de las plantas (izquierda: 100m^2) y patrones de la comunidad (derecha: 2,5km^2) en una tundra biome (ejemplos: Nome, Alaska y Kevo, Laponia)

structure lies in arctic topography that causes small-scale variety in permafrost phenomena such as frost boils, polygons, pingos and palsa, and in soil moisture, soil types, snow coverage, bird nests, lemming burrows, musk-ox dunghills and, consequently, in vegetation patterns (Thannheiser 1988). Microclimatic conditions also differ considerably in this mosaic of ecosystems with open irradiation.

2.2 Boreal zone

This microclimatic differentiation is of less importance in woodlands, such as in the second zone of boreal coniferous forests. These areas have even fewer species than tundra sites because of the greater amount of shadow caused by trees and

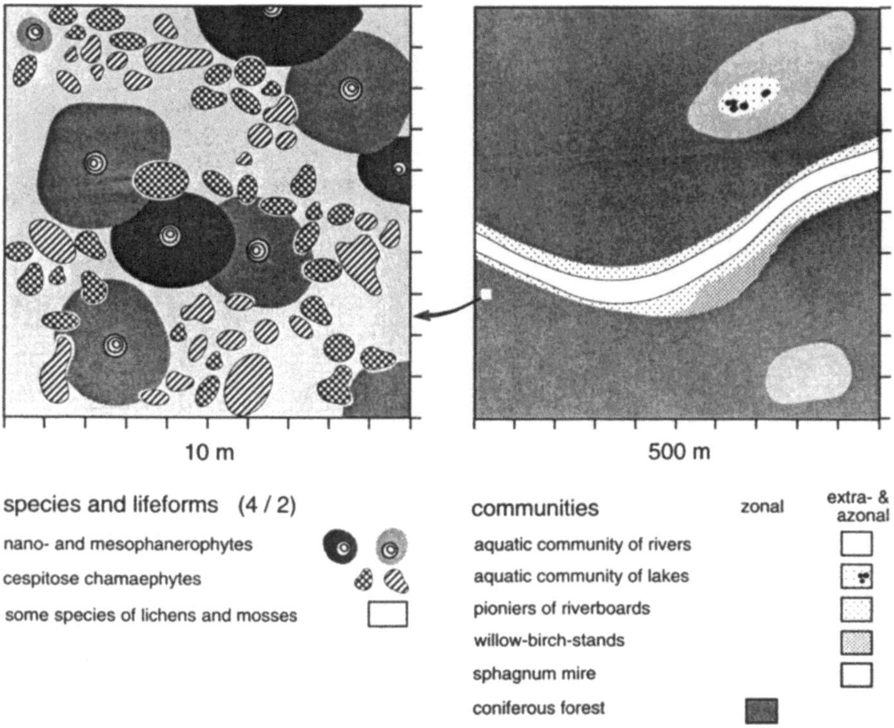

Fig. 4. Illustration of plant distribution (left: 100 m^2) and community patterns (right: 2.5 km^2) in boreal coniferous woodlands (models: Wilhelmina, Sweden and Kitwanga, British Columbia)

Ilustración 4. Ilustración de la distribución de las plantas (izquierda: 100m^2) y patrones de la comunidad (derecha: 2,5km^2) en bosques boreales de coníferas (ejemplos: Wilhelmina, Suecia, Kitwanga, British Columbia)

plants (Fig. 4). This phenomenon involves not only vascular plants but lichens and mosses as well. Of course, they represent one of the most specific characteristics of the zone. But they also occupy homogenous areas in co-association with different coniferous communities.

However, beta diversity remains rather high because the same coniferous communities are broken up by deciduous birch and willow stands as well as by mires and aquatic formations. Such sites may be characterized by permafrost interruptions or discontinuous permafrost. Furthermore, the incidence of disturbances such as fire, storms and forest pests increases towards warmer latitudes. As a consequence, the mosaic of different patterns becomes more variable due to coexisting regeneration processes of different sizes (Fig. 4): this mosaic ranges from partial crown dieback to gaps to cohort mortality (Böhmer & Richter 1996).

With regard to human impact, oligotrophic boreal coniferous stands suffer most from species decline due to acidification caused by sulphur and heavy metal

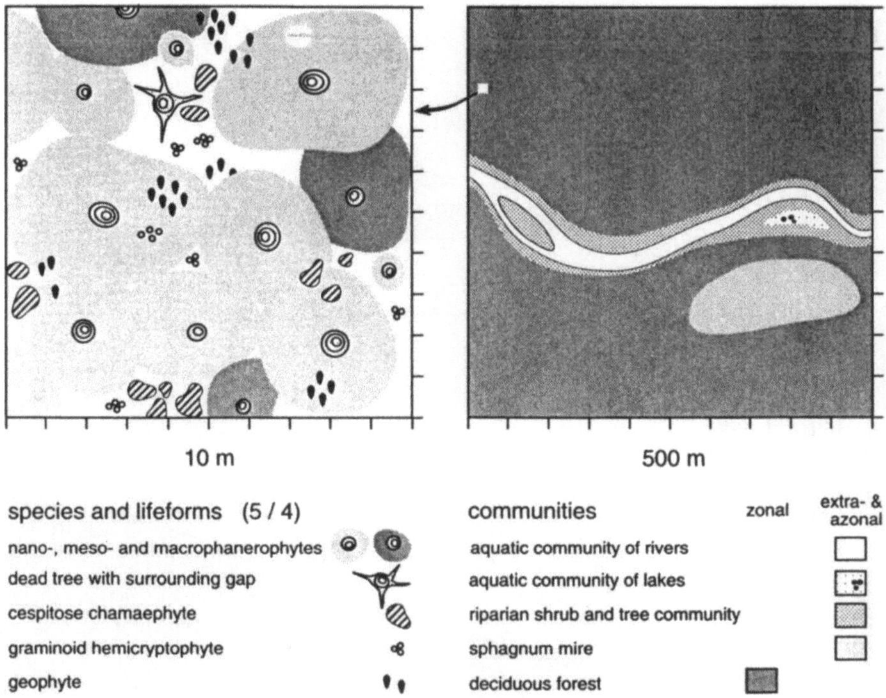

Fig. 5. Illustration of plant distribution (left: 100 m^2) and community patterns (right: 2.5 km^2) in temperate deciduous woodlands (model: Gößweinstein, Franconia, Germany)
Ilustración 5. Ilustración de la distribución de las plantas (izquierda: 100m^2) y patrones de la comunidad (derecha: 2,5km^2) en bosques caducos templados (ejemplo: Gößweinstein, Franconia, Alemania)

pollutants. This is extremely obvious with the apocalyptic-looking "technogenetic deserts" in the vast areas surrounding the Pechenganikel and Severonikel smelter complexes on the Kola Peninsula (Kozlov *et al.* 1993). Another type of human influence – increased forest clearing – must also be considered. As in the case of disturbances in arctic ecosystems, the effects of direct irradiation favour the invasion of adventive pioneers that originate in warmer regions. This applies to roadside communities as well as to eutrophized stands in neighbouring agricultural sites. But this phenomenon of growing species richness should be explained in connection with the third (temperate woodland) zone that encompasses nemoral woodlands and the austral regions of the subantarctic.

2.3 Temperate woodland zone

The majority of forests in temperate climates feature a low level of alpha diversity (Fig. 5). Some North American coniferous and deciduous forests constitute an exception and exhibit much greater diversity. This difference is due to migration

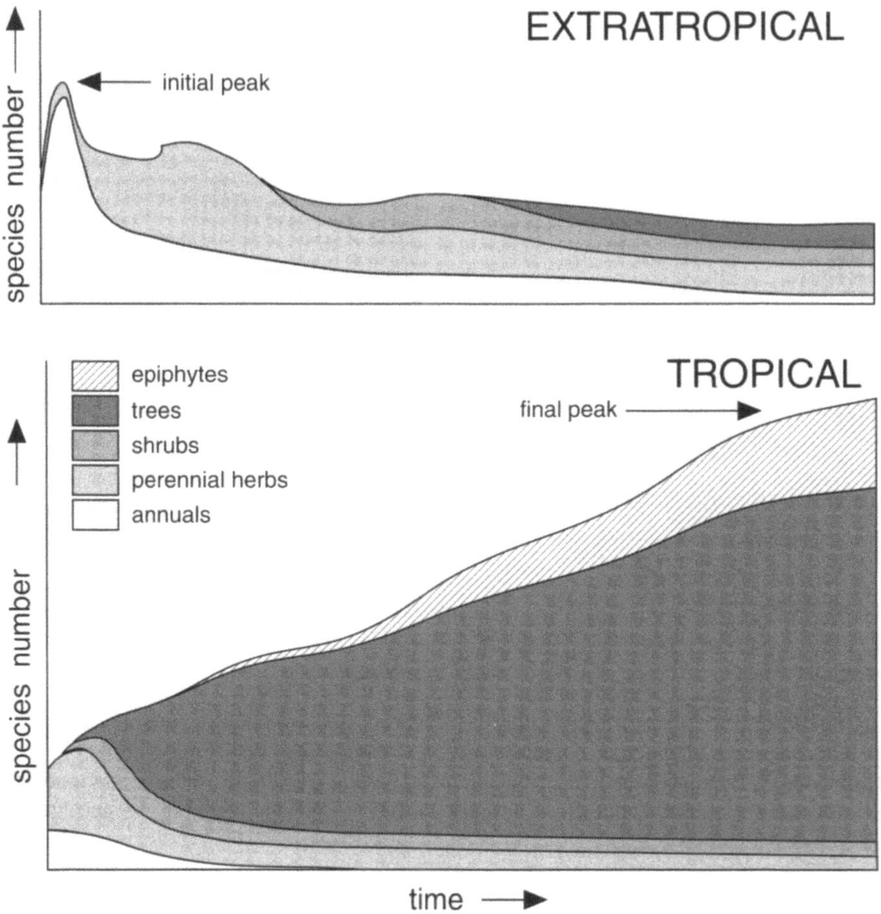

Fig. 6. Diagram of the development of species richness during secondary succession in abandoned fields in Central Europe (above) and in the tropics after shifting cultivation (below)

Ilustración 6. Diagrama del desarrollo de la riqueza de especies durante la sucesión secundaria en campos abandonados en Europa Central (arriba) y en los trópicos después del cultivo alterno (abajo)

caused by climate change in the course of the ice ages. During that period, mountain ranges such as the Alps and Pyrenees formed a barrier that kept plants out. By contrast, North America is dominated by meridional mountain ranges which made floristic migration much easier there. As a result, the species factor in North America can be five times as high as at equivalent and comparable sites in Europe. For example, it is possible to find 15 different species of conifers per hectare in the Klamath Mountains of northern California, whereas in the forests of Central Europe, four to five tree species per hectare would be considered a large number.

Besides the prevailing natural conditions, each of the more or less temperate regions is characterized by a sizeable floristic enrichment caused by human intervention. In Europe, this resulted from the large number of spontaneous invaders since the time of neolithic migration. In North America and at the southernmost tip of South America, a similar phenomenon occurs with the transfer of weeds that accompany seed imports.

When herbicide impact is low, the cultivation stage and the first fallow stage during secondary succession in abandoned fields or vineyards indicate a high level of floristic diversity which is due to a relatively high number of spontaneous annuals (Fig. 6, upper diagram). The level of species richness is lower during later stages of abandonment due to the decrease in annuals and herbs that are superseded by shadow-giving shrubs and trees. As a consequence, high diversity levels in the temperate zones of Central Europe can be conserved only when a variety of land uses that involve limited chemical input are allowed to exist (Richter 1997).

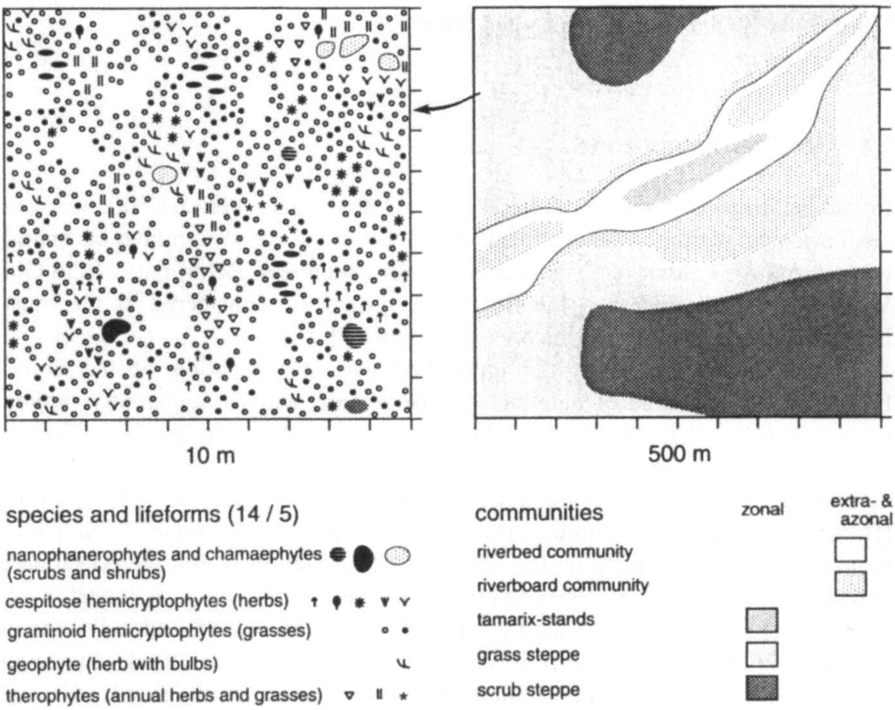

Fig. 7. Illustration of plant distribution (left: 100 m^2) and community patterns (right: 2.5 km^2) in steppe grasslands (models: Kaskelen, Kazakhstan and Badlands National Park, South Dakota, USA)

Ilustración 7. Ilustración de la distribución de las plantas (izquierda: 100m^2) y patrones de la comunidad (derecha: 2,5km^2) en praderas y estepas (ejemplos: Kaskelen, Kazakhstan y Badlands National Park, Sur Dakota, EE.UU.)

2.4 Temperate steppe zone

Even though the fourth zone – steppe grasslands and temperate shrublands – typically has a sparser human population, humans still have a very strong impact in such areas. Alpha diversity is apparently richer in such areas than in most temperate woodlands (Fig. 7). Because there are so few untouched steppe regions in the world, it is difficult to calculate the degree of phytodiversity in natural stands. To date, floristic heterogenity seems to be greater in North American prairies than in similar ecosystems in Eurasia. Once again, this might be the result of migration during the ice age, equivalent to the one mentioned earlier.

On the other hand, Eurasian steppes include sites that have been completely degraded by hundreds of years of overgrazing and by dry farming. Today, a large portion of these areas have been abandoned but there has been no real increase in valuable invaders. The recent loss of a large portion of their idiochorophytes and archaeophytes was due to desertification or, to be more precise, the dehumidification of the upper soil. This problem is well-known from the devastation of the area around Lake Aral and by the disaster provoked by the "Dust Bowl" in the midwestern United States.

2.5 Mediterranean zone

In the subtropical zone, it is necessary to distinguish between the western meditarranean region of the continents where the climate is determined by humid winters and the eastern region which has humid summers. Being part of the Old World, the Mediterranean – the fifth zone – must be considered as the most important genetic pool from which weeds originated.

Prior to man's influence, the untouched natural vegetation of the South European unit consisted of pine and oak woodlands with an apparently limited species variety (Fig. 8). On the other hand, open semi-arid scrub in French garrigues, Spanish tomillares and Greek phrygana reveal the highest level of alpha diversity in all of Europe, with sometimes more than 100 species within 100 m^2 (Bergmeier 1994). Furthermore, northern Mediterranean and Illyrian vegetation in mountainous areas seem to be the hot spots of speciation in Europe (Gaston & David 1994). Akeroyd & Heywood (1994) list the Mediterranean as the site with seven of the nine centres of plant diversity and endemism to be found in Europe.

Man's impact in the form of cleared forests and agricultural expansion supported the evolution of a large number of annual dicots for thousands of years. Today, many of those ruderal and segetal therophytes and steppe grasses must also be regarded as neophytes that threaten the highly endemic and diverse original flora of Mediterranean ecosystems in California, central Chile, South Africa and southwestern Australia.

Large-scale development of phytodiversity in the Mediterranean is just the opposite for precisely that reason. Although the flora of southern Europe

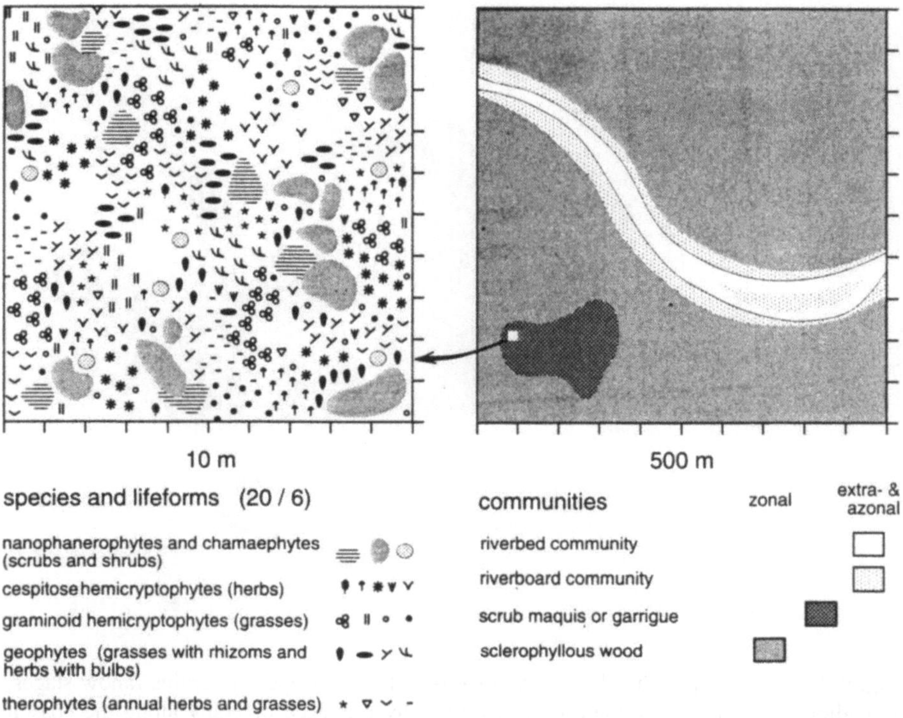

Fig. 8. Illustration of plant distribution (left: 100 m^2) and community patterns (right: 2.5 km^2) in a highly heterogeneous semiarid Mediterranean garrigue. In contrast to the species-rich type of garrigue in the left figure, the one on the right shows a typical homogenous distribution of evergreen forest communities (models left: Mani, Pelopones and Almeria, Andalusia; right: Azrou, central Morocco)

Ilustración 8. Ilustración de la distribución de las plantas (izquierda: 100m^2) y patrones de la comunidad (derecha: 2,5km^2) en una garriga mediterránea semiárida y muy heterogénea. En contraste con el tipo de garriga rico en especies que se observa en la ilustración de la izquierda, la de la derecha muestra una típica distribución homogénea de comunidades de bosques siempre verdes (ejemplos a la izquierda: Mani, Peloponeso, y Almería, Andalucía; a la derecha: Azrou, centro de Marruecos)

has gained a lot of species as the result of man's influence, equivalent environments on other continents are suffering from the mass invasion of the very same newcomers. Furthermore, in the Mediterranean of the Old World, dense assemblages of cultivated land in different stages of fallow intermingled with less impacted semi-natural sites often artificially increase beta diversity (Fig. 9; Carl & Richter 1989). In contrast, the recently colonized Mediterranean parts of the New World do not show similar signs of renaturalization arising from abandonment. Here agro-industrial cultivation as well as extensive pasturing contribute to the decline of these areas' former community richness. For example, natural Californian formations like soft or hard chaparral, coastal scrub, valley grasslands or vernal pools disappear or receive homogeneous structures as a result of human influence.

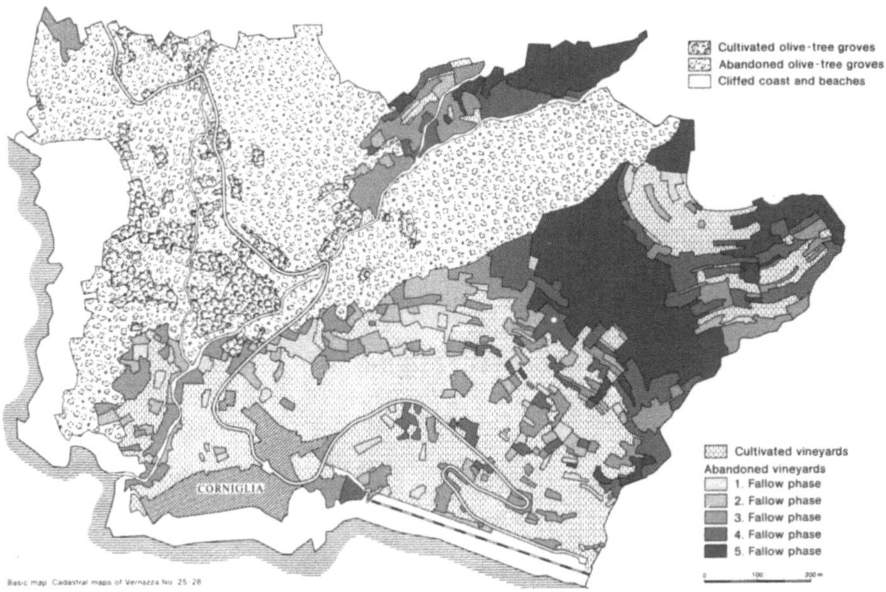

Fig. 9. Land-use map with cultivated vineyards and olive groves at various fallow stages, providing an example of a high level of artificially produced diversity along the coastal slope of the Ligurian Apennine in the Cinque Terre region of Italy
Ilustración 9. Mapa de uso de la tierra con cultivo de viñedos y olivares en varias unidades barbechadas que ofrece una muestra de un alto nivel de diversidad producida artificialmente a lo largo de las lomas de la costa del Apenino Ligur en la región de Cinque Terre en Italia

2.6 Subtropical woodlands along the eastern edges of continents

The sixth zone – the subtropical eastern strip of continents that has at least summer rains or even no seasonal droughts – must be considered as a transitional zone between the temperate and tropical zones. Consequently, elements of both parts meet, creating a high level of diversity especially in tree species. This is particularly true of the southeastern parts of the United States, Brazil, China and South Africa. In the case of the coastal plains of the North American Atlantic transitional zone, the neotropical connection is expressed by epiphytes which include mostly species of *Tillandsia*. On the other hand, the relatively large number of woody lianas in the transitional zone of South Africa evidences a relationship to the paleotropical kingdom (Fig. 10).

However, the subtropical eastern rim of Australia seems to have a lower level of diversity in trees. This might be the result of the continent's isolated location and its predominance of arid regions. Furthermore, many *Eucalyptus* species cause soil acidity that renders problems for herbaceous competition. And finally, most Australian soils are extremely lacking in nutritional properties that favour rather homogenous communities of specialists. In any case, the impact of

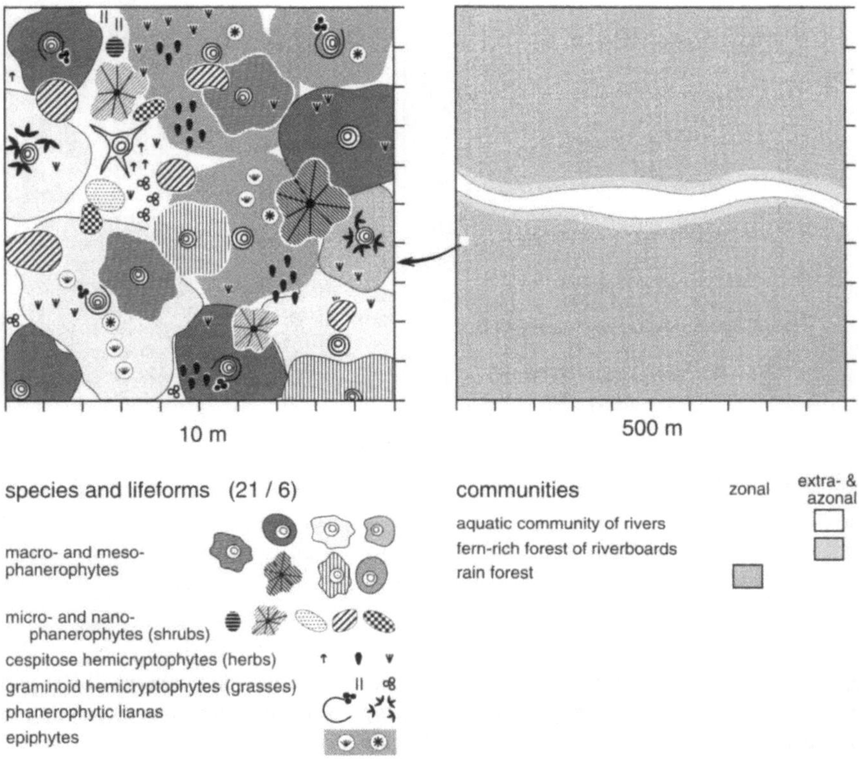

Fig. 10. Illustration of plant distribution (left: 100 m^2) and community patterns (right: 2.5 km^2) in perhumid subtropical types of vegetation along the eastern rim of a continent. The left figure points out a highly diverse South African rain forest with lianas (models: Tsitsikamma, Cape Province and Cape Vidal, Natal), whereas the community figure shows a homogenous assemblage of an East Australian *Eucalyptus* rain forest (Blue Mountains)
Ilustración 10. Ilustración de la distribución de las plantas (izquierda: $100m^2$) y patrones de la comunidad (derecha: $2,5km^2$) en tipos de vegetación subtropical perhúmeda a lo largo del borde este de un continente. La imagen de la izquierda muestra una selva tropical surafricana altamente diversa con lianas (ejemplos: Tsitsikamma, Provincia del Cabo y Cabo Vidal, Natal), mientras que la imagen de la comunidad presenta un grupo homogéneo de una selva tropical de *Eukalyptus* en el este de Australia (Blue Mountains)

clearing land and agricultural degradation in humid and semihumid subtropical transition zones generally lead to reduced diversity. Sugar cane fields are the best example of this!

2.7 Hot arid zone

The seventh zone comprises the arid transition from subtropical to tropical deserts. An interesting aspect has to be mentioned here: In terms of flora, small

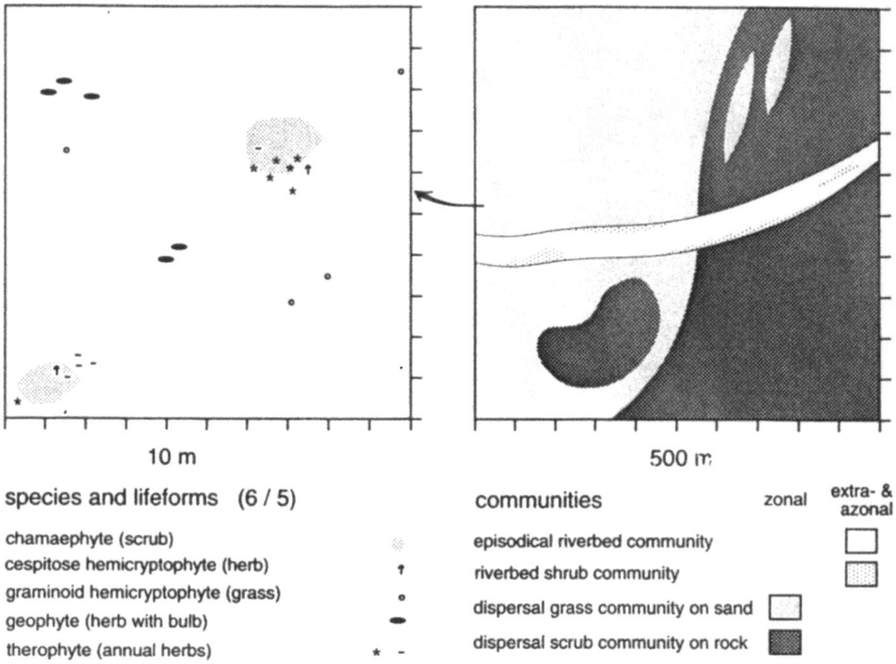

Fig. 11. Illustration of plant distribution (left: 100 m^2) and community patterns (lower part, right: 2.5 km^2) in a semi-desert, pointing out the type of plant coverage as depending on substrate (models: Tozeur, southern Tunis and Mhamid, southern Morocco)
Ilustración 11. Ilustración de la distribución de las plantas (izquierda: 100m^2) y patrones de la comunidad (abajo, derecha: 2,5km^2) en un semidesierto, indicando el tipo de cobertura vegetal dependiente de substratos (ejemplos: Tozeur, sur de Túnez, y Mhamid, sur de Marruecos)

deserts seem to be much more diverse than large deserts. Jürgens' paper (same volume) refers to the very highly developed endemism of the small Nama Karoo region. In this area, the Mesembryantemaceae family appears to be more species rich than the entire range species of all Saharo-Arabian families together. And even though most deserts in the southwestern part of the US are not real deserts, each of these semi-deserts – such as the Mojave, Sonora and Chihuahua – contains more species than the Saharan Province which is much larger. On addition, the very narrow coastal desert of Atacama and Peru has very scarce plant distribution but still contains some 1,000 species – nearly the same number as the much larger Sahara and Australian Desert.

Furthermore, the coastal Atacama in northern Chile displays a highly differentiated speciation at adjacent sites that have a pronounced fog-induced scrub and cacti vegetation due to the paleoecological isolation of the same stands over a long period of time (Richter 1995). Besides paleo-effects of endemism, the degree of diversity in deserts depends on the heterogeneity of the substrate (Fig. 11) that influences nutritional, hydrological and microclimatic conditions. In addition, the amount of therophytes and geophytes determines the varying

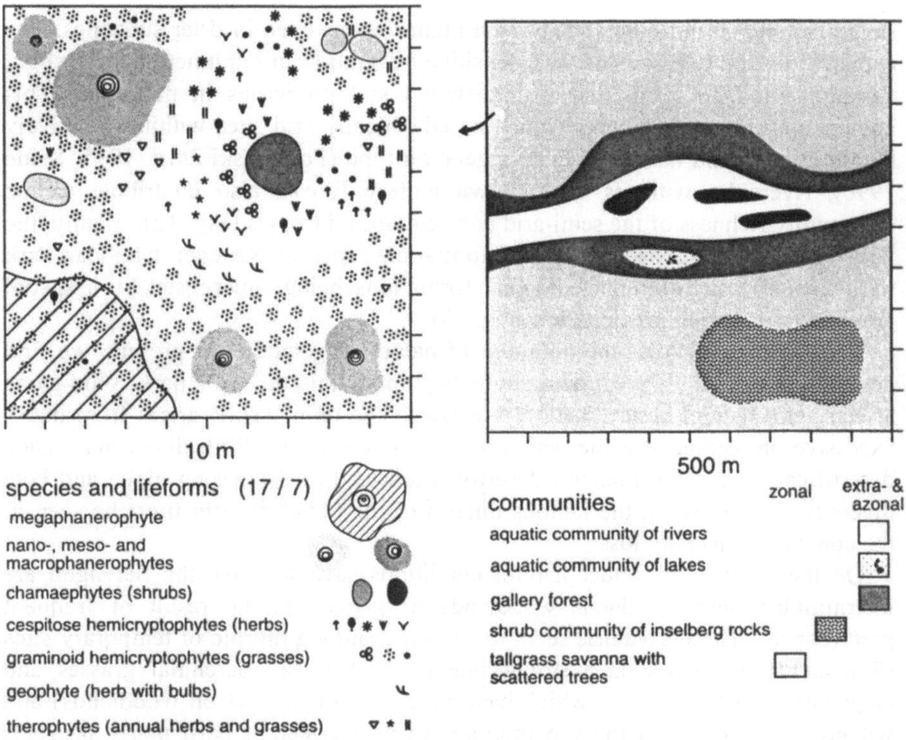

Fig. 12. Sketch of plant distribution (left: 100 m^2) and community patterns (right: 2.5 km^2) in a savanna with scattered tree stands (models: Mazunga/Southern Zimbabwe and Bukalo/Caprivi in Namibia)

Ilustración 12. Esbozo de la distribución de las plantas (izquierda: 100m^2) y patrones de la comunidad (derecha: 2,5km^2) en una sabana con una población de árboles diseminados (ejemplos: Mazunga/sur de Zimbabwe y Bukalo/Caprivi en Namibia)

diversity patterns in plant composition. Finally, an influx of Mediterranean weeds into irrigated locations, especially into palm groves, has to be mentioned here. In contrast, alpha diversity in most semi-deserts suffers from the effects of overgrazing.

2.8 Tropical savanna and deciduous forest zone

The transition from subtropical to tropical zones is not very obvious in warm deserts. Actually, it should be pointed out that warm deserts on the whole cause a much greater break between subtropical and tropical flora than the eastern transitional zone does. Because of this gap between the poles, the input of inner-tropical taxa is much greater in the eighth zone – the savannas, including dry tropical forests – and leads to a dramatic increase in alpha diversity as humidity levels increase.

Beta diversity (if not others) can reach higher levels even in drier subzones. The explanation: the reactions of very sensitive vegetation in conjunction with slight variations in land forms and with different soil properties in particular. This applies not only to inselbergs which are edaphically arid sites within a relatively humid climate and have a distinct vegetation type (Porembski *et al.* 1995, Seine 1996), riversides with its well known gallery forests also contribute to the community richness of the semi-arid and semihumid tropics (Fig. 12). Finally, the very obvious dense patch work of formation types at Krueger National Park results mostly of different geological formations being superposed by a spatial gradient of changing precipitation (Fig. 13).

The same map shows the optimum of natural vegetation. At present, we find devastation caused by overgrazing by wild game within the same park. Wild game affects semi-natural desertification processes. The term "semi" arises from man's excessive protection of game which leads to overgrazing. Both direct man-made desertification and semi-natural desertification lead to decreasing alpha and beta diversity. In any event, the famous images of the Sahel disaster must be seen in the context of diversity loss.

On the other hand, under natural conditions savannas like the Serengeti are intermingled with deciduous woodlands or groves as the result of frequent perturbances. These savannas seem to form a changing mosaic of temporary sites (Sinclair 1979). There is a strong interaction between perennial grasses and elephants, giraffes (both of which have an important impact on woodlands) and wildebeests. Fire has a more pernicious, overall effect. It particularly prevents young trees from growing up into the canopy. This results in uneven age distribution and tree populations may become highly unstable. Decennial or secular fluctuations of this type lead to the hypothesis that alpha and beta diversity may also vary within the same ecosystem.

2.9 Tropical rain forest zone

The relative decline in diversity caused by human influence becomes even more dramatic in the ninth zone tropical rain forests. Tropical heterogeneity on undisturbed sites is supported primarily by the extreme diversity of trees, climbers (Gentry 1982), and, especially in the Neotropics, epiphytes (Ibisch 1996). There is a striking difference between lowlands and areas at higher elevations: The lowlands on the one hand exhibit tree megadiversity but a relatively limited epiphytism that harbours only a few larger examples of vascular plants and a limited number of bryophytes and lichens. On the other hand, at higher levels we find a reduction in tree heterogeneity in combination with epiphyte megadiversity including many smaller vascular plants, mosses and lichens (Richter 1996). Epiphytic diversity patterns depend on the niche distribution of phorophytes, too (i.e. the host): Being drier and having more sunlight, the crowns are much richer in vascular epiphytic species and individuals than the trunks with predominant bryophyte mats in the more humid but obscure lower canopy (Ter Steege *et al.* 1989). Therefore, species and growth-form distribution of epiphytes must

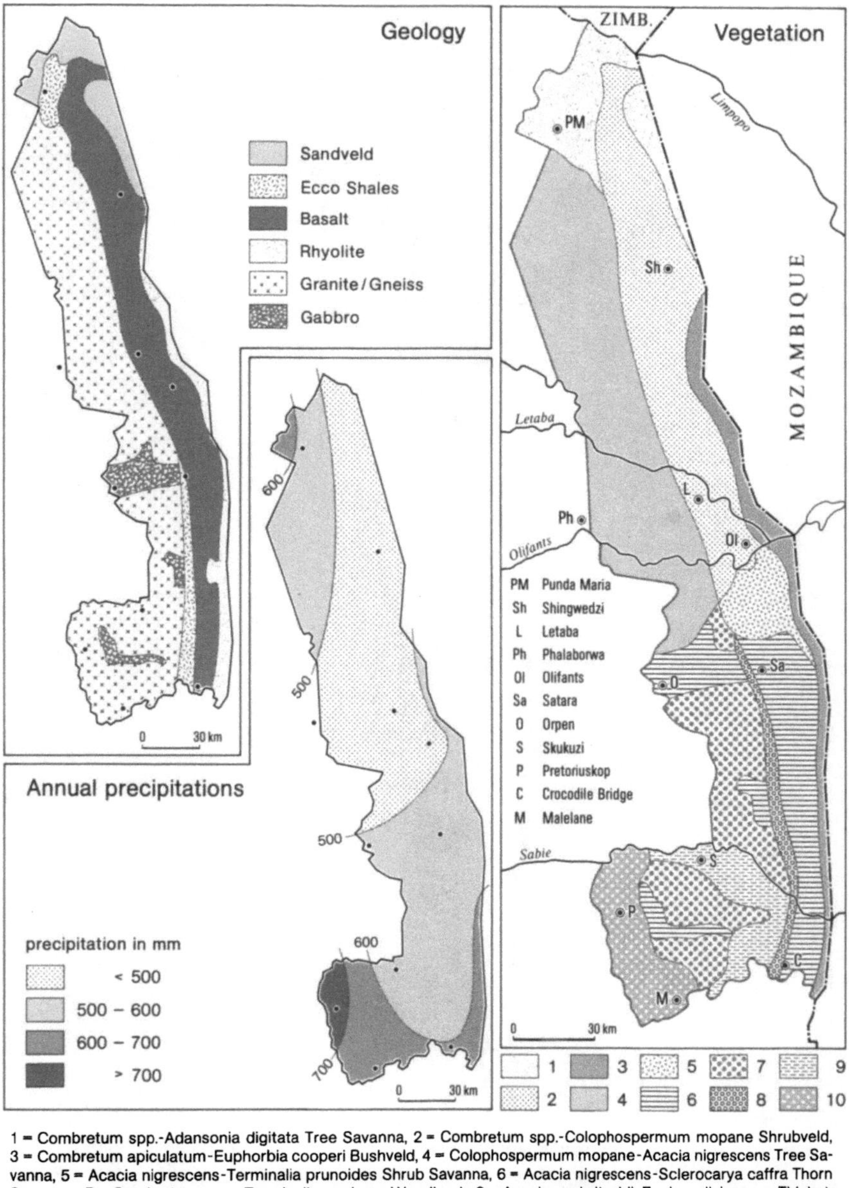

1 = Combretum spp.-Adansonia digitata Tree Savanna, 2 = Combretum spp.-Colophospermum mopane Shrubveld, 3 = Combretum apiculatum-Euphorbia cooperi Bushveld, 4 = Colophospermum mopane-Acacia nigrescens Tree Savanna, 5 = Acacia nigrescens-Terminalia prunoides Shrub Savanna, 6 = Acacia nigrescens-Sclerocarya caffra Thorn Savanna, 7 = Combretum spp.-Terminalia sericea Woodland, 8 = Acacia welwitschii-Euclea divinorum Thicket, 9 = Acacia nigrescens-Combretum apiculatum Thicket, 10 = Combretum apiculatum-Dichrostachys cinerea Bushveld

Fig. 13. Distribution of main plant communities in Krueger NP, South Africa (based on Gertenbach 1983, simplified) in relation to precipitation and geological features as an example for a highly differentiated complex of ecosystems which is the result of the superposition of humidity changes on soil properties

Ilustración 13. Distribución de las principales comunidades de plantas en el Parque Nacional Krueger, Suráfrica (basada en Gertenbach 1983), reducida según las precipitaciones y características geológicas como ejemplo de un complejo de ecosistemas altamente diferenciado que resulta de la superposición de cambios de humedad en las propiedades del suelo

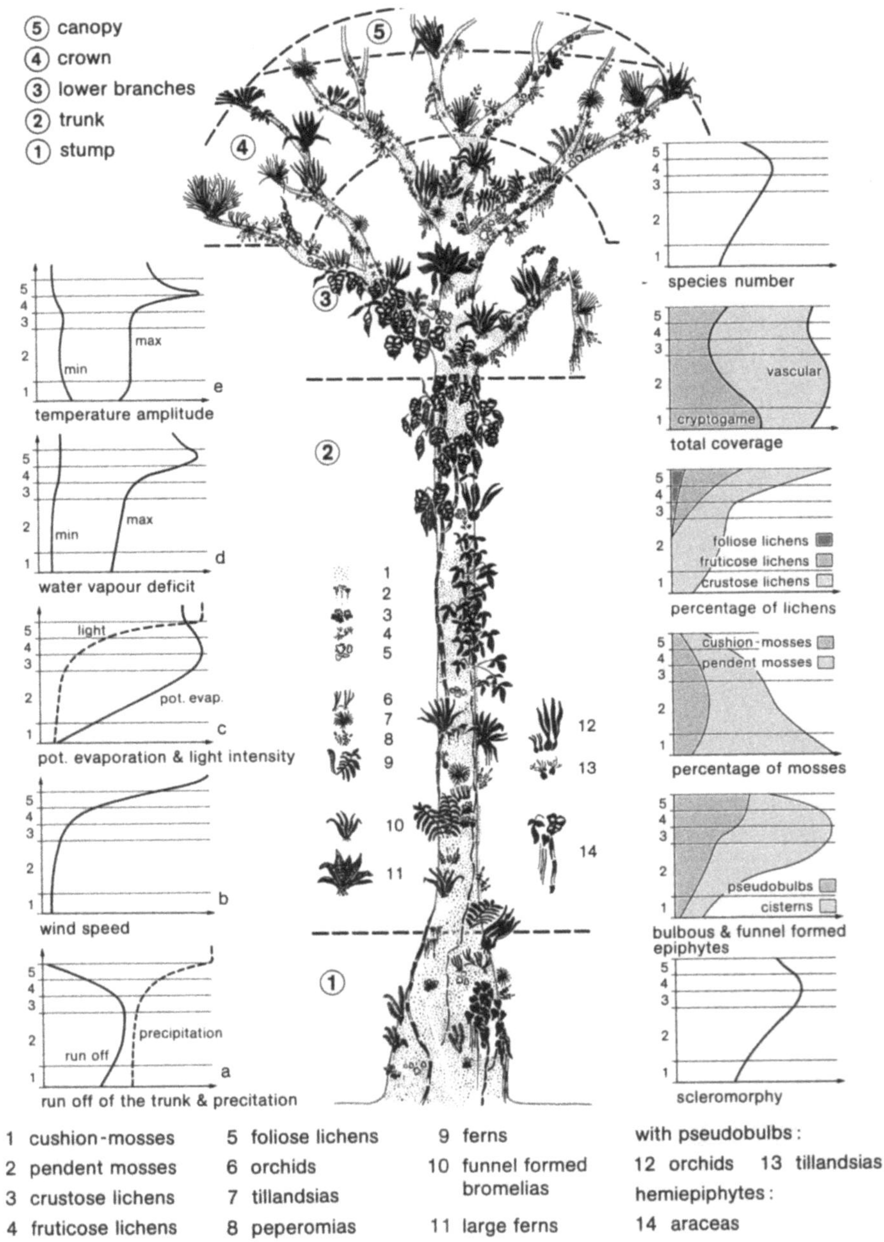

Fig. 14. Scheme of epiphyte zonation on phorophytes which is determined by microclimatic differentiation (l), including the relative spatial distribution of variable groups of epiphytes (r). Based on measurements and type of plant distribution at Finca Irlanda, Chiapas; strata according to Johannson (1974)

Ilustración 14. Esquema de la distribución por zonas de epífitas en forofitas, que viene determinada por la diferenciación microclimática (1), incluyendo la relativa distribución espacial de grupos variables de epífitas (r). Basado en medidas y tipos de distribución de plantas en Finca Irlanda, Chiapas; estratos según Johannson (1974)

be viewed as a multi-layered complex of variable ecological environments within the same forest stand (Fig. 14).

In the tropics deforestation that provokes the loss of diversity is caused by a population explosion in conjunction with population shifts - as is the case at the eastern slopes of the Andes. Of course, in light of the fact that we are sitting in highly artificial Central Europe which has a completely altered landscape and profits from the Third World's low wage levels, we are certainly not in a position to condemn population flux and its destructive effects.

But the very situation in their part of the world seems to be the reason why scientists in Europe feel responsible for the development of sustainability concepts for the tropics that are based on the principle of providing the best imitation of natural plant formations. Their ideas include the cultivation of a variety of trees in dense orchards that have different canopy layers. And if there is a need for annual field crops, all open arable land should at least be situated within a patchwork of tree stands, as it is the case of traditional rice landscapes in southeastern Asia. Tropical agriculture should also strive for high alpha and beta diversity.

2.10 High mountains

In the multizonal complex of high mountains, the greatest species and community richness is situated mainly around timberline ecotones in mountains of the middle latitudes - regardless of whether they are humid or rather arid (Richter 1996). An extreme variety of formations such as dense forest and krummholz stands, heath dwarf shrub, alpine meadows and bogs, fellfields can be found in such areas. And species numbers are high in most of these communities due to the fact that their small microclimatic conditions vary on only a small scale.

This highly developed niche diversity is not comparable to niche diversity in most tropical high mountain areas. For example, the Altiplano type of topography dominates in the páramos of the northern Andes and in the so-called "afro-alpine" tussocks. This topography has less pronounced landforms because the last glaciation had less impact in those regions. Besides this limited gamma diversity, páramos might be influenced by fires set by man or grazing cattle that enhance uniformity at least on beta turnover level (pers. comm. J. Fjeldsa, Copenhagen). With regard to alpha diversity, unintentionally imported plants from Europe might have a suppressive effect on the area's original flora. Today, most parts of the Andean páramos or puna grasslands are affected by "neophytes" such as *Rumex acetosella*, *Trifolium repens*, *Anthoxanthum odoratum* and *Holcus mollis*.

Since high mountains in the middle latitudes are generally not the subject to human influences, an undulating change pattern or altitudinal gradient of diversity seems to be characteristic for such areas, whereas the curve for tropical high mountains displays a slight bulge (Fig. 16). Its most prominent peak generally coincides with the belt where maximum precipitation occurs. And since

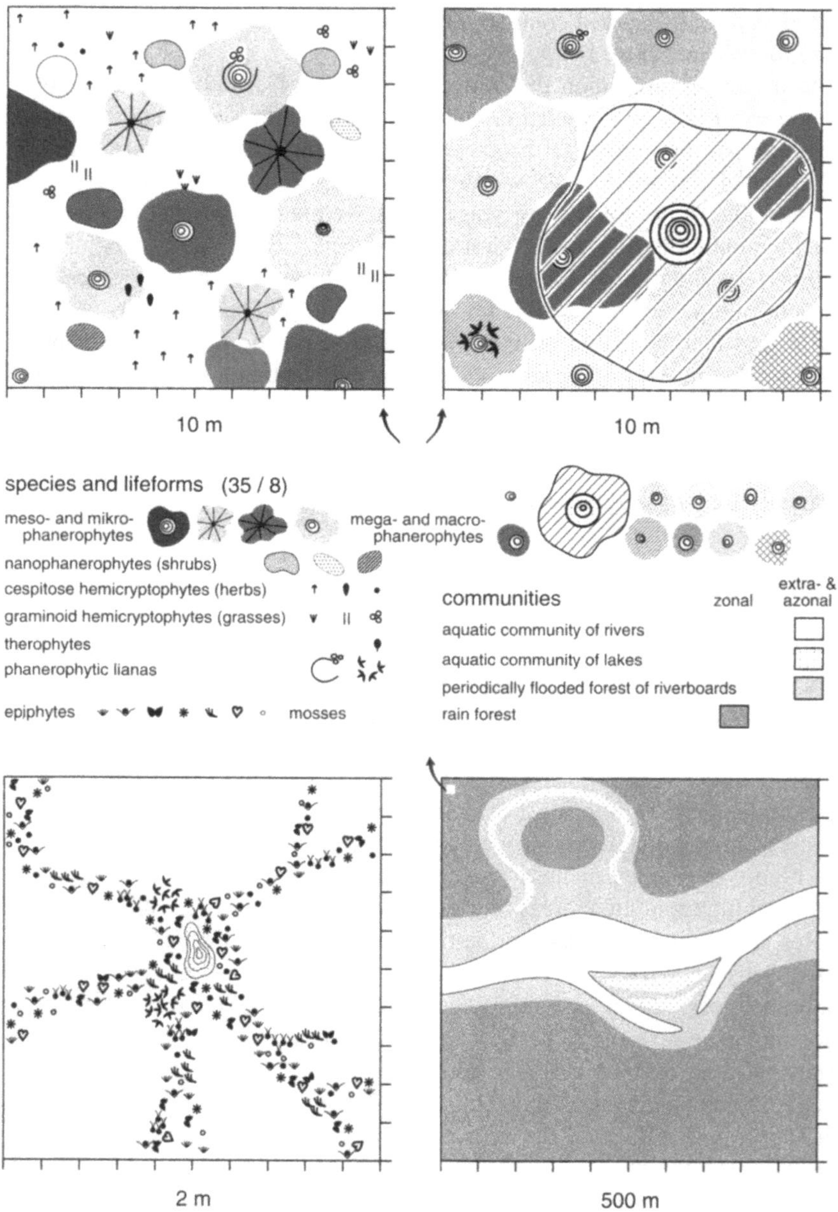

Fig. 15. Illustration of plant distribution (epiphyte patterns at lower left: 4 m^2, pattern of lower strata upper left and pattern of canopy strata upper right: 100 m^2) and community patterns (lower right: 2,5 km^2) in an epiphyte-rich tropical rain forest (models: Huixtla, Chiapas and Zamora, southern Ecuador)

Ilustración 15. Ilustración de la distribución de las plantas (patrones de epífitas arriba a la izquierda: 4m^2, patrones de estratos más bajos arriba a la derecha y patrones de estratos de copas de árboles abajo a la izquierda: 100m^2) y patrones de la comunidad (abajo a la derecha: 2,5km^2) en una selva tropical rica en epífitas (ejemplos: Huixtla, Chiapas, y Zamora, sur de Ecuador)

this belt is only to be found in humid tropical mountains - at rather low levels (Lauer 1976) - the greatest diversity is also located there. According to Gentry (1995), a strong correlation exists between number of tree species and altitude. Witte (1994) and Wolf (1993) demonstrate exponential curves for beta turnover of epiphytes, including bryophytes and lichens (Fig. 16). This is true at least for the hot spot of global diversity which is apparently located at the Pacific side of the Andes in Colombia and northern Ecuador.

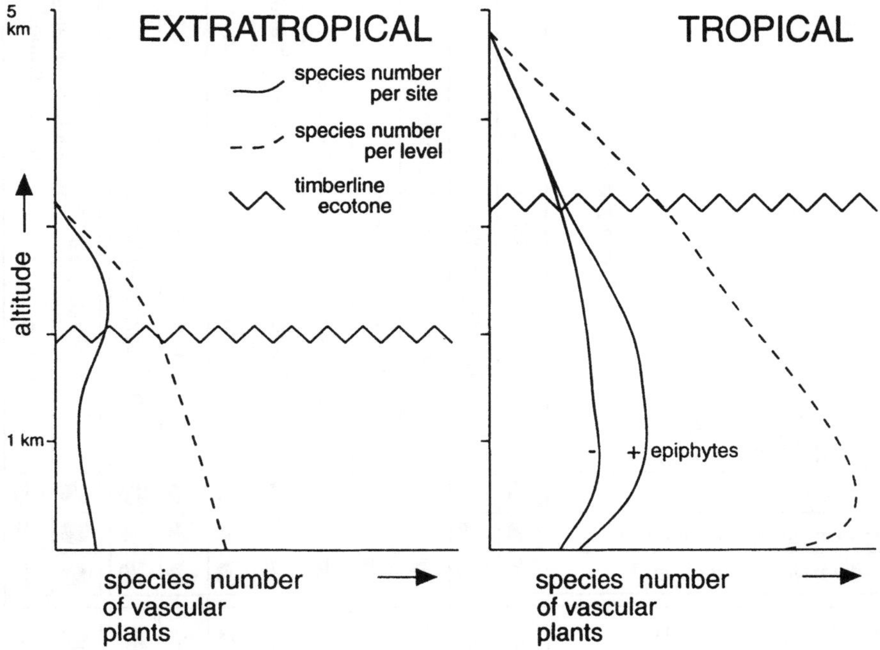

Fig. 16. Altitudinal trends displayed by alpha diversity (number of species per site) and gamma diversity (number of species per level) in high mountains in the humid middle latitudes and humid tropics. The illustration of the tropics shows numbers of species per site without (-) and with (+) inclusion of vascular epiphytes

Ilustración 16. Tendencias altitudinales desplegadas por la diversidad alfa (número de especies por sitio) y la diversidad gamma (número de especies por nivel) en la montañas altas de las latitudes medias húmedas y trópicos húmedos. La ilustración de los trópicos muestra números de especies por sitio sin (-) y con (+) inclusión de epífitas vasculares

3 Conclusions

In conclusion, we should return to the statement mentioned in the introduction that "biodiversity is an answer without question". Of course, this is not really true. First of all, biodiversity means that there are many answers rather than just one single answer. Two of them could be rather important for future environmental development planning and are demonstrated as a (preliminary and very rough) synopsis of this presentation in Fig. 17:

- Every ecozone displays its own specific diversity patterns
- In every ecozone the degree of diversity in natural systems differs from the degree of diversity of man-made ecosystems, generally on a diametrical basis

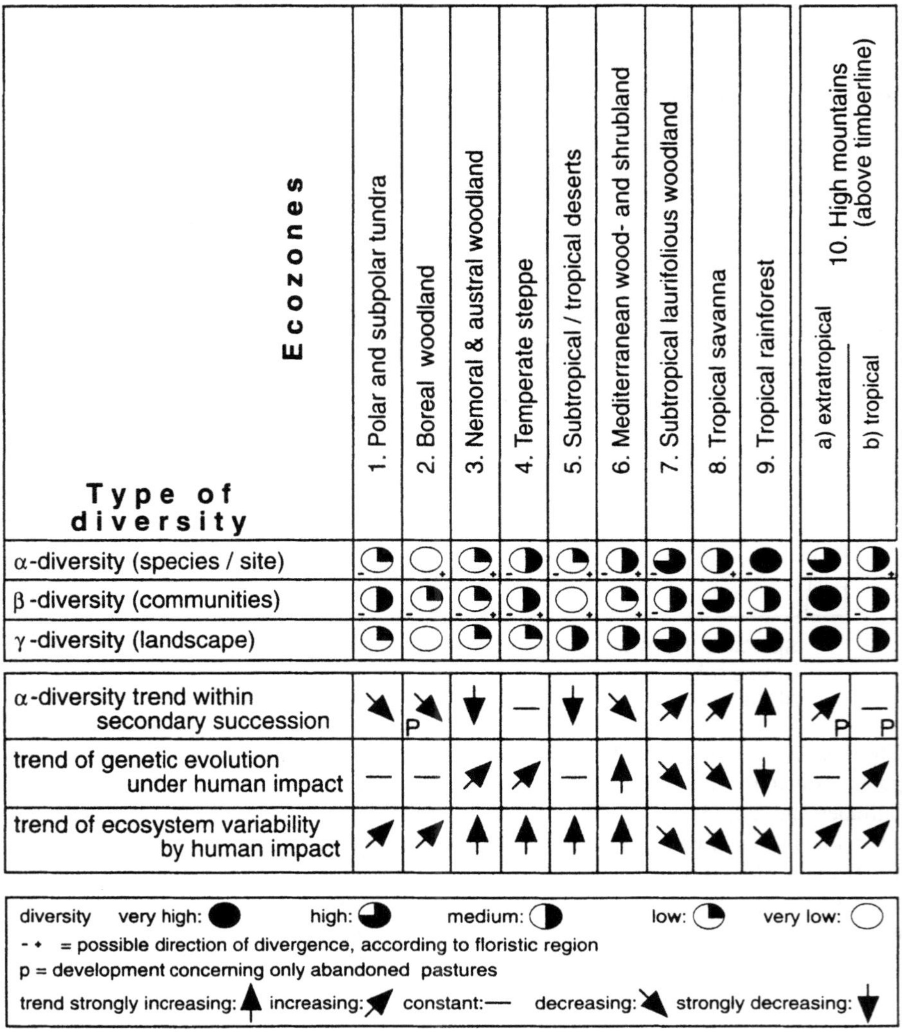

Fig. 17. Hypothetical evaluation of various diversity parameters under mean site conditions and an assessment of diversity change in the course of secondary succession and through human impact in different ecozones

Ilustración 17. Evaluación hipotética de varios parámetros de diversidad bajo condiciones medias de situación y una evaluación de cambios de diversidad en el curso de la sucesión secundaria y a través del impacto humano en diferentes ecozonas

The latter fact is rather obvious from the secondary succession illustrated by Fig. 6: The extremely high diversity of trees and epiphytes in natural tropical forests lead to increased species richness after abandonment - just in opposite of the situation in Central Europe or the Mediterranean.

A few more answers can be presented as fixed rules: Man's clearing of areas where highly diverse tree-, hemi-epiphyte- and epiphyte-communities exist, as it is the case in the tropics, provokes a lower level of phytodiversity. If an area is dominated by a rather uniform tree community, as is the case in the middle latitudes, clearing encourages greater phytodiversity. In such areas, the lack of over-shadowing trees favours the invasion of different herbs and shrubs. Where there is no upper canopy, diversity within the herb and shrub layer depends on the degree of coverage. In such cases, density once again provokes homogeneity as a result of competition and uniform microclimatic conditions.

And finally, questions regarding biodiversity do indeed exist - but scientists should not give all the answers away.

Economists, for example, may ask whether a high diversity is more profitable than low diversity. Of course it is because the chances of finding valuable plants are greater in environments with high species richness. But this argument should produce scepticism, too, because societies that believe solely in economic growth generally tend to destroy nature - by, for instance, harvesting plants of special interest rather than allowing their national "owners" to carefully prospect them. Therefore, although taxonomists might be able to identify plants of medicinal value, they should also take care of chemical and pharmaceutical corporations by not telling them where they can exactly find them!

Scientists raise questions too: High diversity may guarantee more genetic material for tests and cultivation, just as it may provide more samples for phytoindication or phytomonitoring. But in the first two cases, we should once again take a critical view of the economic value of such research. And in the second two cases which involve the ecological aspects of indication and monitoring, it can be said with certain reservations that modern field studies tend to concentrate more on spontaneous flora than on natural vegetation.

For some of us, there is a philosophical question that might be even more important than all of the practical questions with an economic, political and scientific backdrop: Why does man believe that he is more important than the whole of most remaining species communities? What is the reason of his arrogance? We seem to be important from an anthropocentric point of view - but it would be more appropriate if we were to consider ourselves a part of nature and therefore less important.

References

Akeroyd JR, Heywood VH (1994) Regional overview: Europe. In: WWF & IUCN (eds) Centres of plant diversity I, pp 39-54

Barthlott W, Lauer W, Placke A (1996) Global distribution of species diversity in vascular plants towards a world map of phytodiversity. Erdkunde 50: 317-327

Bergmeier E (1995) Die Höhenstufung der Vegetation in Südwest-Kreta (Griechenland) entlang eines 2450 m-Transektes. Phytocoenologia 25: 317-361

Böhmer J, Richter M (1997) Regeneration of plant communities - an attempt to establish a typology and zonal system. Plant Research and Development 45: 74-88

Carl T, Richter M (1989) Geoecological and morphological processes on abandonned vine-terraces in the Cinque Terre (Liguria). Geoökodynamik 10: 125-158

Edlund SA, Taylor B (1989) Regional congruence of vegetation and summer climate patterns in the Queen Elizabeth Islands, N.W.T., Canada. Arctic 42: 3-23

Gaston KJ, David R (1994) Hotspots across Europe. Biodiversity Letters 2: 108-116

Gentry A (1988) Changes in plant community diversity and floristic composition on environmental and geographical gradients. A Missouri Bot Gard 75: 1-34

Gentry A (1995) Patterns of diversity and floristic composition in neotropical montane forests. In: Churchil SA *et al.* (eds) Biodiversity and conservation of neotropical montane forests, pp 527-539

Gertenbach WPD (1983) Landscapes of the Krueger National Park. Koedoe 26: 9-121

Green RH (1979) Sampling design and statistical methods for environmental biologists. New York

Harper KA, Kershaw PR (1996) Natural revegetation on borrow pits and vehicle tracks in shrub tundra, 48 years following construction of the CANOL No. 1 Pipeline, N.W.T., Canada. Artic and Alpine Research 28: 163-171

Holmen K (1957) The vascular plants of Peary Land, North Greenland. In: Meddelelser om Gronland 124/9. Kopenhagen

Ibisch P (1996) Neotropische Epiphytendiversität - das Beispiel Bolivien. Archiv naturwissenschaftlicher Dissertationen 1. Wiehl

Johansson DR (1974) Ecology of vascular epiphytes in West African rainforest. Acta Phytogeographica Suecica 59: 1-136

Kozlov MV, Haukioja E, Yarmishko VT (eds.) (1993) Aerial pollution in Kola Peninsula. St. Petersburg

Lauer W (1976) Zur hygrischen Höhenstufung tropischer Gebirge. In: Schmithüsen J (ed.) Neotropische Ökosysteme, Biogeographica VII, pp. 169-182

Margalef R (1994) Dynamic aspects of diversity. J Veg Sci 5: 451-456

Ouellet H (1990) Avian zoogeography in the Canadian Arctic Islands. In: Harington CR (ed.) Canada's missing dimension 2, pp. 516-545

Petrovsky VV (1988) Vascular plants of Wrangel Island. Magadan (in Russian)

Porembski S, Brown G, Barthlott W (1995) An inverted latitudinal gradient of plant diversity in shallow depressions on Ivorian inselbergs. Vegetatio 117: 151-163

Rannie WF (1986) Summer air temperature and number of vascular species in Arctic Canada. Arctic 39: 133-137

Richter M (1995) Klimaökologische Merkmale der Küstenkordillere in der Region Antofagasta. Geoökodynamik 16: 283-332

Richter M (1996) Klimatologische und pflanzenmorphologische Vertikalgradienten in Hochgebirgen. Erdkunde 50: 205-237

Richter M (1997) Allgemeine Pflanzengeographie. Stuttgart

Rohde K (1992) Latitudinal gradients in species diversity: the search for the primary cause. Oikos 65: 514-527

Rohde K (1996) Rapoport's Rule is a local phenomenon and cannot explain latitudinal gradients in species diversity. Biodiversity Letters 3: 10-13

Sinclair ARE (1979) Dynamics of the Serengeti Ecosystem. In: Sinclair ARE & Norton-Griffiths M (eds.) Serengeti - Dynamics of an ecosystem, pp. 1-30

Seine R (1996) Vegetation von Inselbergen in Zimbabwe. Archiv naturwissenschaftlicher Dissertationen 2. Wiehl

Stevens GC (1989) The latitudinal gradient in geographical range: how so many species coexist in the tropics. American Naturalist 133: 240-256

Stevens GC (1996) Extending Rapoport's rule to Pacific marine fishes. J Biogeogr 23: 149-154

Takhtajan A (1985) Floristic regions of the world. Berkeley, Los Angeles, London

Ter Steege H, Cornelissen JHC (1989) Distribution and ecology of vascular epiphytes in lowland rain forest of Guyana. Biotropica 21: 331-339

Thannheiser D (1988) Eine landschaftsökologische Studie bei Cambridge Bay, Victoria Island, N.W.T., Canada. In: Mitt Geogr Ges Hamburg 78: 3-51

Witte HJJ (1994) Present and past vegetation and climate in the Northern Andes (Cordillera Central, Colombia): A quantitative approach. Amsterdam

Whittaker RH (1972) Evolution and measurement of species diversity. Taxon 21: 213-251

Whittaker RH (1975) Communities and ecosystems. New York, London

Wolf JHD (1993) Diversity patterns and biomass of epiphytic bryophytes and lichens along an altitudinal gradient in the Northern Andes. A Missouri Bot Gard 80: 928-960

Fluctuations, chaos and succession in a living environment

Frank Klötzli
Institute of Geo-Botany, ETHZ, Zürich, Switzerland

Abstract.
Thesis 1. Biodiversity is not stable at any taxonomical or organizational level. Thus, biodiversity impacts not only the non-taxonomical but also the modificational level (for example, invasive plants may undergo specific modifications and eco-physiological adaptations before speciation takes place in a new niche).
Thesis 2. Biodiversity fluctuates in terms of number and taxa. It is impossible to predict the "ascent," temporary to permanent disappearance or "descent" of a given taxon (species).
Thesis 3. The number of a given population is also subject to chaotic fluctuation. Since we know little about the seed bank or other cryptic living parts of a plant, we cannot precisely analyse the long-term state of a given population.
Thesis 4. Unforeseeable fluctuations in certain site factors (such as water, nutrients and warmth) produce fluctuations in species combinations, especially in indicator species in ecotonal areas or the boundaries between any two sites. As a result, such boundaries or areas may display fluctuations among species groups, depending on the changes in site conditions (such as the water or nutrient regime).
Thesis 5. Species and species groups appear and disappear at a fast rate. In other words, the combination of visible plant species you find on a given vegetated spot will be different from year to year.
Thesis 6. Any change may occur rather rapidly, regardless of whether there is no visible reason or whether a site factor has undergone an obvious change. And the speed and quality of these changes are clearly chaotic and unforeseeable. It is not possible to draw a conclusion or make a prediction even after several years or even when a partial knowledge of what is going on in the way of environmental changes exists. There are detectable tendencies which are often caused by unpredictable weather conditions, on management, on animal numbers or pathogenic influences. But in our latitudes, distinct, directed changes can be recognized with a degree of certainty following the passage of five to 12 years' time. Change can be expected to take place more rapidly in more arid tropical countries or in gaps in a rain forest.
Thesis 7. 1) Unforeseen changes in local biodiversity are far greater than changes that occur between drier and wetter periods of up to six years. **2)** The suitability of many plant species changes, i.e. as indicator species or in terms of their regular occurrence in certain types of savanna (plant communities). **3)** The number of plant species may be plentiful one year and then scarce to non-existent the next, a fate that affects the populations of large areas (in our case > 500 km^2), due perhaps to pathogenic factors that are intrinsic to certain savanna types. **4)** However, the presence of a certain number of permanent species can be used to determine the type of savanna in an area. However, quantitative fluctuations among certain grasses in various types of savannas may lead to changes in physiognomy.

Thesis 8. Any analysis of a given type of vegetation can be viewed as the sheet for one specific day on a tear-off calendar. The population of a species group changes as unforeseeably as clouds, with unpredictable patterns, speeds and directions. Coupled with this is biodiversity which fluctuates irregularly within any unit of area, space or time.
Thesis 9. Quantitative values for biodiversity are therefore in no way constant in any given site or time. The larger the surface, the more constant the value used to depict the level of biodiversity will be (see, e.g., Convoy & Noon 1996 regarding scale). Biodiversity may change considerably at a given site due to warming or other influences, just as it may change completely in areas that experience only slight changes in their site conditions.
Thesis 10. In summary, the extent, temporal pattern and distribution of change cannot be foreseen. These factors are completely chaotic. Only long-term studies provide a sounder foundation.

Fluctuaciones, caos y sucesión en un entorno vivo

Resumen.
Tesis 1. La biodiversidad no es estable a ningún nivel taxonómico u organizador. Por consiguiente, la biodiversidad no sólo afecta el nivel no-taxonómico sino también el nivel de modificación (por ejemplo, plantas invasoras pueden sufrir modificaciones específicas y adaptaciones ecofisiológicas antes de la especición en un nuevo nicho ecológico).
Tesis 2. La biodiversidad fluctúa en términos de número y taxa. Resulta imposible predecir el "ascenso", desaparición temporal a permanente o "descenso" de un taxon dado (especie).
Tesis 3. El número de una población dada también está sujeto a fluctuación caótica. Como sabemos poco sobre el banco de semillas u otras partes vivas escondidas de una planta, no podemos analizar con precisión el estado a largo plazo de una población dada.
Tesis 4. Fluctuaciones impredecibles en ciertos factores locales (tales como agua, nutrientes y calor) producen fluctuaciones en las combinaciones de especies, en particular en especies indicadoras de áreas ecotonales o los límites entre cualquiera de ambos lugares. Como resultado, tales límites o áreas pueden mostrar fluctuaciones entre grupos de especies dependiendo de los cambios en las condiciones locales (tales como el régimen de agua o nutrientes).
Tesis 5. Especies y grupos de especies aparecen y desaparecen a gran velocidad. Dicho de otro modo, la combinación de especies de plantas visibles que uno encuentra en una determinada área cubierta de vegetación será distinta de un año a otro.
Tesis 6. Cualquier tipo de cambio puede producirse con bastante rapidez, independientemente de que no existan razones visibles o de que un factor local haya experimentado un cambio obvio. Y la velocidad y cualidad de estos cambios son claramente caóticas e impredecibles. No es posible sacar una conclusión o hacer una predicción, ni siquiera después de varios años o cuando existe un conocimiento parcial de lo que está sucediendo en forma de cambios medioambientales. Hay tendencias detectables, causadas a menudo por condiciones climatológicas impredecibles, sobre el manejo, el número de animales o influencias patógenas. Pero en nuestras latitudes, cambios precisos, orientados, pueden reconocerse con un cierto grado de certeza siguiendo el transcurso de un periodo de tiempo de 5 a 12 años. Se puede esperar que en países tropicales más áridos o en claros en una selva tropical se produzcan cambios con mayor rapidez.
Tesis 7. 1) Los cambios imprevistos en la biodiversidad local son mucho mayores que los cambios que se producen entre periodos más secos y más húmedos de hasta seis años. **2)** La adaptabilidad de muchas especies de plantas cambia, esto es, como especies indicadoras o en términos de su aparición regular en ciertos tipos de sabana (comunidades de plantas). **3)**

El número de especies de plantas puede ser muy abundante un año y al año siguiente reducirse hasta la inexistencia, destino que afecta a las poblaciones de extensas áreas (en nuestro caso > 500 km²), quizá debido a factores patógenos intrínsecos a ciertos tipos de sabana. **4)** No obstante, la presencia de un cierto número de especies permanentes puede servir para determinar el tipo de sabana en un área. Fluctuaciones cuantitativas entre ciertas gramíneas en varios tipos de sabana pueden, sin embargo, llevar a cambios en la fisonomía.
Tesis 8. Cualquier análisis de un tipo dado de vegetación puede considerarse como la hoja de un día específico de un almanaque. La población de un grupo de especies cambia de manera tan impredecible como las nubes, con patrones, velocidades y direcciones que no se pueden pronosticar. Conectado con esto, la biodiversidad fluctúa irregularmente dentro de cualquier unidad de área, espacio o tiempo.
Tesis 9. Los valores cuantitativos de la biodiversidad no son por lo tanto en absoluto constantes en ningún lugar o tiempo dados. Cuanto mayor sea la superficie, más constantes serán los valores usados para representar el nivel de biodiversidad (véase, p. ej., la escala de referencia de Convoy & Noon 1996).

La biodiversidad está sujeta a cambios considerables en un lugar determinado debido a calentamiento u otras influencias; del mismo modo que puede cambiar completamente en áreas que experimentan sólo ligeros cambios en sus condiciones locales.
Tesis 10. En resumen, la extensión, patrón temporal y distribución del cambio no pueden predecirse. Dichos factores son totalmente caóticos. Únicamente los estudios a largo plazo proporcionan una base sólida.

1 Introduction

Is there any stability at all in our environment? Even when we consider virgin or natural sites we find no stability in the way of constancy. Any "stable" natural ecosystem fluctuates around an imaginary point of equilibrium (see e.g., Schulze & Mooney 1983, Walker 1995, van der Maarel 1996).

Heraclit put it in a nutshell when he said "panta rhei" – all is flowing. Species, communities of organisms are flowing – sometimes so rapidly that you could compare them to a river. No species in a given area at a given point in time will be the same when revisited one or several years later.

Of course, this all depends upon how and how often a fluctuating or flowing system is monitored. And it also depends upon the observers and the way they look at it, with the eye of a forester, agronomist or conservationist. "Panta rhei" can be quite relative – especially to the eye of a geologist.

A point of reference, regarding time and place are needed to verify flow or fluctuation. Observations normally focus on the behaviour of selected permanent plots of forest, wetlands or grasslands (compare Heywood & Watson 1995, Kaule 1991, Pickett 1989, 1991, Krahulec *et al.* 1990, Agnew *et al.* 1993, Glenn-Lewin & van der Maarel 1992). Given our aims, we are very fortunate: So-called permanent plots were established for various purposes by older colleagues and myself all over Switzerland (appr. 15,000 plots) starting in the thirties, with most plots being marked in the sixties and seventies. These time series may be analysed for a period of at least twenty to twenty-five years. A period of this length enables us to assess tendencies, trends and fluctuations. Furthermore, the same person handled the vast majority of all plots or special investigation areas, in most cases with the help of watchful co-workers, assistants and students.

The following theory starts by considering fluctuating occurrence at the species level and continues on to fluctuations among groups of organisms, and finally to plant communities in wetlands, grasslands and forests, where diversity and ecosystem functions (e.g., stability) are linked in different ways (McNaughton in Schulze & Mooney 1993; further concepts in e.g. Heywood & Watson 1995, Walker 1995, Vitousek *et al.* 1995, Solbrig 1994, Holdgate 1996). All time series will prove that diversity of a given plot or site is not stable. Diversity is relative and depends on various site factors and in some yet undetermined way on species and time as well. However, organisms themselves are not stable, which Thesis 1 illustrates with a recent example.

Thesis 1.

Biodiversity is not stable at any taxonomical or organizational level. Thus, biodiversity impacts not only the non-taxonomical but also the modificational level (for example, invasive plants may undergo specific modifications and eco-physiological adaptations before speciation takes place in a new niche).

This first theory is related to a botanical and rather typical island story (compare Primack 1995, Heywood & Watson 1995). Due to their isolation, organisms on most islands have or have had their own unique evolution. Any new plant or animal introduced to an island has always met with an ecosystem that was unprepared for that newcomer. Thus feral goats on the island of Hawaii became a problem because they fed on native vegetation, especially above the timberline. As a result of their feeding habits, they eventually became responsible – together with man – for the disappearance of the famous Hawaiian Silversword (*Argyroxiphium sandwicense*), a tall candle-like Asteraceae plant, a life form called "giant caulescent rosette." Today, they exist mostly on Maui. These plants are no longer found on the saddle between Mauna Loa and Mauna Kea.

Approximately 60 years ago, a European weed called "Great Mullein" (*Verbascum thapsus*) was introduced to Hawaii. It also escaped to the saddle area where it adapted to site conditions that differed from those in its native European countries. Great Mulleins growing in the subalpine forests and woodland islets on the saddle look very much like they do in Europe. They are however taller and sturdier. Their growth pattern from seedlings to mature candles is different. They

Fig. 1a. *Agyroxiphium sandwicense* to compare with Fig. 1b (Locality: Haleakale Crater, Mani; Photo Sunder)
Ilustración 1a. *Agyroxiphium sandwicense* a comparar con la ilustración 1b (Localidad: Haleakale Crater, Mani; Foto Sunder)

reach heights of up to more than 2m, at which state their stems lignify. But above the timberline, in the puna-like vegetation on lava scree, they have adapted to the niche once held by the old Silversword. Under these circumstances, their typical appearance has changed to a type of life form approaching the "giant caulescent rosette," and their candle-like stem is sturdier and thicker, primarily due to fasciation (Fig. 1a, b).

Based on these findings, we may conclude:

1. The adaptation of life forms follows certain patterns, that are based on long-established forms which have improved viability
2. The morphological biodiversity of a given system tends to amalgamate introduced organisms, bringing them into line with an "approved" physiognomy

3. Organism biodiversity may change in number, especially when new species are introduced and they exhibit a tendency toward new speciation
4. Given ecosystem structures are more stable than combinations of organisms in a given system. The types of changes described at points 2 and 3 often occur very rapidly, depending upon the effect of stresses, which include man
5. The *Verbascum* has not undergone a similar modification in its natural Old World area (more details in Klötzli 1993b, 1994)

Fig. 1b. *Verbascum thapsus* near Pu`u Kalempeamoa, ≈ 2,900 m, strongly fasciated
Ilustracion 1b. *Verbascum thapsus* cerca de Pu`u Kalempeamoa, ≈ 2.900m, fuertemente aplanado

Thesis 2.

Biodiversity fluctuates in terms of number and taxa. It is impossible to predict the "ascent," temporary to permanent disappearance or "descent" of a given taxon (species).

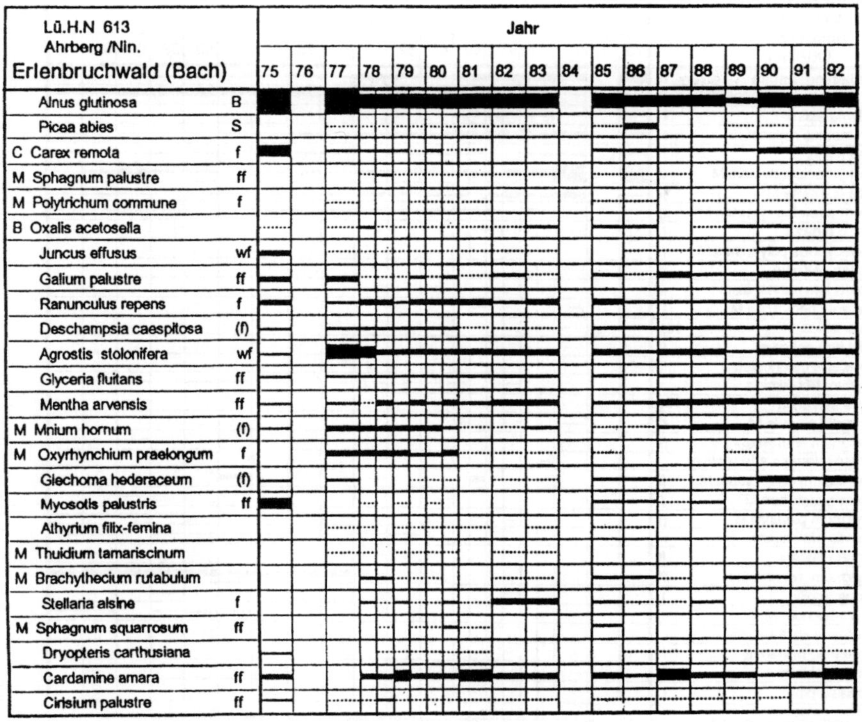

Fig. 2a. Black alder carrforest, near Garlstorf, Nordheide
Ilustración 2a. Bosque de alisos negros, cerca de Garlstorf, Nordheide

Thesis 2 is based on findings obtained from research conducted on the wetlands of the Lüneburger Heide, a famous heathland in northwestern Germany. But it also applies to other areas in Europe and elsewhere where permanent plots have been observed for 20 to 30 years. (Klötzli 1992, 1993a)

Example one examines an alder carrforest which has been monitored for more than 20 years (all examples 1975-96). The level of species fluctuation in this stable forest is rather low. There have however been some newcomers, and other species have vanished for no obvious reason. A windbreak caused by a storm in 1986 led to better light conditions. As a result, the system underwent more obvious changes which then subsided in the course of the following ten years (Fig. 2a).

Example two – a spring basin – is also located in a protected area that is subject to little human influence. The storm of 1986 snapped trees here as well. Wild boar wallow in this wetland and feed primarily on rhizomatous plants. There is a clear tendency towards a more open spring fen and increased local biodiversity (different light and nutrient conditions within the same basin and different degrees of disturbance) (Fig. 2b).

Our third example is a shrub bog that illustrates changes in nutrient conditions and displays a certain local tendency to evolve into a raised bog. Here too, there

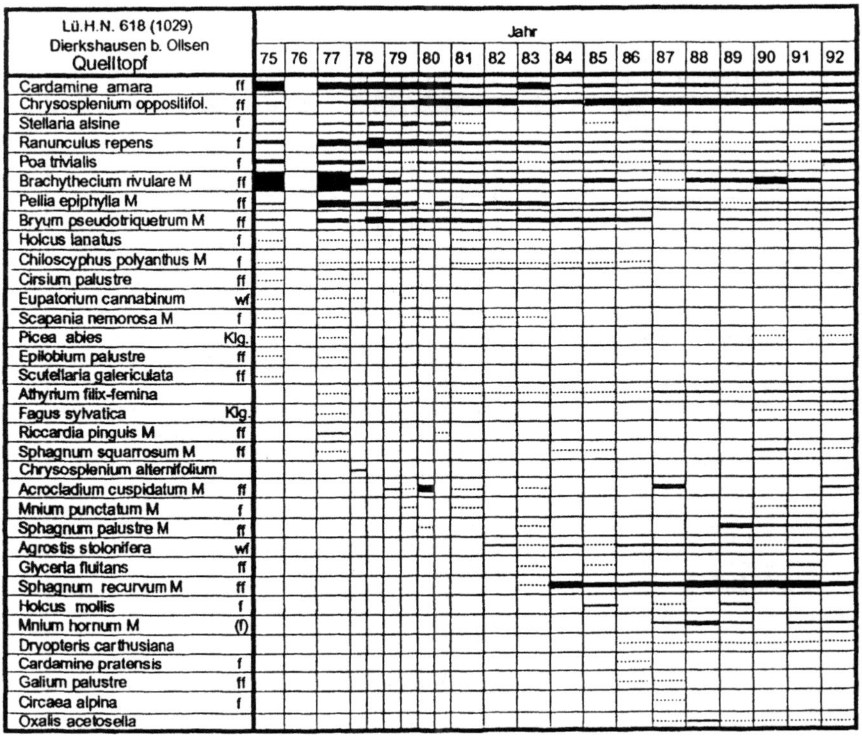

Fig. 2b. *Cardamine amara* spring fer, near Wesel, Nordheide
Ilustración 2b. *Cardamine amara*, cerca de Wesel, Nordheide

are fluctuations which cannot be explained by any known event. These fluctuations can be chaotic, triggered by intrinsic, possibly pathogenic reactions in the ecosystem (see e.g., van der Maarel 1996, "patch dynamics", "carousel model" etc. for more references; compare also Schulze & Mooney 1993, Walker 1995) (Fig. 2c).

Fortunately, there are always a number of species which are always present and thus allow us to make statements regarding the status and conditions of a given site. But the number of fluctuating, emerging or disappearing species cannot be predicted from one year to the next. Biodiversity on a given spot or site is never stable, none of those fluctuations can be foreseen and are, in a way, "chaotic." Only time series of normally more than 10 years allow us to characterize trends or disturbances, using a combination of statistical tests (Wildi & Orloci 1996, Watson 1995).

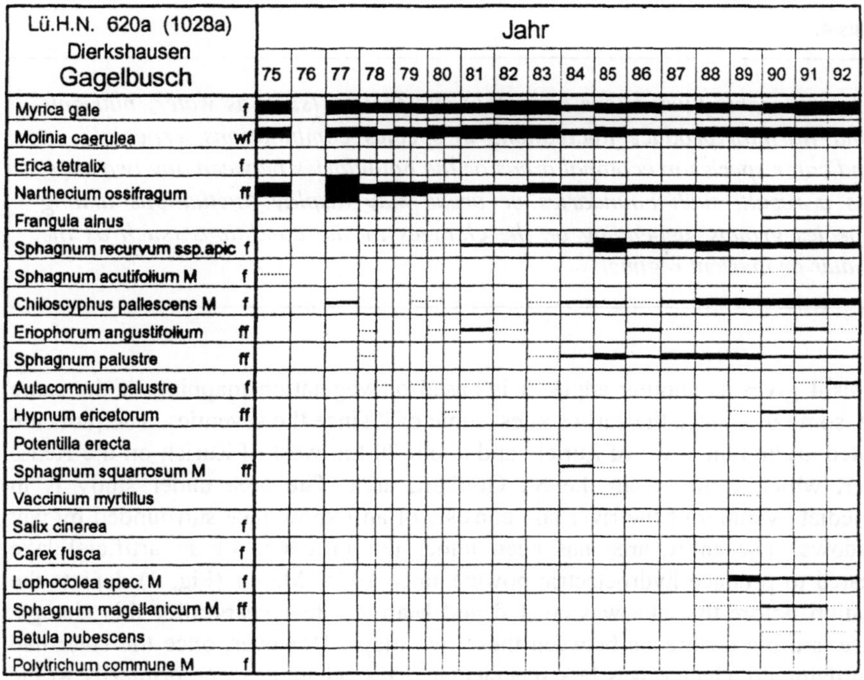

On all tables which show the fluctuations of all species since 1975, the width of the bars are depicting the quantity of a given species in a given year (in percentage cover value)

En todos los gráficos que muestran las fluctuaciones de todas las especies desde 1975, el ancho de las barras representa la cantidad de una especie dada en un año determinado (en porcentaje con valor de cobertura)

cover value (valor de cobertura)
> 75 %
50-75 %
25-50 %
5-25 %
< 5 %
some shoots (algunos retoños)

Fig. 2c. *Myrica* shrub bog, near Dirkshausen, Nordheide (all northern part of Lüneburger Heide)

Ilustración 2c. Pantano de arbustos de *Myrica*, cerca de Dirkhausen, Nordheide (todas las zonas al norte del Lüneburger Heide)

Thesis 3.

The number of a given population is also subject to chaotic fluctuation. Since we know little about the seed bank or other cryptic living parts of a plant, we cannot precisely analyse the long-term state of a given population (See, e.g., Bernhardt 1991 regarding seed bank problems.).

Thesis 4.

> *Unforeseeable fluctuations in certain site factors (such as water, nutrients and warmth) produce fluctuations in species combinations, especially in indicator species in ecotonal areas or the boundaries between any two sites. As a result, such boundaries or areas may display fluctuations among species groups, depending on the changes in site conditions (such as the water or nutrient regime).*

Thesis 4 gives us another angle. It is based on vegetation mapping of the same area, some of it since the early sixties, some of it since the seventies. The plots are located around the town of Zurich, and about 20 km west of Zurich on the Reuss River, which comes from the St. Gotthard area. The area under study is in immediate vicinity of the river and consists of an oxbow lake surrounded by wet meadows. The entire area has been under the influence of an artificial lake (created to produce hydroelectric power) since 1975. Map 1 (Fig. 3) depicts the situation before the lake was created and Map 2 a few years after this change. There is a big difference between these two maps. However, once the river had been dammed, change was less spectacular. Now, change is due primarily to the influence of the precipitation and temperature regimes on the water table and nutrient regime (seepage from fertilized areas). These changes are true environmental changes and naturally influence species occurrence and therefore species groups and plant communities, which in turn change the boundaries between these communities, but without exerting much influence on the central parts of these different units as shown on the map (compare statement of Bakker *et al.* 1996 to link vegetation mapping and permanent plot analysis). This facilitates a direct comparison of such states. To summarize: Boundaries between different ecosystems are by no means stable, especially in ecotonal areas species whose boundaries also shift due to locally changing environmental conditions (Klötzli & Zielinska 1995; compare also Klötzli 1992).

These considerations lead us to our next Thesis:

Thesis 5.

> *Species and species groups appear and disappear at a fast rate. In other words, the combination of visible plant species you find on a given vegetated spot will be different from year to year.*

All species able to establish patch dynamics have their own individual spatial dynamics. This temporal pattern changes every year, thus provoking shifts in species composition on ist "pathway" in a given area, moving species out and bringing species back into the area according to the "carousel model" of van der Maarel (1996). Fluctuations in local biodiversity are coupled with this process. In the case of true successions species replacement also occurs on an individual basis (more details in Karhlec *et al.* 1996, Agnew *et al.* 1993).

Thesis 6.

Any change may occur rather rapidly, regardless of whether there is no visible reason or whether a site factor has undergone an obvious change. And the speed and quality of these changes are clearly chaotic and unforeseeable. It is not possible to draw a conclusion or make a prediction even after several years or even when a partial knowledge of what is going on in the way of environmental changes exists. There are detectable tendencies which are often caused by unpredictable weather conditions, on management, on animal numbers or pathogenic influences. But in our latitudes, distinct, directed changes can be recognized with a degree of certainty following the passage of five to 12 years' time. Change can be expected to take place more rapidly in more arid tropical countries or in gaps in a rain forest.

Another example confirms Thesis 6. Global warming and other aspects of "global change" have already left their mark on vegetation, especially in ecotonal areas such as transitional areas between our summer-green deciduous forests and more warm temperate evergreen broad-leaved forests. The southern-most part of Switzerland (southern Ticino) provides an example of this, especially in winter. More and more evergreen species, shrubs and trees, including the Japanese (or Burma) palm, are spreading on mesic but warm and extremely insulated slopes. In light of meteorological events, this change is "logical" and unprecedented. These species were introduced into the area around 1670, but were kept in gardens and never fully escaped cultivated areas. Obviously, there is a strong causality between higher winter temperatures and damp summer weather – conditions similar to their native areas which include southern Japan and southwestern China – and the invasion of evergreen broad-leaved species in such areas. As a result, there has been a rather rapid shift in biodiversity in that area since about 1970, which has changed the complete physiognomy and structures of these forests, resulting in a true succession, that shows the difference in fluctuations and successional trends (Klötzli *et al.* 1996).

As mentioned above, changes in warmer areas – such as savannas – are progressing more quickly and in an easily detectable way. When we compare change during periods of drier and wetter weather with change that has no obvious cause, we may draw in light of Thesis 7 the following conclusions

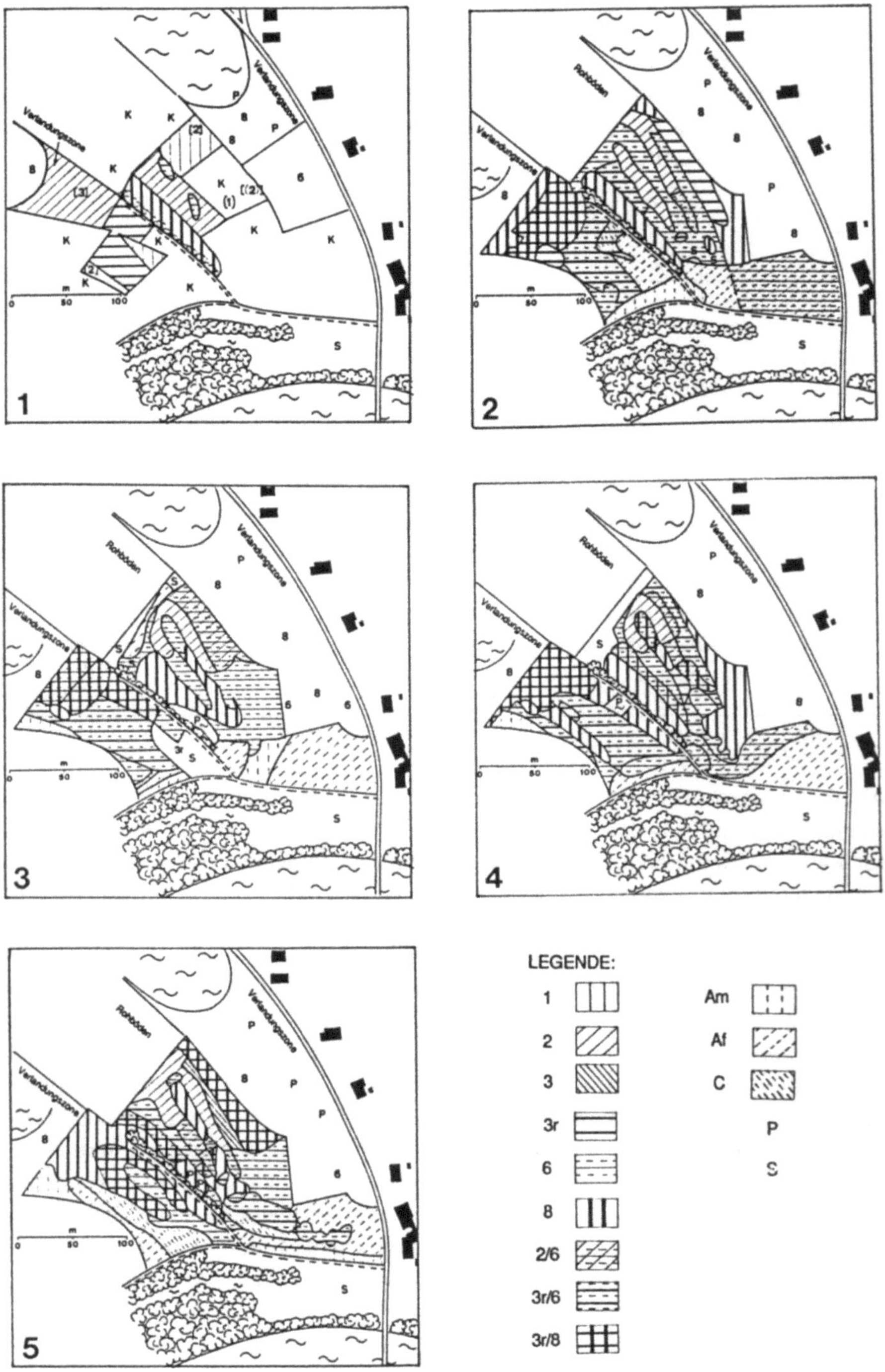
1
2
3
4
5
Verlandungszone
Rohböden
m
0
50
100
LEGENDE:
1
2
3
3r
6
8
2/6
3r/6
3r/8
Am
Af
C
P
S

regarding changes in biodiversity in savanna plant communities (Klötzli *et al.* 1995; studies from 1955- and 1974-1994):

Thesis 7.

1. *Unforeseen changes in local biodiversity are far greater than changes that occur between drier and wetter periods of up to six years.*
2. *The suitability of many plant species changes, i.e. as indicator species or in terms of their regular occurrence in certain types of savanna (plant communities).*
3. *The number of plant species may be plentiful one year and then scarce to non-existent the next, a fate that affects the populations of large areas (in our case > 500 km^2), due perhaps to pathogenic factors that are intrinsic to certain savanna types.*
4. *However, the presence of a certain number of permanent species can be used to determine the type of savannan an area. However, quantitative fluctuations among certain grasses in various types of savannas may lead to changes in physiognomy.*

In conclusion we present Theses 8 through 10:

Legend to Fig. 3.

1,2,3	different *Molinia* meadows from dry to moist
3r	inundation basins with small sedges
6	macrophorbe meadow
8	tall sedge (*Carex*) fen
2/6	respective
3r/6	mixtures, transitions,
3r/8	mosaiques
Am, Af, C	different fertilized meadows from mesic to moist
P	reed (*Phragmites*)
S	disturbed area

Fig. 3. Wetland plant communities at "Stilli Rüss", an oxbow pond, 15km to Zurich, and their fluctuations in a year before damming the river Reuss, and in three different years after damming the river. ((→%)). Density of hatching symbolizes humidity of a given plant community, wide hatching relatively dry, dense relatively wet sites of the vegetation units

Ilustración 3 Comunidades de plantas en las tierras húmedas del "Stilli Rüss", un brazo muerto del río, a 15 km de Zurich, y sus fluctuaciones en un año antes de embalsar el río Reuss y en tres años diferentes después de embalsar el río. ((→%)). La densidad del sombreado simboliza la humedad de una comunidad de plantas dada: un sombreado espaciado indica lugares relativamente secos de las unidades de vegetación y uno denso lugares relativamente húmedos de las mismas

←

Thesis 8.

Any analysis of a given type of vegetation can be viewed as the sheet for one specific day on a tear-off calendar. The population of a species group changes as unforeseeably as clouds, with unpredictable patterns, speeds and directions. Coupled with this is biodiversity which fluctuates irregularly within any unit of area, space or time.

Any inventory is like a sheet on a tear-off calendar. The data it provides might be of use at a later date when conducting a time series. But one single inventory offers only very arbitrary facts that may change within less than a year, even without human intervention (compare Grubb 1984).

Thesis 9.

Quantitative values for biodiversity are therefore in no way constant in any given site or time.
The larger the surface, the more constant the value used to depict the level of biodiversity will be (see, e.g., Convoy & Noon 1996 regarding scale).
Biodiversity may change considerably at a given site due to warming or other influences, just as it may change completely in areas that experience only slight changes in their site conditions.

The following example illustrates such a case in European forests. The average number of species in central European forests has generally declined over the past thirty years, particularly in ecotonal and changing areas, due to changes in temperature, nutrients and water, even without deliberate human influences (around 15% to > 30%). But even so-called stable systems of plant communities exhibit large fluctuations (Klötzli 1995, Klötzli *et al.* 1996).

Thesis 10.

In summary, the extent, temporal pattern and distribution of change cannot be foreseen. These factors are completely chaotic. Only long-term studies provide a sounder foundation.

However, we must have the opportunity to look behind the processes, fluctuations, trends and even intrinsic pathogenic stresses. We must find ways of assessing threshold values that influence the physiognomy of such ecosystems and the changes in their biodiversity. More sophisticated time series might reveal more secrets of biodiversity and unforeseeable fluctuations in biodiversity (compare e.g. Picket 1989, 1991).

These statements force us to examine some political aspects.

When observing the changing environment, there can be no doubt that there is a strong threat to biodiversity which is evidenced by the extinction of many organisms, a process that continues to accelerate (Primack 1995, Heywood & Watson 1995, Holdgate 1996, partly on the basis of WLMC 1992, McNealy 1992, WRI *et al.* 1993).

For this reason, the "Rio Convention" (on biodiversity) urges signatory states to place more emphasis on conservation. However, the political consequences involve conservation, agriculture, forestry and recreational activities. According to recent findings – to be politically correct – we have to know much more about the dynamics of biodiversity. Only then will we be able to do our best for conservation within the boundaries set by political constraints. We have to improve our knowledge of biological processes throughout the world, particularly if we want to safeguard rare species in a fluctuating environment. We have investigated the effects that man's activities have on protected areas and their organisms in agricultural regions. The same applies to forestry and how management methods influence biodiversity. We must also know more about how recreational activities affect ecosystems and what the limits of these activities have to be in or near protected areas (on development, assessment and threat to biodiversity see, e.g., Wilson 1995).

Our political structures require us to decide within a comparatively short period, only one to four years, on the boundaries and form of a future conservation area. Unfortunately, the rules of nature are not taken into account in such processes; change and fluctuation are comparatively slow in our latitudes. Organisms in lower latitudes extend their ranges faster than organisms in higher latitudes. As a result, biodiversity also fluctuates more rapidly in lower latitudes. Conservation decisions have never taken such factors into consideration. And a normal (European) legislative period is also far too short in terms of natural – fluctuating or chaotic – changes in biodiversity.

For this reason, the Rio Convention must go beyond mere recommendation status, it must be more than an element in our national education and information systems (examples in Aragno *et al.* 1992, see Grose *et al.* 1995 for a list of relevant institutes), it must be made part of the law of any nation ready to safeguard its patrimony.

References

Agnew, ADQ, Collins, SL & van der Maarel, E (eds) (1993) Mechanism and processes in vegetation science. J Veg Sci 4: 145-278

Aragno, M *et al.* (eds) (1992) Education and sciences for monitoring biodiversity. Internat Sympos Basel 1992, UNESCO and Swiss Acad Sci Bern

Bakker, JP, Olff, H, Willems, JH, Zobel, M (1996) Why do we need permanent plots in the study of long-term vegetation dynamics. J Veg Sci 7: 147-156

Bernhardt, KG, (1996) Die Samenbank und ihre Anwendung im Naturschutz. Verh Ges Ökol 20: 883-888

Convoy, MJ, Noon, BR (1996) Mapping of species richness for conservation of biological diversity: Conceptional and methodological issues. Ecol Applic 6: 763-773

Glen-Lewin, DC, van der Maarel, E (1992) Patterns and processes of vegetation dynamics. In: Glenn-Lewin, DC, Peet, RU, Vebleu, TT (eds) Plant succession: theory and prediction. Chapman & Hall, London. pp 11-59

Grose, K, Howard, ES, Thierry, C (eds) (1995) A sourcebook for conservation and biological diversity information. IUCN, Gland / CH

Grubb, PJ (1984): Some growth points in investigative plant ecology. In: Corley, JH, Golley, FB (eds) Trends in ecological research for the 1980's. Plenum Press, New York. pp 51-74

Holdgate, M (1996) The ecological significance of biological diversity. Ambio 25: 409-416

Kaule, G (1991). Arten- und Biotopschutz. Ulmer, Stuttgart

Klötzli, F (1992) Die Einstellung des Naturschutzes auf die Rahmenbedingungen in der Schweiz. NNA (Norddeutsche Naturschutz-Akademie) Ber 5: 62-64

Klötzli, F (1993a) Grundsätze ökologischen Handelns. In: DVGW-LAWA Koll Ökologie und Wassergewinnung. DVGW-Schriftenreiher Wasser 78: 9-24

Klötzli, F (1993b) Dornpolster und Kissenpolster - zwei divergierende Adaptionen. Diss Bot 196: 155-162.

Klötzli, F (1994) Vegetation als Spielball naturgegebener „Bauherren" (am Beispiel von Verbascum in der „Puna" Hawai'is). Phytocoen 24: 667-675

Klötzli, F (1995) Projected and chaotic changes in forest and grassland communities. Preliminary notes and theses. Ann Bot 53: 225-231

Klötzli, F, Zielinska, J (1995) Zur inneren und äusseren Dynamik eines Feuchtwiesenkomplexes am Beispiel der "Stillen Rüss" im Kanton Aargau. Schriftenreihe Veg Kde, Sukopp-Festschrift 27: 267-278

Klötzli, F, Lupi, C, Meyer, M, Zysset, S (19959 Veränderungen in Küstensavannen Tanzanias. Ein Vergleich der Zustände 1975, 1979 und 1992. Verhandl Ges Ökol 24: 55-65

Klötzli, F, Walther, GR, Carraro, G, Grundmann, A (1996) Anlaufender Biomwandel in Insubrien. Verhandl Ges Ökol 26: 537-550

Krahulec, F, Agnew, ADQ, Agnew, S, Willems, JH (eds.) (1990) Spatial processes in plant communities. Academia, Praha

van der Maarel, E (1996): Pattern and process in the plant community: Fifty years after A.S. Watt. J Veg Sci 7: 19-28

McLaughlin, A, Minean, P (1995) The impact of agricultural practices on biodiversity: Agric Ecosyst Environ 55: 201-212

McNeely, JA, Miller, KR *et al.* (1992): Conserving the world's biological diversity. WRI Publ, Baltimore, MD

Pickett, STA (1989) Space for time substitution as an alternative to long-term studies. In: Likens, GE (ed) Long-term studies in ecology. Springer, New York, pp.110-135

Pickett, STA (1991) Long-term studies: past experience and recommodations for the future. In: Risser, PE (ed) Long-term ecological research. Wiley, Chinchester, pp 71-88

Plachter, H (1991) Biologische Dauerbeobachtung in Naturschutz und Landschaftspflege. Lauf Sem Beiträge 7: 7-29

Raver,PH, Wilson EO (1992): A fifty-year plan for biodiversity surveys. Science 258: 1099-1100

Schulze, ED, Mooney, HA (1993) Biodiversity and ecosystem functions. Ecol Stud 99. Springer, Berlin, Hamburg, New York

Solbrig, OT (1994): Biodiversität. Wissenschaftliche Vorschläge für die internationale Forschung. MAB, Bonn

Thompson, JN (1996) Evolutionary ecology and the conservation of biodiversity. Trends Ecol Evol 11: 300 - 303

Vitousekl, PM, Loope, L, Andersen, H (eds) (1995) Islands. Biological diversity and ecosystem function. Ecol Stud 115. Springer, Berlin

Walker, B (1995) Conserving biological diversity through ecosystem resilience. Cons Biol 9: 747-752

Walter, H, Breckle, SW (1984) Ökologie der Erde 2. Die tropischen und subtropischen Zonen. G. Fischer, Stuttgart

Wildi, O, Orloci, L (1996) Numerical exploration of commmunity patterns. A guide to the use of MULVA-5, 2nd edition. SPB Acad Publ, Amsterdam

Wilson, EO (1995) Der Wert der Vielfalt. Die Bedrohung des Artenreichtums und das Überleben der Menschen. Piper, München

World Conservation Monitoring Center, Cambridge (UK) (1992) Global biodiversity. Status of the earth's living resources. Chapman & Hall, New York

WRI, IUCN, UNEP, FAO, Unesco (1993) Global biodiversity strategy. WRI-Publ. Baltimore, MD

Landscape diversity - a holistic approach

Hartmut Leser[1], Peter Nagel[2]
[1] Institute of Geography, University of Basel, Switzerland
[2] Institute of Nature-, Landscape- and Environmental Conservation, University of Basel, Switzerland

Abstract. The term landscape ecosystem designates a higher rank within the hierarchy of ecosystems which focuses on the structural, functional and historical peculiarities of a certain locality while the actual landscape constitutes its spatial representative. With regard to diversity, ecosystems are characterized by their special biodiversity and geodiversity. Consequently, landscape ecosystems also have a characteristic diversity of living and abiotic elements and systems. Site-specific abiotic elements, complex units and processes form the primary framework for the composition and dynamics of organisms and, at least in part, the densities and activities of human populations. Nature and culture are specifically linked, mutually effective and interdependent in such landscape ecosystems. Geodiversity and biodiversity are thus subsets of the methodologically more complex model of landscape diversity. The role of landscape diversity is discussed with regard to the objectives of conservation. The link to development issues is given special consideration by taking into account the sustainable use of natural resources.

Diversidad del paisaje: un enfoque global

Resumen. El término *ecosistema del paisaje* designa una categoría superior dentro de la jerarquía de ecosistemas que se orienta hacia las peculiaridades estructurales, funcionales e históricas de un determinado lugar, mientras que el paisaje actual constituye su representación espacial. Respecto a la diversidad, los ecosistemas se caracterizan por su especial biodiversidad y geodiversidad. Por lo tanto, los *ecosistemas del paisaje* tienen también una diversidad característica de sistemas y elementos bióticos y abióticos. Elementos abióticos específicos de un lugar, complejas unidades y procesos forman el principal entramado para la composición y dinámicas de los organismos y, al menos en parte, las densidades y actividades de las poblaciones humanas. En tales *ecosistemas del paisaje,* naturaleza y cultura suelen aparecer generalmente unidas y ser mutuamente efectivas e interdependientes. Geodiversidad y biodiversidad son por tanto un subgrupo del modelo metodológicamente más complejo de la diversidad del paisaje. El papel que juega la diversidad del paisaje es discutido con respecto a los objetivos de conservación. Se concede especial consideración a la relación con temas de desarrollo, teniendo en cuenta el uso sostenible de los recursos naturales.

1 Introduction

The theme of this article is an old topic: the holistic approach to the science of ecology which encompasses both the theoretical and the practical approach, and how it affects the need for interdisciplinary and transdisciplinary studies. The original aim of the founders of ecology and ecosystem sciences was to achieve a holistic approach. Over the years, it has – with few exceptions – been split up into several disciplinary sectors and has consequently lost its overall view to an increasing degree. Let us investigate the problems of the holisitic approach and how they can be solved at least to a certain extent. Landscape diversity will have to be discussed in relation to other types of diversity. In addition, this paper deals with the contribution that diversity research makes to the environmental sciences against the backdrop of environmental and development policies. It also discusses the aims and purposes of conservation strategies. Brief examples of practical experience with regard to development and environment are provided to help illustrate the theories.

2 Holism: reviving the roots of ecosystem science

In ecology's early years as a scientific discipline (Möbius, Haeckel last century, Dahl, Hesse, Tansley, Thienemann, Elton, Friederichs at the beginning of this century) and even later (Tischler, Kühnelt, MacArthur, Odum, Wilson, Remmert, Troll, Ellenberg, Walter, to list only a few), its most obvious new aspects were holism with regard to theory, and pluridisciplinarity with regard to methodology. The geographers Carl Ritter and Alexander von Humboldt at the turn of the 19th century have to be considered the main forerunners of modern ecology. For them, the analysis of the "entirety of a landscape" ("Totalcharakter einer Erdgegend," A. von Humboldt) was the objective of their scientific discipline. This "entirety of a landscape" meant the totality of linkage and the interrelation of geoscientific, biological and anthropological factors with regard to structural, functional, spatial and temporal patterns. This approach has been taken up and further developed by Ratzel, Waibel, Passarge, Gradmann, Troll, Neef, Schmithüsen and others, and finally resulted in the landscape ecosystem approach (see Naveh & Lieberman 1994, Leser 1997).

During the second half of this century, and mainly during the last 10 to 20 years, the tendency for the holistic approach to split into mono-disciplinary or mono-sectoral approaches has increased considerably. Nowadays, this seems to be the standard approach (or more precisely, set of approaches) to ecology and related disciplines. Practical and methodological problems are the primary – although not sole – reasons for this sectoral view and the splitting and reductionism in scientific theory and practice (for example, one would have to have an equally in-depth knowledge of the abiotic and biotic aspects of soil, the chemistry of a certain pollutant leaking through the soil layer, and the taxonomy and life cycle of a spring tail used for the biomonitoring of soil pollution). Other reasons for this situation include a lack of willingness and/or practical opportunities to pursue interdisciplinary or transdisciplinary studies.

In nature conservancy, this sectoral approach in diversity studies has led to some strange products. To illustrate: The factors "diversity" and "endangering" of organisms are often no longer regarded as parts of a holistic approach. Instead, they are taken as isolated, separate factors which therefore suddenly take on overwhelming importance. For example, artificial ponds on the tops of hills in a sandstone area where no standing water bodies of any kind would naturally exist over a longer period are designated to be "extremely valuable ecosystems" for the simple reason that the number of species has increased (the alpha diversity of aquatic organisms in this case) and rare or endangered species for which official red data exist have come to the area and established themselves there. We do not know how many ecologists would accept such isolated criteria for nature conservancy and the assessment of ecosystems for conservation purposes. But we have to discuss at least the goals and purposes of conservation strategies by also taking into account development aspects.

3 Landscape ecosystems: their role in ecosystem science

Let us take the example of a Mopane woodland in the Zambezi Valley between Mozambique and northern Zimbabwe (Fig. 4). Its key features include its particular structure, very few species (low alpha-diversity) and low beta-diversity. This lack of species applies not only to plants but extends to animals, and to insects and birds in particular. The area has an alluvial sandy soil. This particular type of ecosystem is the product of the area's combination of climate, soil and history – geological history as well as the history of the last 5000 years of human settlement. Up until about 15 years ago, the area was sparsely populated and the people living there were restricted to areas along of non-persistent streams. This Mopane woodland is thus a unit of organisms, geological background, geographical features, man and his activities, and time. All this together forms the landscape ecosystem which is dominated by the tree species *Colophospermum mopane* with its sclerophyllous leaves and large but shallow root system. The entire landscape ecosystem is however very complicated and complex and can be understood only when we regard it on a regional level and take into account related biological, cultural and geoscientific aspects.

Landscape ecosystems (or geoecosystems, cf. Troll 1939, Huggett 1995) are thus true ecosystems, and most current definitions of ecosystems fit landscape ecosystems well. However, they hold a special ranking within the scale hierarchy of ecosystems. They range in size from about a dozen square metres to sections of an entire continent. Accordingly, spatial dimension is a particular consideration. The attribute landscape underscores its link with the structural, functional and historical (geological and human time-scale) peculiarities which are specific to a locality (Leser & Rodd 1991, Naveh & Lieberman 1994, Leser 1997). Taken on its own, the term landscape is designation for the spatial representative of a landscape ecosystem.

4 Diversity as an inherent component of ecosystems

A primary forest is more diverse than a plantation, a sandy beach is less diverse than a fixed dune area, and the Mopane woodland is less diverse than the adjacent Miombo woodland. The term "diversity" is very often simply a reference to alpha diversity or to the richness of one or several ecosystem components (cf. World Conservation Monitoring Centre 1992). Sometimes the term "diversity" is also used in the sense of a more or less even distribution of subsets among a given number of sets – the so-called Shannon-Wiener diversity which MacArthur and subsequent authors introduced into ecology shortly before 1970 (cf. Whittaker 1972, Nagel 1978). This marked the height of popularity of the view of a strictly positive correlation between the terms stability and diversity, a principle which MacArthur introduced into ecology in the mid-fifties. Although there might be a positive correlation between some kinds of diversity and stability, it is also possible that exactly the reverse situation exists, a fact which was not known 30 years ago.

Different scientific disciplines use the term diversity in different ways, which makes the term even more obscure. One can refer to the totality of ecosystems or to individual, generally conspicuous components, such as the diversity of topographical relief, soil texture, insects, moss flora or the temperature regime. In other words, there are many different kinds of diversity and there are – and this is even more important – many methods and strategies for studying and recognizing them. In any event, diversity is an inherent characteristic of an ecosystem's structural, functional and dynamic parameters.

In the following sections, we concentrate on the three main types of diversity. These types are referred to as geodiversity, biodiversity and landscape diversity (Leser & Schaub 1995) (Figs. 1 and 2).

4.1 Geodiversity

Geodiversity is a component of all earth sciences but it is normally not specifically mentioned. As a result, other scientific disciplines are not always aware of its existence. Although it has not been specifically mentioned, we have heard a lot about geodiversity during this symposium. Examples: the soil catena (the ecological zones and regular sequence of soil types along hill slopes or at increasing distance from a water body, for example), climatic gradients, geomorphological sequence of running water from the spring to the mouth, and so on. Geodiversity is most evident in spatial patterns which indicate a certain geoecological structure. Geodiversity could be defined as the diversity of non-biotic systems – in other words, the structure, function and dynamics of the abiotic parts of ecosystems, such as the geosphere, climosphere and hydrosphere.

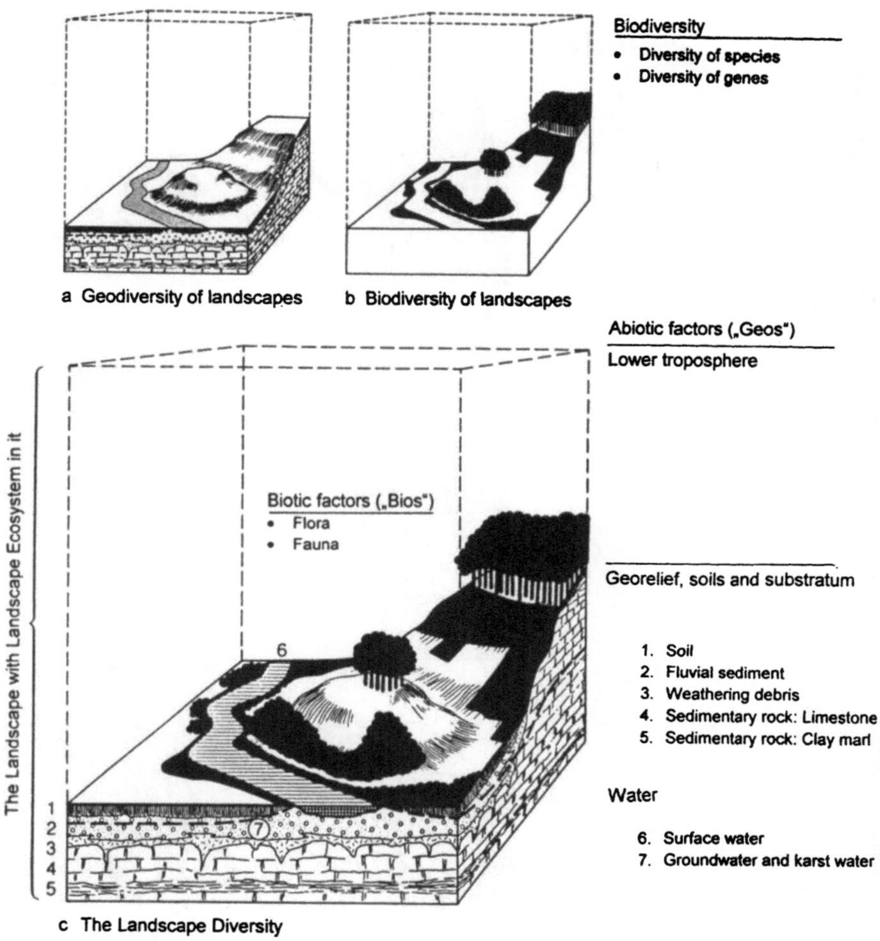

Fig. 1. Landscape and different types of diversity in landscape ecological systems
Ilustración 1. Paisaje y tipos diferentes de diversidad en sistemas ecológicos del paisaje

And once again, spatial scale and dimension are of particular importance. Although there are techniques and methods which apply exclusively to the study of geodiversity, the latter comprises only part of a more comprehensive model. Geodiversity constitutes part of the framework for biodiversity. At the same time however, geodiversity and biodiversity interact with and are mutually dependent on one another, and are part of what we call landscape diversity.

4.2 Biodiversity

Biodiversity seems to be the most important type of diversity because this term is most often used to not only at symposiums like the present one but also in connection with environmental law and international conventions. The frequent

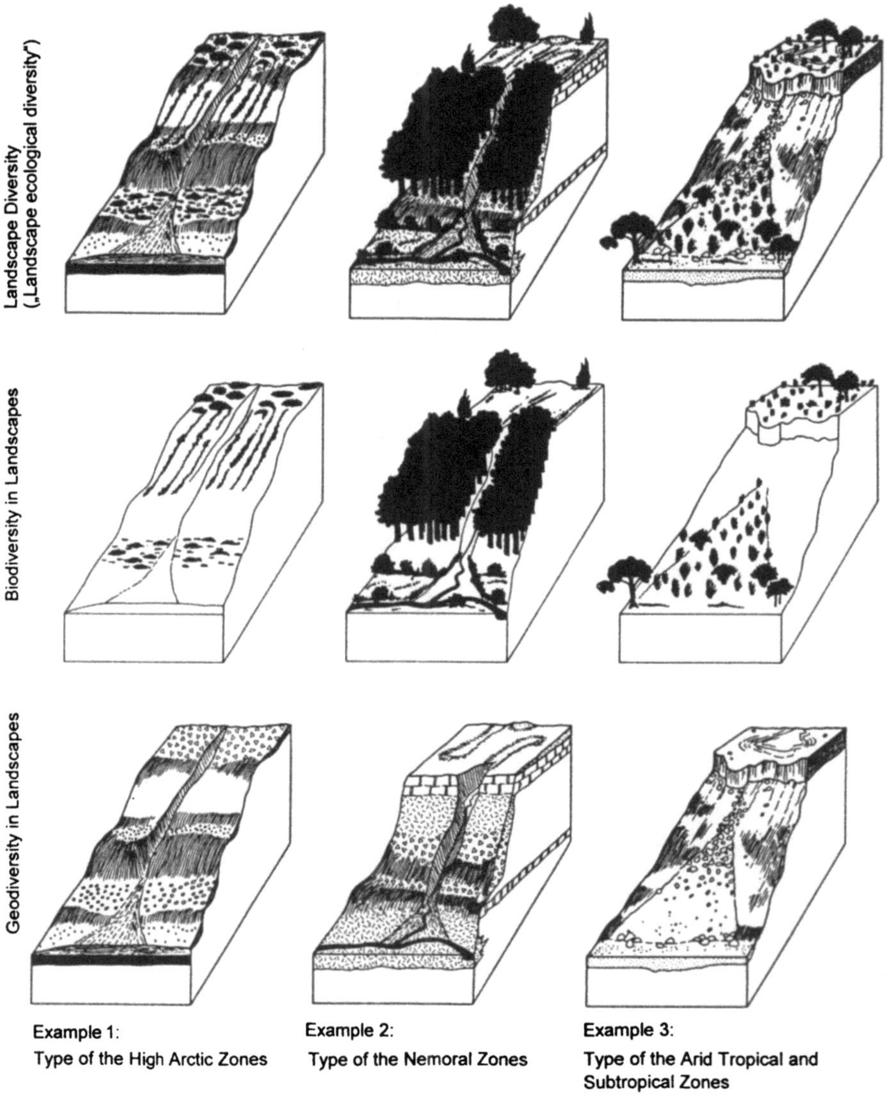

Fig. 2. Types of landscape diversity: Examples from different ecozones
Ilustración 2. Tipos de diversidad del paisaje: ejemplos de ecozonas diferentes

use of the term in scientific and non-scientific discussions is perhaps just a consequence of such international conventions. Dealing with biodiversity nowadays almost certainly guarantees the allocation of funds, at least in some countries. Which is fine. However, there is not just one kind of biodiversity. The term is subject to a variety of definitions among a wide range of life sciences. Richness of life forms, species richness, genetic diversity, and diversity in relation to habitat are just a few examples (cf. Wilson 1988, Solbrig 1994). In the context of biodiversity studies however, the spatial dimension is frequently ignored.

It is quite surprising that the "space factor" is sometimes missing because species conservation in the field, for example, usually necessitates habitat conservation. Although everyone would accept that species conservation is normally not possible without habitat conservation, the focus and the discussion are not on the spatial dimension but on alpha diversity. Furthermore, microbiological and genetic approaches as opposed to macrobiological or organismic approaches are currently gaining importance. This means that there is the danger that biodiversity studies may slip beneath the "topos-specific" dimension (= "topisch", a term of the "Theory of geographical dimensions", see Leser 1997, p.198-199, 202-205) and lose their direct relation to and importance for conservation and other environmental approaches. Similarly to geodiversity, biodiversity can be defined as the diversity of biotic systems; in other words, the structure, function and dynamics of the biotic parts of ecosystems such as populations, species, communities and life forms. However, the link to the spatial dimension also has to be considered which is normally not done. Although there are techniques and methods that apply exclusively to the study of biodiversity, which is only part of a more comprehensive model: Biodiversity and geodiversity are mutually dependent and are part of what we call landscape diversity.

4.3 Landscape diversity

The term landscape diversity refers to a particular position on the spatial scale ("topos-specific" dimension or larger, see explanation above) and consists of a diversity of interacting abiotic and biotic systems, each characterized by a specific geodiversity and biodiversity. Although they are often ignored, man's activities are also important factors in systems. Local people grow up in a particular area, live in it, have a certain impact on it and are also influenced by the biotic and abiotic characteristics of this particular landscape. A mutuality of effects of non-biotic and biotic systems on the one hand and of human society and its techniques (= culture) on the other hand actually exists and varies from locality to locality (see Fig. 3). This combined mutuality exhibits itself in the diversity of landscapes. Landscape diversity reflects the structural, functional, spatial and temporal patterns of landscape ecosystems (cf. Schulze & Gerstberger 1993). It is more than just the sum of its parts. From this point of view it is evident that geodiversity and biodiversity are just two components of the more comprehensive landscape diversity (see definition in Leser 1994, Leser & Schaub 1995).

The term "ecological diversity" came into widespread use about 25 years ago but remained restricted mainly to biological diversity (Pielou 1975). It has been suggested that the term "ecodiversity" be defined as "the total diversity of a region which is a combination of bio- and geodiversity" (Barthlott *et al.* 1996, p. 320). "Ecodiversity" could be used as a generic term that contains many different individual "diversities." There is however just one type of diversity of landscape ecosystems. "Ecodiversity" does not make it clear from the start that it refers to

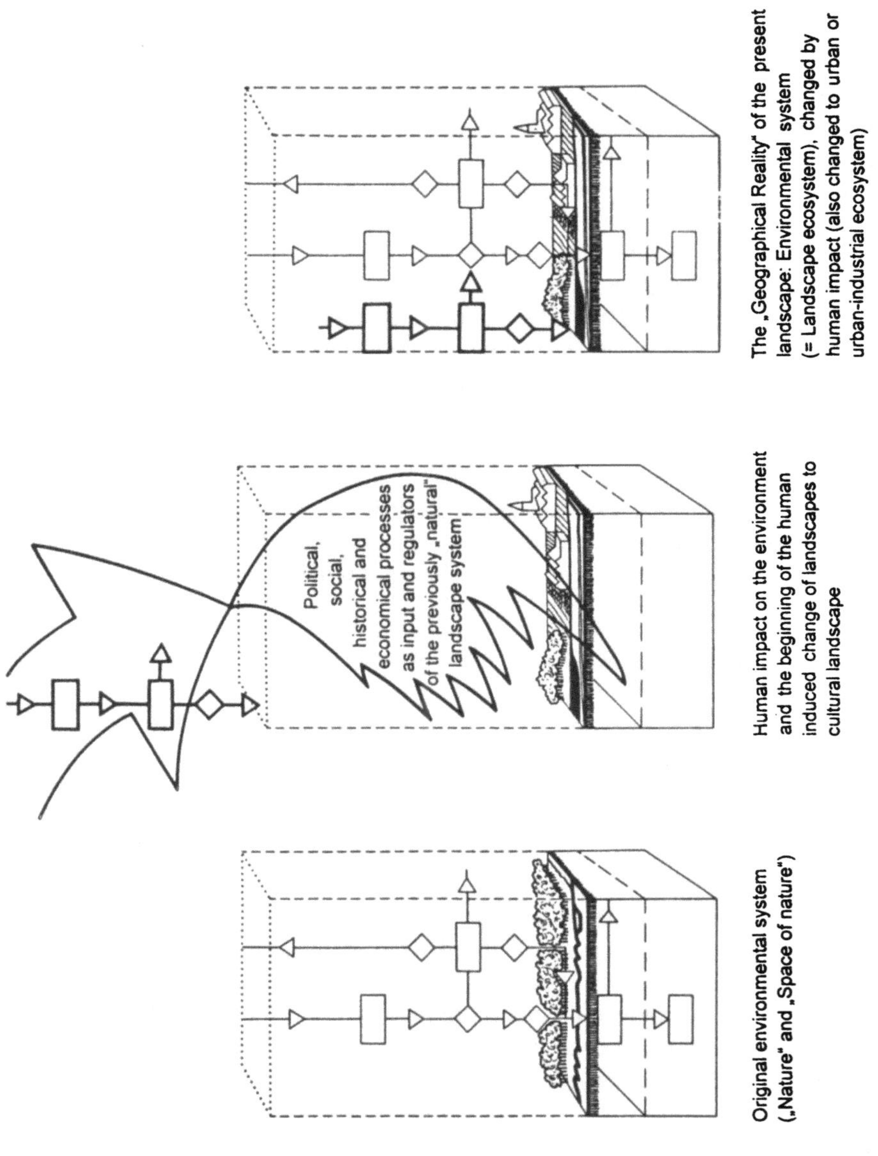

Fig. 3. The spatial approach of landscape research and different types of landscape diversity: A time-scale problem

Ilustración 3. El enfoque espacial de investigación del paisaje y diferentes tipos de diversidad del paisaje: un problema de escala temporal

the diversity of a particular sector of a landscape ecosystem. In other words, it conceals the "regional" aspect. Scientific approaches have only recently begun to consider human beings to be a factor that is reciprocally connected with the environment within the framework of ecosystems. The necessary consideration of man as a structural and functional element in systems is not as clearly a part of the term "ecodiversity" as it is of the term "landscape diversity."

Barthlott *et al.* (1996) consider the term landscape diversity to be reserved for denoting gamma diversity. Whittaker (1972, p. 231) defines gamma diversity as "richness in species of a range of habitats (a landscape, a geographic area, an island)." He does not use the term landscape diversity at all. Gamma diversity is pure biodiversity – generally a "list of species for geographic units" (Whittaker 1972, p. 232) – and therefore is not to be confused with 'landscape diversity'. In view of the fact that the term *Landschaftsvielfalt* (landscape diversity – not to be confused with the 'richness in landscapes' of a specific area) has been used with a signification that is at least similar to 'ecodiversity' for far more than a century, we would not favour its proposed replacement. We would instead prefer the terms 'landscape ecological diversity', 'landscape diversity' or 'landscape ecodiversity'. The latter wording has been suggested by Naveh (see Naveh & Lieberman 1994, p. 1-8).

5 The holistic approach compared to the sectoral approach

The conclusions drawn from the above definition of landscape diversity are also, but not exclusively, a call to intensify studies on geodiversity and biodiversity with regard to the spatial dimension and to incorporate culture-related sciences to a greater degree into ecosystem studies. For example: A risk assessment on the effects of certain chemicals that are used in the African savanna to control Tsetse flies which transmit disease to humans and cattle also has to take into account socio-cultural aspects such as the local population's needs and constraints in relation to its environment, its right not to have to suffer from a particular disease and its right to development (cf. Nagel 1995).

Although earth sciences and biological sciences proclaim interdisciplinarity in research, a proper analysis readily reveals that different disciplines often work in isolation in terms of their methodology and objectives, and have – at most – only the study area in common. Sometimes the aim is a common objective that is situated between two or more disciplines. However, each discipline will still use its own methodology. A holistic approach calls for generating new ideas and findings by practicing true cooperation between several disciplines. These include not only biosciences and geosciences but human sciences as well. One possible example: Biosciences and geosciences (physical and human geographical disciplines) could accept and apply one another's respective theories and methods which in turn could result in new theories and methods to the benefit of our shared environment.

6 Aims and objectives of conservation strategies

Conservation originally operated at the biodiversity level, although it soon became evident that species conservation in the field is possible only by means of habitat conservation and, consequently, the conservation of whole ecosystems. From biodiversity's position within the model of landscape diversity one would infer the necessity to focus on the landscape ecosystem level when referring to conservation. This type of 'biosphere reserve' concept explicitly integrates man's non-adverse activities; in other words: those that do not affect the sustainability of the system.

Looking at how diversity is dealt with in (most parts of) the scientific community (although not in most parts of the general public or the political sector), the aim of species conservation has evolved from the preservation of a large number of species to the preservation of the 'landscape ecosystem-specific' composition of biota. Consequently, the aim of conservation is to preserve or restore (by supporting or starting the regeneration of) the respective ecosystem to a state that is as close as possible to its original structure, function and dynamics. This means that especially the relation between its inherent dynamic and the extrinsic natural dynamic is not hampered. A landscape ecosystem can never be taken as an isolated unit of nature plus culture; it must be regarded within the context of the more comprehensive, higher ranks of landscape ecosystems all the way up to global level. This means that the specific needs of conservation issues can be considered together with the needs of current and future human generations in a balanced manner (also in light of a balanced spatial distribution of land-use patterns). The use-directed objective of environmental science activities including conservation, has been aptly defined as sustainability (Brundtland *et al.* 1987: Sustainable development meets the needs of the present without compromising the ability of future generations to meet their own needs). In certain cultures (if not in all), ethical reasons are also accepted as grounds for conservation.

Efforts to preserve historical land-use features such as orchards, artificial pond and lake areas or certain heathland areas in Central Europe stem from and are accepted by local human society. The preservation of these man-made changes in the earth's surface does not at all mesh with the aim of conservation formulated above. Although in such cases biodiversity and geodiversity are generally both artificially maintained to a high degree, this approach is even contradictory to sustainability. One may accept this type of conservation strategy (i.e. conservation of cultural landscapes, cf. Droste *et al.* 1995) but should then clearly differentiate between this work-intensive approach and the sustainability-oriented conservation or restoration of landscape ecosystems.

7 Landscape diversity, development and environmental ecology

It is quite easy to demonstrate the advantages of a holistic landscape diversity approach in conservation. With regard to the recognition and establishment of nature reserves, it is indisputable that characteristic landscapes with all their geo-elements and bio-elements have to be taken into consideration, without however ignoring the needs of the local human population. When dealing with conservation decisions that are based on, for example, the concept of dispersal centres, it is not just a centre of endemism or a biodiversity centre that has to be considered: The area's vegetation history or even its entire landscape history, together with its climatic and geographic history background, must also be taken into account (Müller 1973a, b). Regarding development, pollution and environmental protection, the guarantee that disease control will have no adverse effect on the sustainability of natural resources is not enough: Right from the start of disease-control programmes aimed at, for example, trypanosomosis or malaria, we also have to take into account the characteristics of the relevant landscape ecosystems with regard to the question of what happens after a successful control (Nagel *et al.* 1995).

Fig. 4. Mopane woodland with dominant *Colophospermum mopane*: northern Zimbabwe, Zambezi valley, dry season aspect (1987)
Ilustración 4. Bosque de mopane donde predomina el *Colophospermum mopane*: en el norte de Zimbabwe, valle del Zambeze, aspecto en la estación seca (1987)

Many parts of the Mopane woodlands in the Zambezi Valley in northern Zimbabwe shown in Fig. 4 have been turned into cotton fields in the last fifteen years (like Fig. 5). The question arises, what are the consequences of this change for the country and the region when this unique type of woodland is used for a particular type of agriculture. Dealing with these aspects, for example, with regard to risk assessment from an environmental point of view, we have to check if there are no other landscape ecosystems in the entire region or beyond which are more suitable for growing cotton. Apart from the fact that in the tropical and subtropical zones cotton belongs to that group of cultivated plants that require the most pesticides, the local soil does not have enough nutrients or water capacity to grow cotton on a long-term basis. Furthermore, the frequency and amount of rain in the region is not ideal for cultivating cotton. Therefore, sustainability is not guaranteed. So we have to look beyond this region for landscape ecosystems where cotton can be grown on a sustainable basis. Taking into account ecophysiological requirements, traditional cotton production and the current possibilities of world-wide trade, it appears that the most suitable cotton-producing areas in the tropics are only those areas with deep, well-drained soil and high temperatures during the cultivation period (geomorphologically and pedologically suitable sections of semi-arid and humid savanna) and certain temperate regions' steppe landscape ecosystems.

Fig. 5. *Isoberlinia* dominated woodland, transformed into cotton fields: northern Ivory Coast (1991)
Ilustración 5. Bosque donde predomina la *Isoberlinia* transformado en campos de algodón: en el norte de Costa de Marfil (1991)

8 Summary and conclusions

The original approach in ecology as well as in geography was characterized by holism. At present, one can observe an increasing splitting into disciplinary and sectoral approaches. This occasionally results in strange views such as considering a large number of species in a single area to be of great value per se, irrespective of whether their presence is autochthonous (native) or anthropogenic.

Landscape ecosystems are true ecosystems of and beyond the "topos-specific" dimension (in general, a minimum of a few dozen square metres; see explanation above). The term underscores a focus on the peculiarities of a certain locality and the consideration of explicitly abiotic, biotic and human factors. The term landscape denotes spatial representative of landscape ecosystems.

Diversity is an inherent component of ecosystems. The current discussion often ignores the fact that there are many different kinds of "diversity", such as species richness or relief diversity. The introduction of the Simpson and the Shannon-Wiener indices into ecology led to a discussion of the importance of stability in ecosystems. Today we know that diversity and stability are not necessarily linked unidirectionally and are sometimes not linked at all. The current ecosystem model is a dynamic one and is based on the assumption that mostly unpredictable dynamic (= chaotic or stochastic forces within certain limits) govern the structural and functional parameters of ecosystems.

The term geodiversity designates the diversity of abiotic systems, such as soil catena or climatic gradients. It is applied as a matter of course in all earth sciences and, without direct mention, in most field ecology studies. Biodiversity means the diversity of biotic systems such as species richness or richness of life forms. Currently, the trend is towards an increase in genetic and microbiological studies and ignorance of the spatial dimension. Landscape diversity comprises geodiversity, biodiversity and man's activities; it takes the spatial dimension into account. It is not a simple cumulation of these factors. Instead, it reflects the structural, functional, spatial and temporal patterns of landscape ecosystems.

A comparison of the holistic and sectoral approaches indicates that a mutual acceptance and consideration of disciplinary methodologies could result in new ideas, strategies and findings.

With regard to the aims and objectives of conservation strategies we therefore call for a shift away from the biodiversity level and towards the landscape diversity level, a shift away from the 'species richness approach' and towards a 'landscape-ecosystem-specific' approach, and a shift towards the preservation of the 'natural dynamics' of ecosystems (human activities not excluded).

In the context of landscape diversity, development and environmental ecology, the sustainable use of natural resources is facilitated by linking development with landscape ecology.

Geodiversity and biodiversity together with cultural issues (such as land-use characteristics) comprise inseparable aspects of landscape ecosystem science. Landscape diversity is not just a cumulation of geological, biological and anthropological factors. They form indispensable parts of landscape diversity. It reflects the number of sets – that is to say, the number of subsystems, of landscape ecosystems. We regard this holistic approach as the most promising one for development and conservation issues. It aims at maintaining and revitalizing the natural dynamics of landscape ecosystems and at guaranteeing the sustainability of natural resources.

References

Barthlott W, Lauer W, Placke A (1996) Global distribution of species diversity in vascular plants: towards a world map of phytodiversity. Erdkunde 50: 317-327

Brundtland GH et al (1987) Our common future. World Commission on Environment and Development. Oxford University Press, New York

Droste Bv, Plachter H, Rössler M (eds) (1995) Cultural landscapes of universal value. Components of a global strategy. G Fischer, Jena, Stuttgart, New York

Huggett RJ (1995) Geoecology. An evolutionary approach. Routledge, London, New York

Leser H (1994) Räumliche Vielfalt als methodische Hürde der Geo- und Biowissenschaften. In: Brunner H (ed) Festschrift für Heiner Barsch. Potsdamer Geographische Forschungen 9. Potsdam, pp 7-22

Leser H (1997) Landschaftsökologie. Ansatz, Modelle, Methodik, Anwendungen. Mit einem Beitrag zum Prozess-Korrelations-Systemmodell von Thomas Mosimann. 4. Auflage. UTB 521, Ulmer, Stuttgart

Leser H, Rodd H (1991) Landscape Ecology - Fundamentals, Aims and Perspectives. In: Esser G, Overdieck D (eds) Modern Ecology: Basic and Applied Aspects. Elsevier, Amsterdam, London, New York, pp 831-844

Leser H, Schaub D (1995) Geoecosystems and Landscape Climate - The Approach to Biodiversity on Landscape Scale. GAIA 4: 212 - 220

McNeely JA, Miller KR, Reid WV, Mittermeier RA, Werner TB (1990) Conserving the world's biological diversity. IUCN, Gland, Switzerland; WRI, CI, WWF-US, World Bank, Washington D.C.

Müller P (1973a) The dispersal centres of terrestrial vertebrates in the Neotropical Realm. Biogeographica 2, The Hague

Müller P (1973b) Amazonische Nationalparks. Ent Zeitschr 83: 57-64

Nagel P (1978) Speziesdiversität und Raumbewertung. Verh dt Geographentag 41 (Mainz 1977): 486-498

Nagel P (1995) Environmental monitoring handbook for tsetse control operations. Margraf, Weikersheim

Nagel P, Erdelen W, Peveling R (1995) Tsetsefliegen-Kontrolle und Landnutzungsdynamik in der Côte d'Ivoire. Verh Ges Ökologie 24: 67-72

Naveh Z, Lieberman AS (1994) Landscape Ecology. Theory and Application. 2nd ed. Springer, New York, Berlin, Heidelberg

Pielou EC (1975) Ecological Diversity. Wiley & Sons, New York, London, Sydney, Toronto

Schulze ED, Gerstberger P (1993) Functional aspects of landscape diversity: a Bavarian example. In: Schulze ED, Mooney HA (eds) Biodiversity and ecosystem function. Springer, Berlin, Heidelberg, New York

Solbrig OT (1994) Biodiversität. Wissenschaftliche Fragen und Vorschläge für die internationale Forschung. Deutsches Nationalkomitee für das UNESCO-Programm „Der Mensch und die Biospäre" (MAB), Bonn und Deutsche UNESCO-Kommission, Bonn (eds). Rheinischer Landwirtschaftsverlag, Bonn

Troll C (1939) Luftbildplan und ökologische Bodenforschung. Z Ges Erdkunde 1939: 241-298

Whittaker RH (1972) Evolution and measurement of species diversity. Taxon 21 (2/3): 213-251

Wilson EO, Peter FM (eds) (1988) Biodiversity. National Academy Press, Washington D.C.

World Conservation Monitoring Centre (1992) Global Biodiversity. Status of the earth's living resources. Chapman & Hall, London

Loss of biodiversity in European agriculture during the twentieth century

Walter Kühbauch
Faculty of Agriculture, University of Bonn, Germany

Abstract. Today, agriculture in Europe can be generally classified as having a low level of biodiversity. This article describes indications of biodiversity in agriculture and of how biodiversity has changed over time. It is highly likely that in the past, over a period of approximately 1,000 years, agriculture in central and western Europe was a matter of biodiversity, due to low-intensity cultivation methods and the principle of self-sufficiency that most farms practised. Biodiversity actually increased between the eighteenth and twentieth centuries due to a growing variety of cultivated crops brought about by the more advanced farming methods used in England and the Netherlands at that time. The loss of biodiversity in recent decades was stimulated primarily by the economic "environment" which was shaped by rising labour costs and relatively low prices for agricultural products which, in turn, accelerated mechanization and specialization in a few products (crops, animals) and put pressure on farmers to increase their yields. At present, loss of biodiversity is apparent in many ways. The European Union is trying to make biodiverse farming attractive to farmers through and farm ecology programmes. A general shift towards more biodiversity is, however, unlikely to occur because the prevailing political, commercial, administrative and economic forces are against it.

Pérdida de biodiversidad en la agricultura europea durante el siglo XX

Resumen. Hoy en día se puede decir que la agricultura en Europa tiene, en téminos generales, un bajo nivel de biodiversidad. Este artículo describe algunas indicaciones de la biodiversidad en la agricultura y de cómo ha cambiado la biodiversidad con el paso del tiempo.

Es muy probable que en el pasado, durante un periodo de 1,000 años aproximadamente, la agricultura en el centro y oeste de Europa fuera un tema de biodiversidad, debido al uso de métodos de cultivo de baja intensidad y al principio de autosuficiencia practicado por la mayoría de los agricultores.

La biodiversidad se incrementó realmente entre el siglo XVIII y el siglo XX por la creciente variedad de cultivos que se originaron en ese periodo a raíz del empleo en Inglaterra y Holanda de métodos de producción agraria más avanzados.

La pérdida de biodiversidad en las últimas décadas fue estimulada en primer lugar por el "entorno" económico, determinado por dos factores: por un lado, una mano de obra que resulta cada vez más costosa y, por otro lado, los precios relativamente bajos de los productos agrícolas que, a su vez, provocaron la acelerada mecanización y especialización en unos pocos productos (cultivos, animales), llevando a los agricultores a ampliar su

cosecha. Esta pérdida de biodiversidad se puede observar actualmente de muchas maneras. La Unión Europea está intentando hacer más atractiva para los agricultores la agricultura biodiversa y fomentando los programas de cultivos ecológicos. No obstante, no es probable que se produzca un cambio general hacia una mayor biodiversidad, ya que las fuerzas dominantes a nivel político, comercial, administrativo y económico están en contra.

1 The basis for biodiversity

Talk about the "loss of biodiversity" in European agriculture would seem to indicate that there was a time when biodiversity existed. Before delving into this subject, it is necessary to discuss possible yardsticks for "biodiversity in agriculture." Fig. 1 shows a number of criteria that can be used to recognize biodiversity in agriculture.

- Number of crop species used in single farms or in regions respectively
- Subdividing arable land into individual fields
- Size of individual fields
- Percent area and distribution of land that is not used with crops (hedges, bushes, and trees, balks etc.)
- Intensity and frequency of farm management treatments (fertilization, pesticide application, harvesting)
- Number and species used in animal husbandry, per farm, per hectar, in regions
- Diversity of non agricultural vegetation including weeds
- Diversity of companion plants
- Number of individual farms in regions and countries
- Others

Fig. 1. Indicators to recognize biodiversity in agriculture
Ilustración 1. Indicadores para identificar la biodiversidad en la agricultura

One criterion is the number of crop species used in crop rotation on individual farms or in regions, counties and the like: The larger the number of crop species used, the more likely biodiversity is to occur because each single crop not only depends on certain environmental conditions but also exerts an influence on them. Water, nutrient and light requirements, and the physical and biological conditions of the soil which root growth influences, create a specific kind of microclimate and environment for each specific crop. This in turn opens niches for other organisms which naturally include crop antagonists in the flora and fauna domains.

For instance, the disappearance of several weeds such as *Cameline alyssum*, *Galium spurium* and *Silene linicola* which were once common where flax (*Linum usitatissimum*) grew can be traced back to when flax cultivation was discontinued. When hemp crops, *Cannabis sativa,* disappeared, *Orobanche romosa* disappeared as well. The same happened with buckwheat, *Fagopyrum esculentum*, which has also disappeared now that an old kind of corn, *Fagopyrum tataricum*, is no longer grown.

Farmers know, for instance, that each crop species has its "own" spectrum of weed species. Accordingly, it is virtually impossible to find a rape seed field, *Brassica napus* var. *napus*, that does not contain the weed species *Stellaria media*. In contrast, *Veronica* species appear in only 30 % of all rape seed fields, but in nearly all winter cereal fields. This demonstrates that the more different crops are used, the more niches and biodiversity occur.

Taking this into account, such a simple task as subdividing large fields into smaller plots may support biodiversity.

In this context it is clear that the distribution and percentage of land not under agricultural cultivation (hedges, brush woods, verges) will increase biodiversity. In other words, uncultivated land broadens and stabilizes the ecosystem because nature always establishes and stabilizes ecosystems through numerous organisms that compete, support and supersede each other in a very complex manner. This in turn means biodiversity.

Since intensive and frequent treatment of agricultural crops (fertilizer and pesticide application; the number of harvests per season in grassland and forage production) is strictly intended to support one single crop, many accompanying organisms eventually lose their ability to compete and may even disappear.

For instance, many weed species disappear simply as a result of more effective or deeper tillage, or when summer fallow is abandoned. Species of this type are called geophytes and include deep-rooting plants and plants which form onions such as *Allium vineale*, *Neslia paniculata* and *Polycnemum arvense*.

On the other hand, nutrient and pH levels can be raised to uniformly high levels through the intensive application of fertilizers. At the same time, endemic plant species that have adapted to low nutrient conditions over a long period of evolution may lose their competitiveness, and the greater part of their population may disappear within just a few years. Many of them are mini-species that are not able to compete in soil with high nutrient levels and are considered endangered species today. Others, such as weed grasses, are definitely invigorated by the intensive use of nitrogen fertilizer and naturally create new weed problems.

In grasslands, frequent cutting suppresses the flowering of most grassland species and hence reduces the amount of food and habitat available to insects.

The number and kind of animals used in livestock husbandry is also a measure for biodiversity because all livestock needs certain kinds of feedstuffs (= crops) and management systems in order to be productive. For example, hog raising farms that want to generate a profit can feed only a few crop species – mostly wheat, barley, maize and/or a mix of concentrates in which cereal grain is the major component. When climatic and soil conditions are also taken into account, the number of crops that are both economical and appropriate for pig production is then reduced in many instances to no more than two or three. And, as a further example, intensive beef production in Europe is predominantly based on one single crop, maize silage.

The number and kind of different plant species in rural areas outside crop fields and in crop fields (weeds) provide a direct measure of biodiversity in agriculture. It takes an expert, of course, to use this measure. On the other hand, a simpler method involves charting the change in the number of farms in certain areas or countries over time as a measure of biodiversity: It is generally accepted that the individual farmer is the primary actor on his farm. If this is true, the decreasing number of farmers is very likely an indication of loss of biodiversity.

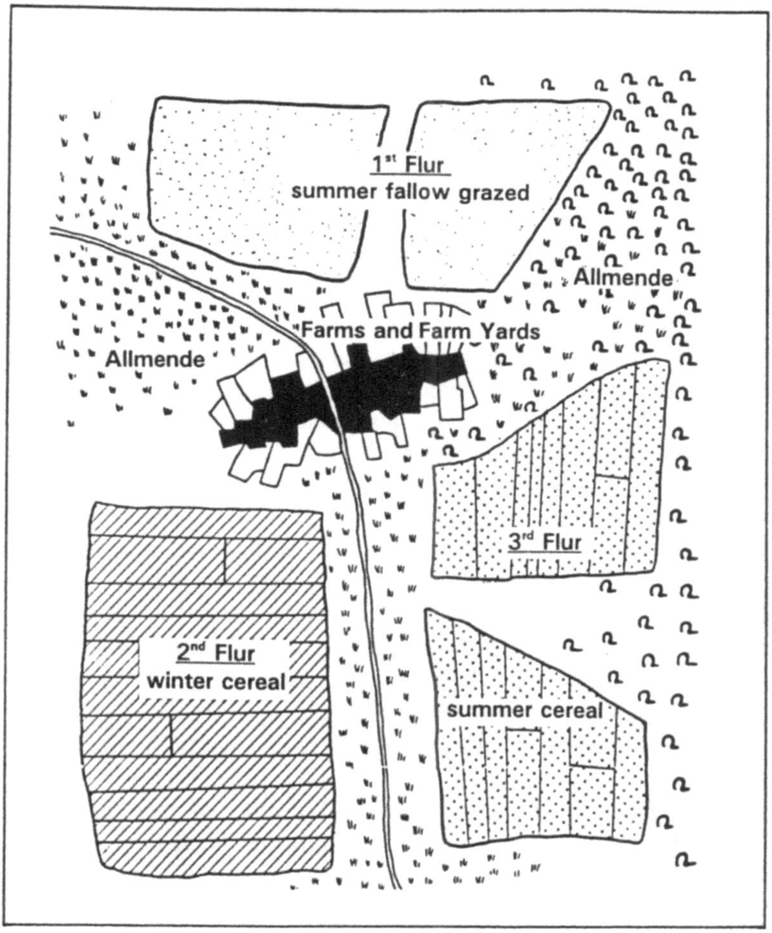

Fig. 2. Three-field farming in central Europe ninth to eighteenth century (Könnecke 1967)
Ilustración 2. Cultivo alterno a tres campos en Europa central del siglo IX al siglo XVIII (Könnecke 1967)

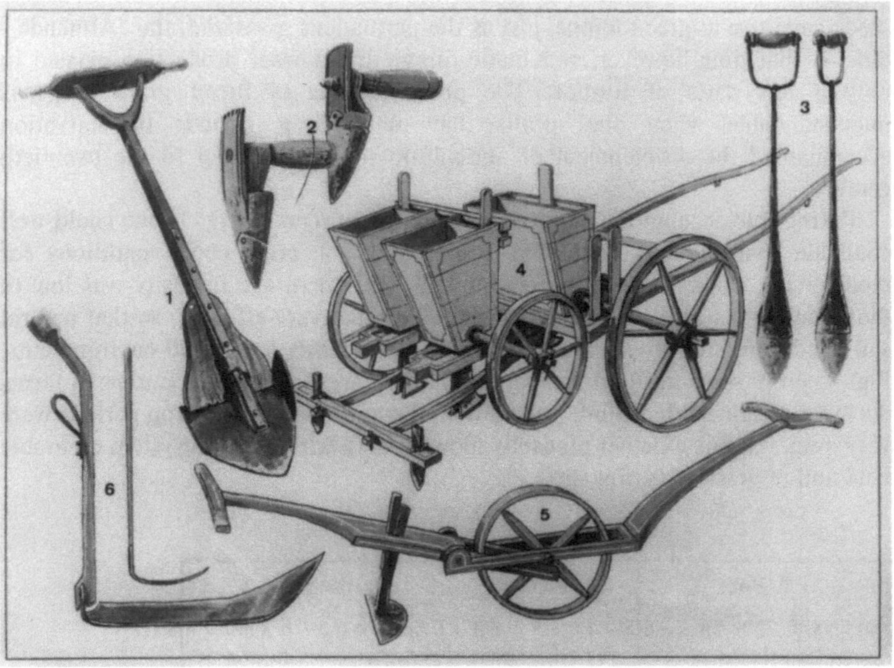

1. Lawn plough to skim grass sod prior to ploughing - 18th century
2. Dibbling tool for beans; women moved across the field on their knees to dig holes - 18th century
3. Dibbling tool to dig holes in a walking position - 18th century
4. Jethro-Tull-mechanic drill (1730) beginning of farm mechanisation
5. Hoe; one man pulling, one pushing - 18th century
6. Hainaut- or Normandy-scythe (held with one hand) and small grain hook

Fig. 3. Farm implements used in eighteenth century Europe - Shell Chemie 1978
Ilustración 3. Utensilios agrícolas utilizados en la Europa del siglo XVIII (Shell Chemie 1978)

2 Change in biodiversity

The methods for gauging biodiversity in agriculture outlined above will now be used to evaluate the past change in European agriculture. The oldest method of systematic agricultural land use in the central and western parts of Europe was a three-field system consisting of summer fallow, winter rye and oats or barley over three consecutive years. During the ninth century, Charles the Great made this crop sequence virtually compulsory throughout his empire. It was used throughout the greater part of central and western Europe until the eighteenth century.

This nearly 100% bred crop rotation survived for so long because cereals were so important for human nutrition. In the fallow section, shattered cereal and weed seeds gave rise to green forage, just as the permanent grassland, the "Almende," did. At that time however, systematic forage and animal production existed in only a few parts of Europe. The predominance of bread grain crops is understandable when one recalls that many long periods of starvation accompanied the development of agriculture in Europe prior to the twentieth century.

Taking into account the biodiversity criteria outlined in Fig. 1, one could well conclude that the old three-field system did not offer good conditions for biodiversity. However, it must be admitted that soil tillage intensity was low in those days. As a rule, almost all activities were not very efficient, so that natural soil conditions still provided a broad variety of fertility levels and environments. Fig. 3 shows some farm implements commonly used on modern European farms during the eighteenth century. In addition, the undisturbed non-crop periods were very long. All this together probably allowed for a fairly rich ecosystem on arable land and in grasslands (Fig. 4).

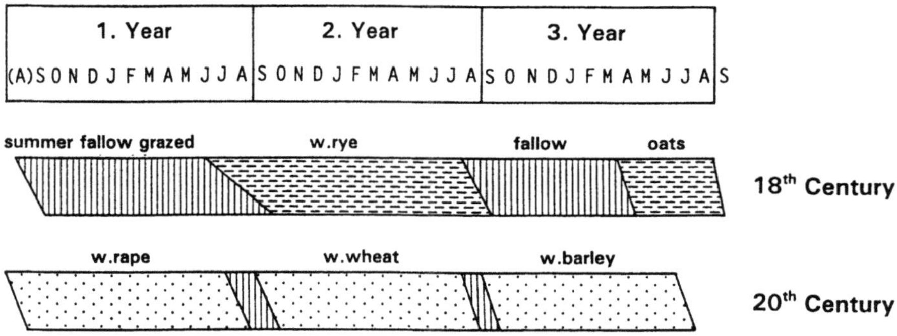

Fig. 4. Loss of biodiversity in farming in central Europe as a result of intensified land-use – see close rotation of crops in the twentieth century

Ilustración 4. Pérdida de biodiversidad en la agricultura en Europa central como resultado de un uso intensivo de la tierra. Véase la estrecha rotación de cultivos en el siglo XX

Agriculture in eighteenth-century central and western Europe underwent a change that is much more deserving of the term "Green Revolution" than any other progress or change in agriculture to date: The summer fallow was replaced by legumes, and by red clover in particular. Not only was this the first time that forage was systematically produced in a deliberate, straightforward manner, it also provided a number of advantages such as (1) all subsequent crops enjoyed an increased nitrogen supply from the remains of N_2-binding legumes; (2) forage production enabled improved animal husbandry which in turn (3) increased the amount of barnyard manure which is a prerequisite for growing demanding crops such as potatoes and beets. It is a well known fact that barnyard manure was the most important fertilizer on almost all European farms until around 1950/60. Red clover not only supported other crops by making a huge contribution to soil fertility, it also (4) enforced broader rotations because of its auto-incompatibility. In other words, at least five years must pass before red clover can be planted again in the same field. As result of the "Green Revolution," European agriculture experienced its broadest biodiversity ever, which lasted until the middle of this century. Table 1 offers some examples of how the number of crops in central and west European agriculture changed between the ninth and twentieth centuries.

Table 1. Rotation and field sequences in Europe from 9th to 20th century

9th	18th/19th century	Early 20th century		End of 20th century			
Year	3-field farming	wide rotation	"Rheinische rotation"	New 3-field farming		New 2-field farming	
				Köln-Bonn area	Schleswig Holstein	no animal farms	pig farms
1	summer fallow	red clover	fodder and sugar beet (with farm-yard)	sugar beet	winter rape	sugar beet	corn
2	rye	oats	winter wheat	winter wheat	winter wheat	winter wheat continue	winter wheat continue
3	oats	winter rye	rye (companion crop)	winter barley continue	winter barley continue	winter barley continue	
4		potatoes	red clover (with farm-yard)				
5		winter wheat	oats etc.				
6		rye					
7		fodder beet					
8		summer barley					
9		rye + red clover companion crop					

However, this beautiful agricultural system was swept away within only a few decades. Economic constraints – first and foremost, the dramatic increase in the cost of labour which rose by about 300% in Germany between 1950 and 1970, while prices for agricultural products rose by only 25% – forced farmers to increase their efficiency and productivity. They did so by specializing in a few of the most productive crops or animals to save labour and reduce machinery requirements, and/or by giving up animal husbandry to, in many instances, concentrate on no more than two or three cash crops. They also resorted to increasing their use of nutrients and pesticides in order to increase and protect their yields.

We must accept that this was the only way for a farmer to survive under the economic constraints prevailing in high tech-countries with high labour costs. Despite their struggle to intensify and specialize, many farmers ultimately had to give up, a fact that underscores the dramatic change that happened within just a few decades. Figs. 5, 6 and 7 give simplified examples for this change (see also Table 1, twentieth century).

During the same period, there was a dramatic increase in agricultural efficiency. Agricultural efficiency, which is measured by the number of people fed by one farmer, is viewed as a success story. One could just as well call it a story of lost biodiversity.

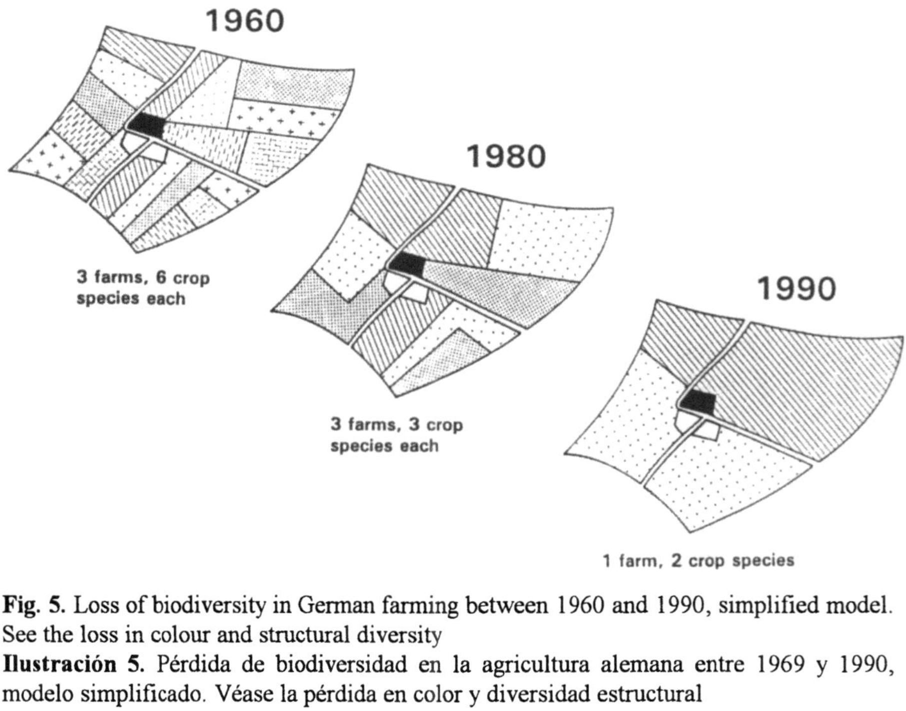

Fig. 5. Loss of biodiversity in German farming between 1960 and 1990, simplified model. See the loss in colour and structural diversity

Ilustración 5. Pérdida de biodiversidad en la agricultura alemana entre 1969 y 1990, modelo simplificado. Véase la pérdida en color y diversidad estructural

- Number of farms in Germany and France

	Germany	France
1960:	1385250	1773500
1990:	665100	1017000

- Percent area of cropland in Germany and France covered by small grain

	Germany	France
1960:	53.0%	38.5%
1990:	54.9%	37.3%

- Percent area of certain small grain species as related to total small grain area in Germany and France

	Germany		France	
	1960	1990	1960	1990
wheat[a]	29.6%	40.6%	53.7%	71.6%
barley[b]	21.0%	41.1%	24.1%	24.4%
rye	31.5%	10.0%	4.0%	0.9%
oats	17.9%	8.2%	18.2%	3.0%

- Percent area of cropland in Germany and France covered by the total area of wheat + barley + rapeseed + corn, 1990

Germany	France
55.6%	47.6%

[a] winterwheat + summerwheat
[b] winterbarley + summerbarley

Fig. 6. Comparative figures for agriculture change in Germany (West) and France 1960-1990
Ilustración 6. Cifras comparativas para cambio de la agricultura en Alemania (Oeste) y Francia 1960-1990

3 Prospects

Papers of this kind usually conclude with a few optimistic words. However there are an overwhelming number of reasons to be pessimistic. For instance, almost all our structures – administrative, political, commercial and economic – are against biodiversity. Even our common sense about what is necessary for life and the quality of life has been gravely distorted. One of our most forward-looking

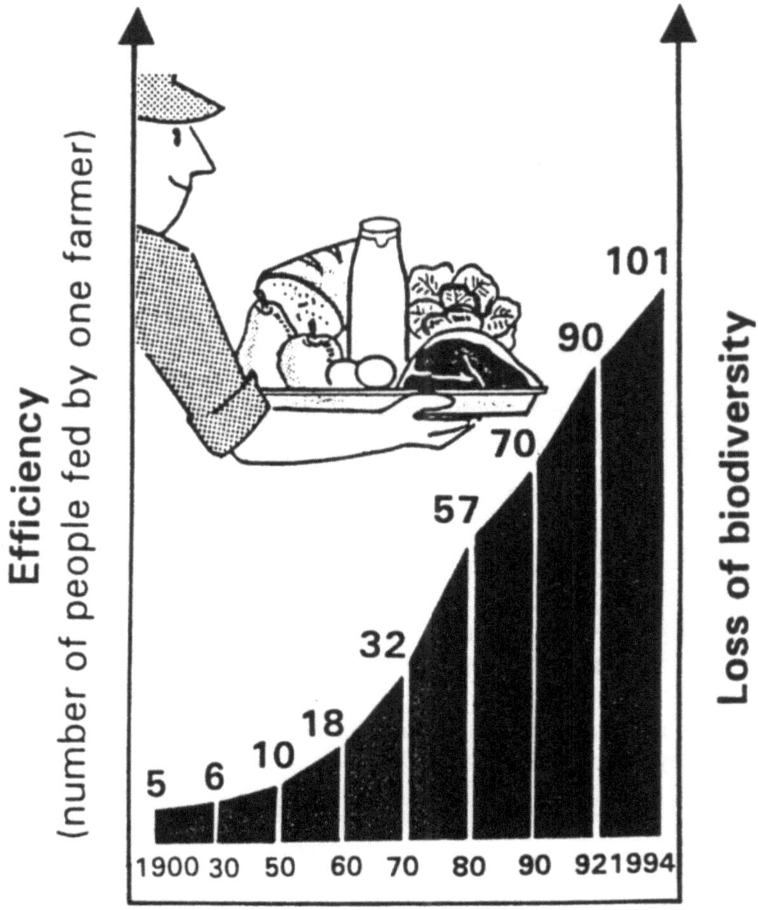

Fig. 7. A story of success?
Ilustración 7. ¿Una historia exitosa?

journalists once stated, "We know the price of everything and the value of nothing" (Horst Stern).

Let me put this in another way to demonstrate that this statement is indeed very applicable to European agriculture: In Germany, for instance, farmers feed a population of 80 million and cover more than 90% of the country's foodneeds. Although agriculture is essential to life, it accounts for less than 2% of the country's gross national product (GNP). On the other hand, products, items, services and even destructive activities contribute much more to the GNP than agriculture does. This in turn places continuous pressure on agriculture – both the business and its practices – especially in high-wage countries. It is possible that organizations such as the Common Market and international agreements such as GATT[1] accelerate the loss of biodiversity more than we would normally assume at first glance. One of the EU's primary administrative activities is to regulate, standardize and harmonize the agricultural business within Europe and in doing so, it faces enormous problems in dealing with regional peculiarities. This goal is just the opposite of "diversity." GATT policy increases the pressure on our farmers to be economically efficient, whereas economic efficiency usually means a loss of biodiversity.

At the same time, European agricultural policy tries in many ways to de-intensify agriculture, reduce surplus production and protect nature. There are numerous programmes to support farmers who reduce fertilization intensity (especially with nitrogen and phosphorus). It also supports organic farming, sustainable agriculture and the like. But the total of all such farm products comprises a maximum of only 10 to 15% of the food market. And the more sustainable methods of food and fodder production in agriculture have a difficult competitive environment to deal with.

As far as biodiversity in European agriculture is concerned, it seems that we will have (1) regions with intensive agricultural production and low biodiversity, (2) nearly bare regions, some of them reforested with medium biodiversity, (3) a small percentage of areas and farms which use so-called sustainable or organic farming methods and have relatively high biodiversity, and (4) some uncultivated islands of natural resources with oligotrophic soils that preserve high biodiversity.

From a biological standpoint, one needs only one single sentence to explain the reason why biodiversity in agriculture (and elsewhere) is endangered: "Homo sapiens" has become too efficient and competitive in a short period of time, and this superior efficiency not only endangers biodiversity, it ultimately endangers man himself.

References

Diercks R (1983) Alternativen im Landbau. E. Ulmer, Stuttgart

Kühbauch W (1993) Intensität der Landnutzung im Wandel der Zeit. Die Geowissenschaften 11: 121-129

[1] General Agreement on Tariffs and Trade

Traditional use of tropical biodiversity in Melanesia

Wulf Schiefenhövel
Max-Planck-Institute of Behavioral Physiology, Andechs, Germany

Abstract. Do members of traditional cultures live in harmony with their environment and, if so, why? Do they classify the living world and, if so, how do they accomplish this without access to natural science, its libraries and databases? There are a number of misconceptions regarding non-literate societies and their view of and relationship with nature.

Like all animal species, humans have always had an impact on the environment in which they live. Yet, the efficacy of their actions was limited, mainly due to a lack of efficient tools. Much more primary forest has been cut down in New Guinea since the introduction of the steel axe than during the times when areas for cultivation had to be cleared with stone adzes.

It is typical for members of traditional societies to have a very sophisticated system of understanding and classifying nature. At first glance it seems surprising and highly improbable that they use classificatory methods that are very similar to our own Linnéan system. From the perspective of evolutionary biology however, a similarity between the various ways of seeing the living world is not all that unlikely. The human brain looks for structure in a very similar way, regardless of the specific culture. Humans are, by their very nature, "natural scientists," even when the object of their study is a plant without any direct material, nutritional or religious significance.

On the basis of their intimate knowledge of the living world in their environment, people in Melanesia have a large variety of methods for utilizing biodiversity – and they have even increased biodiversity. Food plants are usually propagated via vegetative seedlings. During the long history of Melanesia and through the curiosity and skill of the Papuan and Austronesian populations a very rich array of edible species (*Colocasia esculenta*, *Dioscorea alata*, *Ipomoea batatas*, *Saccharum edule* and *officinarum*, *Rungia klossii*, *Abelmoschus manihot* and many others) has been developed which in fact constitutes one of the genetic treasures of our planet.

Animals including fish and other maritime creatures are caught in thoughtful, clever ways that are based on a very precise knowledge of the ecology, biology and behaviour of the respective species. Neolithic horticulture with long fallow periods and mulching techniques has yielded good results. Rather than being mutually exclusive, biodiversity and cultural diversity are natural partners.

Uso tradicional de la biodiversidad tropical en Melanesia

Resumen. ¿Viven los miembros de las culturas tradicionales en armonía con su entorno y, si es así, por qué? ¿Clasifican el mundo en que viven y, si es así, cómo lo hacen sin tener acceso a las ciencias naturales, sus bibliotecas y bases de datos? Existe una serie de conceptos erróneos respecto a las sociedades sin literatura y su visión de y relación con la Naturaleza.

Como todas las especies animales, los humanos siempre han afectado el entorno en el que viven. No obstante, la eficacia de sus acciones era limitada, principalmente debido a la falta de herramientas eficientes. En Nueva Guinea, se han talado muchos más bosques primarios desde la introducción del hacha de acero que durante el periodo en el que las áreas de cultivo tenían que ser despejadas con azuelas de piedra.

Un rasgo característico de los miembros de sociedades tradicionales es tener un sistema muy sofisticado de entender y clasificar la Naturaleza. A simple vista parece sorprendente y muy improbable que empleen métodos clasificatorios muy similares a nuestro propio sistema Linnéan. Desde la perspectiva de la biología evolutiva, sin embargo, no resulta tan improbable una semejanza entre las distintas formas de contemplar el mundo viviente. El cerebro humano busca estructuras de forma muy similar, independientemente de la cultura específica. Los humanos son, por su propia naturaleza, "científicos naturales", incluso cuando su objeto de estudio es una planta sin ninguna importancia directa a nivel material, nutritivo o religioso. En base a su conocimiento profundo del mundo viviente en su entorno, los habitantes de Melanesia poseen una gran variedad de métodos para utilizar la biodiversidad; incluso han aumentado la biodiversidad. Plantas alimenticias suelen estar propagadas a través de plantas de semillas vegetativas. Durante la larga historia de Melanesia y debido a la curiosidad y habilidad de las poblaciones de Papua y Austronesia se ha desarrollado un conjunto muy rico de especies comestibles (*Colocasia esculenta, Dioscorea alata, Ipomoea batatas, Saccharum edule y officinarum, Rungia klossii, Abelmoschus manihot* y muchas otras) que constituye de hecho uno de los tesoros genéticos de nuestro planeta.

Los animales, incluyendo peces y otras criaturas marinas, son capturados mediante métodos cuidadosos e inteligentes basados en un conocimiento muy preciso de la ecología, la biología y del comportamiento de las respectivas especies. La horticultura neolítica con largos periodos de barbecho y técnicas de *mulching* han dado buenos resultados. Biodiversidad y diversidad cultural no sólo no se excluyen mutuamente, sino que son socios naturales.

1 Introduction

The perspective I am taking for this presentation partly reflects cognitive anthropology, or ethnoscience. In this so-called emic approach one tries to see and understand the world, as much as one can, through the eyes and the concepts of the people one is living with – in my case, the Eipo of the highlands of Irian Jaya, West New Guinea, the Trobriand Islanders in the Solomon Sea of Papua New Guinea and other peoples in Melanesia. Scientific insight can be increased by switching perspectives, and in anthropology this would be a switch from the emic to the so-called etic approach. In the latter case, one tries to see and handle data collected in the field according to one's own traditions of obtaining and analyzing information. This process of combining emic and etic perspectives such as indigenous botany and academic botany plus related sciences as they have developed in our own scientific institutions has proven its heuristic and explanatory strength (cp. Berlin 1992, Schultes & von Reis 1995, Cotton 1996) and may help to shed light on how members of Melanesian societies perceive and use nature.

2 Two Melanesian cultures

In the years since 1965, I have conducted most of my research on the island of New Guinea and other surrounding islands. Two of these projects, fieldwork among the Eipo and the Trobriand Islanders, were supported by the DFG (German Research Council) and I am glad to have the opportunity during this presentation in Bonn to express my gratitude for this support in the presence of some of its representatives. The Eipo live just north of the Central Cordillera of Irian Jaya at approximately 140° eastern longitude and 4° 26' southern latitude in climatic and ecologic conditions that are rather typical for the mountainous interior of New Guinea. The Eipo are descendants of Papuan immigrants who arrived on the shores of the main island at least 35,000 (Jorgensen 1994) and probably 50,000 to 60,000 years ago. The people and their culture (Schiefenhövel 1991) are well adapted for life in the island's rugged high mountains which protect them from malaria and other infectious diseases that occur in lower altitudes. Archaeologists have shown that plants and corresponding horticultural techniques were domesticated at least 9,000 years ago in highland New Guinea, making it one of the very early centres of agriculture worldwide.

We have also been conducting interdisciplinary fieldwork since 1982 on one of the Trobriand Islands which have become known through the work of B. Malinowski, one of the founders of modern anthropology. Malinowski (1922, 1935) convincingly demonstrated the complexity of traditional cultures with his descriptions of their techniques for growing food in the coral gardens, getting food from the sea and conducting daring sea voyages and large-scale economic transactions. In contrast to the Papuan Eipo, the Trobrianders are Austronesians; their ancestors arrived in well-designed sea-worthy outrigger sailing canoes some thousand years ago and their language is related to the other Austronesian languages that span more than half the globe between Taiwan and Hawaii in the north, Madagascar in the west, New Zealand in the south and the Polynesian societies of the Pacific that extend as far east as Easter Island.

3 Living close to nature

When we began our fieldwork in 1974, the material and social conditions in which the Eipo in the highlands of West New Guinea lived resembled a neolithic setting (Schiefenhövel 1976). They can therefore be viewed as 'modern models of the past'. From cradle to grave, members of this culture are embedded in nature in a very direct sense. When a child is born in an Eipo village it is usually born onto the grass outside the women's house. *Me delina*, putting the child down, is their term for birth. In this society, post partum life usually starts with direct physical contact with nature – as did its very beginning: Sexual intimacy is usually enjoyed in the garden or some other part of nature outside the village. In Eipo poetry (cp. Hiepko & Schiefenhövel 1987) lush nature, the wilderness of the forest and the growing energy of the open gardens is contrasted with the village – which serves as a metaphor for norms and narrow regulations, as well as for being protected and at home.

Fig. 1. The Eipo place their dead on boughs in the tops of trees. In a second ceremony, the mummified bodies are brought to little huts in the gardens. Sometimes the skull and long bones are transferred to a third location under overhanging rocks
Ilustración 1. Los Eipo depositan a sus muertos sobre las ramas situadas en la cima de los árboles. En una segunda ceremonia, los cuerpos momificados son transportados a pequeñas chozas en los jardines. A veces, el cráneo y los huesos largos son trasladados a un tercer lugar bajo rocas sobresalientes

When a person dies, his body is put onto the boughs of a tree (Fig. 1) from which branches have been removed so that there is space to surround the corpse with a shelter made of bark and leaves. This protects the corpse to a certain degree from the rain. Corpses exposed in this way mummify in the course of several months and are then usually transferred to a secondary place, generally under the roof of a small garden hut. And in cases where there is tertiary exposure, the skull and long bones are kept under overhanging rocks. These customs clearly illustrate the fact that life in societies as traditional as that of the Eipo is virtually embraced by nature.

4 Utilizing nature in the highlands

In the afternoon the Eipo women are coming home, carrying their own weight in garden products plus fire wood, a total of more than 40 kgs. The Eipo are a very small but very powerful version of *Homo sapiens*. They are surprisingly healthy and able to sustain highly demanding physical activities for long periods. In the background one sees fires – a sign of the typical slash-and-burn technique used to clear land for new gardens. Old garden plots which have been allowed to lie fallow and are now overgrown with secondary vegetation are usually used again. We estimate that the fallow period lasts about fifteen years. The Eipo use *Trema tomentosa*, the typical tree in anthropogenous grasslands, as their most important bio-indicator: When the stems have reached a certain circumference the fallow land is considered fertile again and fit for new usage. The process of shifting cultivated areas in such a way ensures sufficient fallow time (cp. the contribution of W. Kühbauch in this volume) and, consequently, virgin forest has to be cleared only occasionally. This principle of re-using old garden plots coincides with cost-benefit considerations: It is much more time and energy consuming to cut down large forest trees – with stone adzes – than to clear secondary growth.

At the start of our fieldwork in 1974, the Eipo were living under neolithic conditions. They had stone adzes and other tools made of stone, wood, bone and teeth. The technology to produce metal was unknown. In the West, there is a tendency to portray traditional peoples as always being in sacred harmony with nature. Unfortunately, this image is far from reality. *Homo sapiens* is a maximizer, alas. The large flightless Moa bird of New Zealand was exterminated by the country's indigenous Maori population (genetically, linguistically and culturally related to that of the Trobriand Islands). When one lives with people in Melanesia one is surprised by how often trees and other plants are injured by senseless blows from adzes or axes and how often animals are brutally treated or killed. The concept of preventing cruelty towards living beings is definitely not a worldwide phenomenon. Certain cultures are able to control some of the factors that are critical to the equilibrium between humans and nature more effectively than others, but it seems that most of the nature-culture balance we see in traditional societies is due to the fact that their members just don't have the technology to do lasting damage. New Guinea provides an example for the scope of change that can be brought about just by replacing stone adzes with steel axes.

Fig. 2 shows the harvest of one day. A woman, her husband and child are about to return to the village, their string bags filled with sweet potatoes *(Ipomoea batatas)*, the staple diet in this part of the highlands, and green leafy vegetables.

Ipomoea batatas was probably introduced less than 300 years ago. There is still some debate whether this new and very successful crop was perhaps introduced from South America much earlier through Polynesian contacts, but this seems doubtful. *Rungia klossii* and *Abelmoschus manihot*, domesticated shrubs with dark green leaves (taxonomic identifications from Hiepko & Schultze-Motel 1981), are the most important source of protein in the diet of the basically vegetarian (by necessity, not by choice) Eipo.

Fig. 2. A woman, her husband and child leaving their garden. Sweet potatoes (in this case a cultivar with a reddish skin) provides most of the carbohydrates needed and leafy greens (wrapped in bundles) most of the protein
Ilustración 2. Una mujer, su esposo y su hijo salen de su jardín. Los boniatos (en este caso una variedad de piel rojiza) proporcionan la mayor parte de los hidratos de carbono necesarios y las hortalizas hojosas (enrrolladas en manojos) la mayoría de las proteinas

Various sweet potato species, that are used, look much the same to us. But even children are able to identify and name the common varieties – one example of how effective a non-formalized system of knowledge transfer can be. Our children, on the other hand, have only a meagre knowledge of similar cognitive tasks such as identifying even the most common trees in our forests – despite the efforts of their teachers.

Taro *(Colocasia esculenta)* was very probably the main staple food before the arrival of *Ipomoea batatas*. Particularly large specimens of taro are grown in special places and with special care. This food is used to honour guests and in sacred ceremonies. The sweet potato is everyday food, taro is served at feasts. Gerd Koch, Klaus Helfrich and Thomas Michel, the anthropologists on our interdisciplinary team, collected names for different cultivars of some main food plants. The results are shown in Table 1. If one accepts the view that *Colocasia esculenta* (and perhaps other *Araceae*) originated from pre-Ipomoean horticulture in this region of the highlands, then it is somewhat surprising that *Ipomoea batatas* has been bred, in less than 300 years, in such way that it has produced almost as many cultivars. Perhaps (this is just a guess on the part of a non-botanist) mutation rates are higher in sweet potatoes than in taro.

Table 1. Cultivars of the most important food plants of the Eipo in the highlands of Irian Jaya. These figures were obtained by three anthropologists in three different villages (from Hiepko & Schiefenhövel 1987)

Species	N of cultivars in 3 different villages:		
	Malingdam	Moknerkon	Talim
Taro (*Colocasia esculenta*)	56	39	65
Sweet potatoe (*Ipomoea batatas*)	34	30	44
Banana (*Musa x paradisiaca*)	21	20	24
Sugar Cane (*Saccharum officinarum*)	17	21	21
bace - Vegetable (*Saccharum edule*)	13	16	12
touwa - Vegetable (*Abelmoschus manihot*)	13	12	13
teyang - Vegetable (*Setaria palmifolia*)	10	18	10
mula - Vegetable (*Rungia klossii*)	5	8	9

from: Hiepko & Schiefenhövel 1987

This table demonstrates that there is a rich variety of other important food plants as well. Knowing that stone-age people were responsible for generating this enormous genetic richness makes one aware of the effectiveness of neolithic plant domestication and of the threat we pose to such cultures through acculturation, logging and the introduction of monocultures, pesticides, herbicides and the like. Other contributors to this symposium have convincingly demonstrated the threats to this genetic richness. I am convinced that besides protecting biodiversity, we must protect cultural diversity as well in order to facilitate the former's survival.

Wild food plants are very important for this population. Fruit stands of *Pandanus brosimos* that grow in the rain forest at elevations above approximately 2,000 m are carried home. The enormous load is carried in string bags and supported by a band around the forehead. *Pandanus brosimos* contributes a little bit of fat (and probably other valuable nutrients) to the diet of the Eipo. Fat is otherwise very scarce in their culinary repertoire. Fig. 3 shows a festive meal being prepared with the red sap of *Pandanus conoideus*, a domesticated species that grows at elevations below 1,600 m – in other words, outside the home territory of the Eipo who get this highly valued and symbolically powerful food from neighbours in lower lying areas in exchange for things that grow well at higher altitudes. The ethnic groups in the isolated mountain valleys of New Guinea, sometimes comprised of only a few hundred in number, do not live an isolated life. Necessity makes them trade with other groups who often live on the other side of the central cordillera. The lowest passes that have to be crossed, barefoot, are about 3,700 m high.

Fig. 3. The red, slightly bitter sap of *Pandanus conoideus* is a rare delicacy
Ilustración 3. La savia roja y ligeramente amarga de la *Pandanus conoideus* es un raro manjar

Meat is highly valued but very scarce because wild animals (all except a few transferred placentalia are marsupials) are small and difficult to hunt or snare, and the meat of the few domesticated pigs that are raised is saved for festive ceremonies (Fig. 4). During their daily walks to and from the gardens and their activity in the gardens, women collect small reptiles and their eggs as well as certain spiders and various kinds of insects and their larvae. These tiny but physiologically probably very important quantities of animal protein are reserved for women and infants. In this quasi neolithic culture, those who need high quality food the most actually get it. In my opinion, nutrition is so precarious in this region that its culture had to find a wise way to distribute the scarce protein resources by creating regulated "ecological niches" – and by defining every flying, crawling or otherwise moving creature as food. Certain rather intensely smelling stinkbugs *(Pentatomidae?)* are also eaten, and are steamed as a spice in leafy foods. Primatologists have provided new insights into the importance of such "bizarre" foods such as ants, termites and the like which represent sources of valuable animal protein. It is very likely that insect food played a prominent role in the human past; the Eipo and other traditional peoples are carrying this nutritive strategy over into our times.

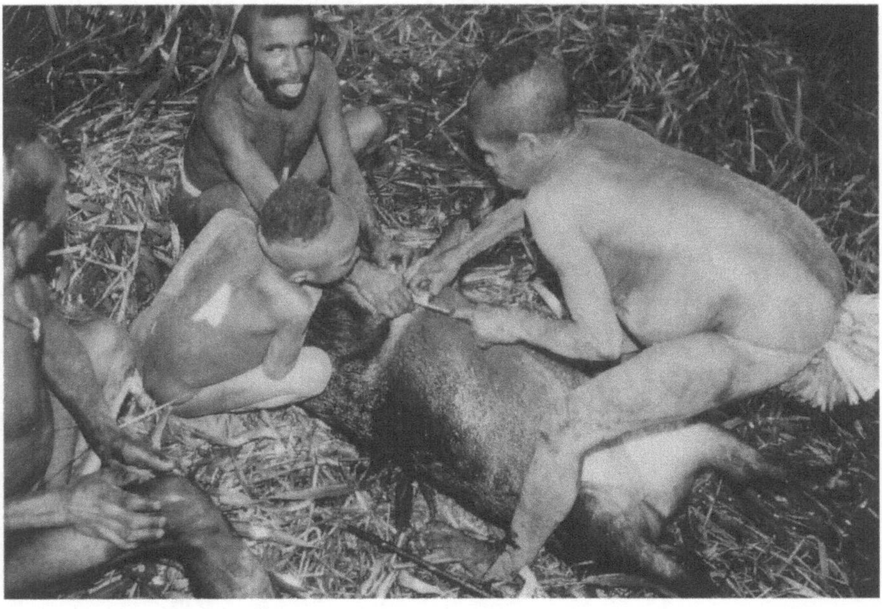

Fig. 4. Pigs are the most important animal in the culture of the Eipo. Women take a lot of care to raise them. Their number is limited and they are killed only at special occasions. Therefore pork contributes very little fat and animal protein to the daily diet

Fig. 4. Cerdos son los animales mas importantes en la cultura de los Eipo. Las Mujeres tienen mucho cuidado para criarlos. Su número está limitado y se los matan solamente en ocasiones especiales. Por eso, a la dieta diaria el carne de cerdo contribuye muy poco de grasa y de proteina de animales

The Eipo hunt arboreal marsupials with bows and (non-poisoned) arrows, sometimes assisted by specially trained dogs. They also use snares to catch these animals. These traps are constructed preferably on horizontal tree stems which form part of a runway. Fig. 5 (from Blum 1979) shows the snare's individual elements which are cleverly combined to achieve maximum effectiveness, and makes it clear that the Eipo, like members of other traditional cultures, not only have a perfect knowledge of how to build such ingenious snares but also have command of a precise linguistic terminology containing functional lexems for all the elements of this construction: *Homo sapiens* and *Homo faber* at their best.

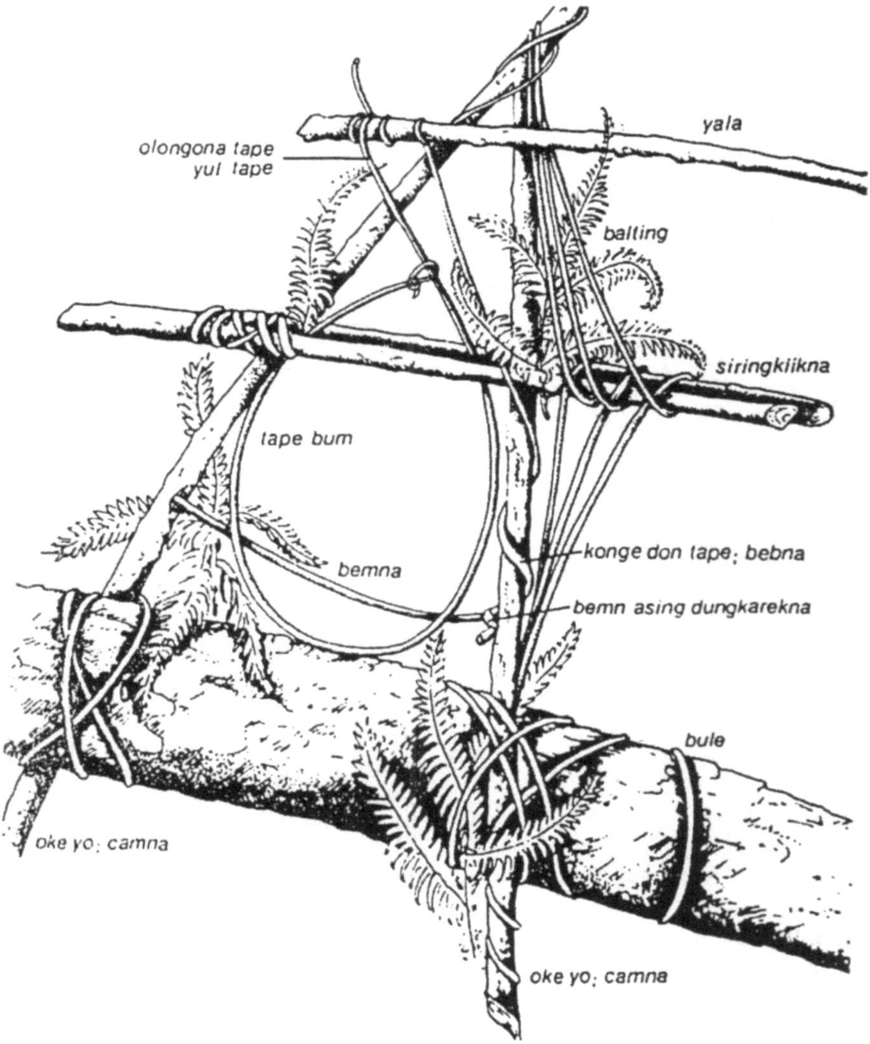

Fig. 5. Snare to catch marsupials (from Blum 1979)
Ilustración 5. Trampa para cazar marsupiales

As is proper for botanists, Paul Hiepko and Wolfram Schultze-Motel (1981), the botanists in our team among the Eipo, collected plants and recorded their names in the local language. Some of the native taxonomy was puzzling, as it appeared that plants were grouped together which did not seem to be related to each other in our Linnean system (for example the seemingly very different species of Saxifragaceae, *Polyosma induta and Polyosma sp.* which are both termed *bensukwe* in the Eipo language). Sometimes Eipo informants provided specific explanations such as: "This plant is the uncle of that one," thereby creating a relationship between plants following the principle of kinship among humans. Ultimately, it turned out that the Eipo were excellent botanists. In those cases where they had claimed a relationship between different plants (which was sometimes not immediately recognized by academic specialists) this relationship existed in our taxonomic system as well. This is certainly no surprise to those familiar with ethnobotany and ethnozoology (cp. Berlin *et al.* 1973), but it is still very powerful proof of the human capacity and propensity to detect and conceptualize structure and order in the living world.

5 Utilizing nature on the islands

Most of the Trobriand Islands are coral atolls with a very limited fauna (especially among terrestrial animals) and a somewhat limited flora as well. In contrast, the biodiversity of the coral reefs is breathtaking. The inhabitants of this region of the Solomon Sea have a number of cultural traits which differ from those found in the Eipo highland society. The Trobrianders are matrilineal and have a structured political system with strongly matrilineally inherited chieftainships. Institutionalized, ritualized forms of competition as well as a very highly developed sense of aesthetics are also characteristic of their culture.

Marine resources are exploited in countless very clever ways. During the night, torches made of dried coconut palm leaves are lit and swung through the air so that they flare up. This light allows the fishermen to see their prey in the shallow water and spear them. *Panulirus ornatus* and *P. punctatus* spiny lobsters form part of the culinary repertoire, as do many other species of shellfish and fish from the various regions of the Trobriand environment. These people know virtually every animal that lives in the coral reef and in the deeper sea; they have given them names, arranged them in a classificatory system and catch them in a special way. It is amazing to see how these people utilize different techniques, including *Derris elliptica*, a fish poison also known in other parts of the world. It was very surprising to witness Trobriand divers using *Derris* in the coral reef because one would expect the sea to dilute the poisonous latex so quickly that it would be ineffective. But the Trobrianders use a very clever method to avoid this: They keep crushed *Derris* roots in their clenched fist and release them under or inside coral stocks so that the poison remains there and is prevented from being watered down. The fish *(Oxycirrhitus?)* – having adapted, also in their colouring, to life in and around these reddish coral stocks – take refuge in their hiding place from the human figure approaching them and are thus incapacitated by the poison. They then float in the water and can easily be collected. A small bucket of fish can be collected in 20 minutes and there is no need to worry about possible bad effects from eating the catch. Almost every species of fish is caught using a special method which was developed on the basis of a precise knowledge of the ecology of the particular animal, its preferred habitats, chronobiology and behaviour – applied ethology in a non-literate society.

This will be made clear by the example of a porcupine fish *(Diodon holacanthus,* Diodontidae*)*. While we were looking at the amazing creature and photographing its dramatic way of changing shape a Trobriand man said, „You must be very careful because there are two very similar fish. You eat the wrong one you die, but this one is the type you can eat, nothing will happen to you." Obviously, he was referring to one of the puffer fish *(*Tetraodontidae*)* which is related to the *Diodontidae*. One of them is the famous *fugu* with its deadly poisonous bile; it is considered a delicacy in Japan and prepared by specially licensed chefs in expensive restaurants.

What was said before about the excellent botanical knowledge of the highland Eipo can be repeated here with regard to marine biology in the Trobrianders' environment. Ecology, form, colour, behaviour and any peculiarity of the creatures in and around the sea are known in detail and utilized to catch them for consumption.

6 Ethnoscientific knowledge – The enigma and danger

Table 2 (from Schiefenhövel & Prinz 1984) lists a number of well-known drugs without which modern medicine would be inconceivable. Modern anaesthesia and surgery, for instance, would not be possible without curare. All these medicines have been taken from the pharmacological thesaurus of traditional cultures, including our own.

All cultures have their own kinds of science. This knowledge has seldom been put into museums or published in journals or books. It was passed on, in an incredibly efficient manner, orally from generation to generation, seemingly without loss and with an occasional addition. For me, one of the very surprising aspects of natural science in traditional societies is the fact that the people know plants and animals even when these species do not lend themselves to being researched or known. Our friends in New Guinea usually also know very insignificant plants which are neither useful, dangerous or remarkable in any other way. *Homo sapiens* longs to collect knowledge, to find structures and to be a natural scientist. We are a curious species and, having a scientific bent, we find hidden secrets in nature and construct our concepts of the world.

Another example. People from India to Melanesia love to chew betel nut *(Areca catechu)*, the main active principle of which is the alcaloid arecoline. It has, apart from being a strong parasympathicomimetic, a mild hallucinogenic effect. The Trobrianders often call it „our beer." But getting this much-sought effect is not that easy. Chewing the seed of the *Areca* nut produces a more or less pronounced astringent effect, but nothing happens to the central nervous system.

Burnt and slaked lime has to be added in order to free the arecoline and turn it into a pharmacologically active form. On the Trobriand Islands this precious substance is produced by collecting certain white corals, burning and then slaking them carefully with a spray of water coming from the mouth of an expert lime maker. To pep up the effect of the released alcaloid and add a culinary quality to their morsel, people everywhere add *Piper betle* or some other peppery plants.

Who on earth discovered all this? Who found out about how to treat *Manihot esculenta* tubers containing harmful levels of prussic acid so that they could be turned into delicious Cassava dishes? Who found out about the hidden powers of tens of thousands of medicinal plants? How did *Homo sapiens* develop all this vast and detailed knowledge before the days of chemistry and pharmacology? I don't know how many plant species are accessible to members of individual cultures in various types of environments, but the numbers must be overwhelming.

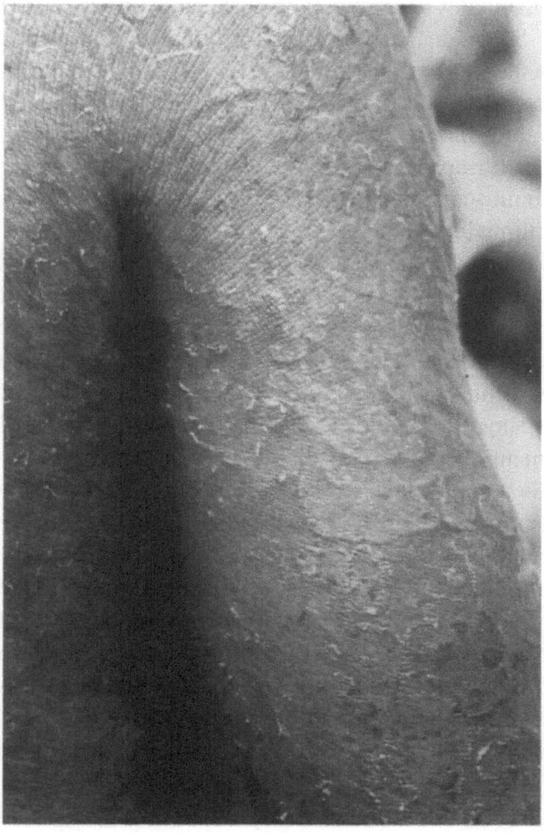

Fig. 6. Many persons in lowland Melanesia suffer from 'ringworm' *(Tinea imbricata)*, a superficial mycosis that is feared because of its stigmatizing effect

Ilustración 6. Muchas personas de las tierras bajas de Melanesia sufren de tiña (*Tinea imbricata*), una micosis superficial temida por sus efectos estigmatizadores

For me, this is one of the most intriguing questions in ethnoscience. It is inconceivable that our own forefathers and the forefathers of others populations set out on a gigantic experiment to test all possible herbal and other substances for possible beneficial and/or harmful effects using neolithic technology. Such intimate knowledge of nature can not have come about by simple trial and error. It must, I am certain, be the product of early humans' neurobiological perception mechanisms and learning abilities. Otherwise it would have taken much, much longer to discover all these active principles.

I believe we have to assume that the brains of animals, and perhaps of mammals in particular, possess certain search and evaluation mechanisms that are able to foster, as it were, pharmacognostic curiosity, trigger specific evaluation processes and store the results in the central nervous system in a way that allows them to be retrieved. There are indications that the *postrema area* in our brain is one of the elements of this type of intelligent (non-esoteric) system that helps us find the right foods and the right cures (cp. Schiefenhövel, in press).

In the low-lying areas of Melanesia many people suffer from *Tinea imbricata* (Fig. 6), a basically harmless, but socially stigmatizing superficial mycosis caused by *Trichophyton concentricum.* Some decades ago my informants said that a certain plant, *Cassia alata,* was the appropriate cure for this skin disease. And indeed, this plant (called seven candle sticks in English) is pharmacologically active, particularly through its main principle, chrysarobic acid. *Cassia alata* was, not surprisingly, integrated into the European pharmacopoeia and used to treat, for instance, *Psoriasis vulgaris.* It would seem that the local people should have every reason to continue using this traditional pharmacon.

There is only one problem in this. Doctors in hospitals, nurses and medical orderlies at aid posts, and teachers and missionaries started handing out Griseofulvin tablets long ago, a very powerful oral antimycotic drug that can cure *Tinea imbricata* in a much shorter time than *Cassia* leaves which must be applied for days or weeks. Papua New Guineans have realized that this new way of regaining a healthy, nice skin is superior to their traditional methods. This is of course true for many other elements of modern medicine as well. There is only one hitch: Griseofulvin is not only a very effective antimycotic but has potentially dangerous side effects as well, particularly for the liver. Rather than risk possible gastrointestinal, hepatic, teratogenic and other consequences from insufficiently controlled doses of this modern medicine, *Tinea* patients should opt for the traditional alternative. But they don't. When one asks people these days about possible usages of the conspicuous *Cassia* with its yellow flowers, one is likely to hear either, „I don't know," or „Our grandparents used to rub the leaves of this plant on the skin of people suffering from 'ringworm', but we don't have to do this any more. The white people have tablets which make our skin healthy much quicker."

Melanesia has one of the highest levels of biodiversity in the world. I would like to repeat: Let us protect biodiversity, but let us not forget about protecting cultural diversity – often the latter will benefit the former.

Table 2. Our modern inventory of powerful drugs contains many herbal medicines from traditional cultures (from Schiefenhövel & Prinz 1984)

Medicinal plants taken from the pharmakopoeia of traditional cultures

Plant	Region	Substance	Usage in Modern Medicine
Physostigma venosum	W-Af	Physostigmine	Parasympathikomimetic
Pilocarpus sp.	S-Am	Physostigmine	Parasympathikomimetic
Atropa belladonna	Eur	Atropine	Parasympathikolytic
Hyoscyamus niger	Euras	Hyoscyamine	Parasympathikolytic
Ephedra sinica	China	Ephedrine	Indir. Sympathikomimetic
Rauwolfia sp.	India	Reserpine	Antisympathikotonic
Digitalis sp.	Eur	Digitoxine a.o.	Glycoside (heart)
Strophantus sp.	Af	Strophantine	Glycoside (heart)
Scilla maritima	Eur	Scillarene	Glycoside (heart)
Strychnos sp.	S-Am	Curare	Muscle Relaxation
Chondodendron sp.	S-Am	Curare	Muscle Relaxation
Erythroxylum coca	S-Am	Cocaine	Local Anaesthetic
Papaver somniferum	Eur	Morphine	Analgetic
		Codeine	Anti-Cough
		Papaverine	Spasmolytic
Cataranthus roseus	Af, M-Am	Vinblastine	Anti-Cancer
		Vinchristine	Anti-Cancer
Colchicum autumnale	Euras	Colchicine	Anti-Gout, Anti-Cancer
Cassia alata	Trop	Chrysarobine	Anti-Mycotic
Cephaelis ipecacuanha	S-Am	Emetine	Anti-Amoebic
Cinchona sp.	S-Am	Quinine	Anti-Malaria

from Schiefenhövel & Prinz 1984.

References

Berlin B (1992) Ethnobotanical Classification. Principles of Categorization of Plants and Animals in Traditional Societies. Princeton University Press, Princeton

Berlin B, Breedlove D, Raven P (1973) General Principles of Classification and Nomenclature in Folk Biology. American Anthropologist 75: 214-242

Blum JP (1979) Untersuchungen zur Tierwelt im Leben der Eipo im zentralen Bergland von Irian Jaya (West-Neuguinea), Indonesien. Reimer, Berlin

Cotton, C. M. (1996) Ethnobotany. Principles and Applications. Wiley & Sons, Chichester etc.

Hiepko P, Schiefenhövel W (1987) Mensch und Pflanze. Ergebnisse ethnotaxonomischer und ethnobotanischer Untersuchungen bei den Eipo, zentrales Bergland von Irian Jaya (West-Neuguinea), Indonesien. Reimer, Berlin

Hiepko P, Schultze-Motel W (1981) Floristische und ethnobotanische Untersuchungen im Eipomek-Tal, Irian Jaya (West-Neuguinea), Indonesien. Reimer, Berlin

Jorgensen D (1994) Pacific Peoples in a Modern World. In: Burenhult G, Rowley-Conwy P, Schiefenhövel W, Hurst Thomas D, White JP (eds) The Illustrated History of Humankind, Vol. 5, Traditional Peoples Today, Harper & Collins, New York, pp 99-121

Malinowski B (1922) Argonauts of the Western Pacific. An Account of Native Enterprise and Adventure in the Archipelagoes of Melanesian New Guinea. Routledge & Kegan Paul, London

Malinowski B (1935) Coral Gardens and Their Magic. A Study of the Methods of Tilling the Soil and of Agricultural Rites in the Trobriand Islands. 2 vols. Reynolds, New York

Schiefenhövel W (1976) Die Eipo-Leute des Berglands von Indonesisch-Neuguinea. Homo 26,4: 263-275

Schiefenhövel W (1991) Eipo. In: Hays TE (ed) Encyclopedia of World Cultures, Vol. II, Oceania. G. K. Hall & Co, Boston, pp 55-59

Schiefenhövel W (in press) Good Taste and Bad Taste - Preferences and Aversions as Biological Principles. In: Macbeth H (ed) Good Taste and Bad Taste, Berghahn, Oxford etc.

Schiefenhövel W, Prinz A (1984) Ethnomedizin und Ethnopharmakologie - Quellen wichtiger Arzneimittel. In: Czygan FC (ed) Biogene Arzneistoffe. Vieweg, Braunschweig, pp 223-238

Schultes RE, von Reis S (1995) Ethnobotany. Evolution of a Discipline. Chapman & Hall, London etc.

Part 4

Consequences for development and environmental policy

Science and technology - Sustainable economics (condensed version)

José A. Lutzenberger
Fundaçao Gaia, Rio Grande do Sul, Brazil

1 Introduction

As human beings we are all concerned with what is happening to the incredible richness of life in all its different forms and functions. This fantastic pageant of organic evolution – a symphonic process that put us here on this planet – is in danger. Since we are all concerned about the demolition of this incredible process, I would like to dwell on one of the factors in this tremendous loss of biodiversity, a factor which is also one of the causes of this negative trend: namely, modern agriculture.

2 Traditional and modern agriculture – a comparison

According to conventional wisdom, modern agriculture is so efficient today that in First World countries such as Germany, the US, Australia, Japan or Canada, two percent of the population can feed the entire population and even produce some surplus for export. In contrast, at the end of the last century about 60 percent of the people had to work hard on the land to feed themselves and the rest of the population. And although few people remember this today, in January 1945, it was the old peasant farmer who saved Germany. Now, if it were true that only two percent of the population were required to feed a country, we wouldn't need to talk about alternatives. If you simply compare the number of people who were called farmers in the past with the number of people who call themselves farmers today, you end up with these figures. But this is an illusion. This calculation is based on a fundamental error.

To explain this, we need to take a look at traditional farming, a holistic look from an overall macroeconomic point of view. The old peasant farmer was a self-contained, autarkic food production and distribution system. He was autarkic because he produced all his own inputs: The traditional peasant produced his own fertiliser using the dung from his animals, he produced his own energy which also came from his draft animals and he produced almost all his own tools or, rather, they were made by the carpenters and blacksmiths and other artisans in his village, who also counted as farming population and – this is important – he delivered most of his food directly into the hands of the consumer. In my national language, in Portuguese, the days of the week are still called Second, Third, Fourth, Fifth, Market (segunda-feira, terça-feira, quarta-feira, quinta-feira, sexta-feira, sábado, domingo), rather than Tuesday, Wednesday, Thursday and so on. This is a leftover from the day when farmers delivered their food to the weekly market.

On the other hand, today's modern farmer is just a small cog in an enormous machinery, in a gigantic technocratic, bureaucratic and legislative infrastructure. Therefore, any activity that contributes to the modern farming system is a part of this apparatus and must be included in the calculation. That means that modern farming starts in the oil fields of Saudi Arabia, Iran, Iraq, Kuwait and other oil producing nations, and can be found in the oil refineries, the chemical industry, the engineering industry and other industries. A modern economist looking at a pesticide or chemical fertiliser factory would call it part of the chemical industry, but in reality it is part of modern farming. A tractor or combine factory is part of the engineering industry for an economist, but in fact it is part of modern farming and should be counted with it. The bank employee who sits in front of a computer and manages credits to farmers and thinks of him or herself as a data controller or a software person is part of modern farming, and his or her man-hours have to be included in this calculation as well. And when the American housewife drives her two-ton car twenty miles to the nearest shopping centre to buy a hundred pounds of pre-cooked, deep-frozen and over-packaged food to put in her deep-freezer at home, she too becomes part of the calculation, along with the factories that processed the food, made the packaging, the freezer and so on.

Based on this calculation, nearly 40 percent of all working hours in a modern economy are dedicated to producing and distributing food.

If we look at the overall picture we will see that our modern-day food production and distribution system is a gigantic apparatus. How did this come about? Industry took everything that is profitable and involves little risk away from the farmer and left him everything that is less profitable and involves high risks.

Seen in this light, modern farming is no more efficient in terms of man-hours than traditional peasant farming. And it may come as a surprise, but more often than not it is also no more productive in terms of yield per acre. A few farmers in Holland may be able to produce 10,000 kilos of wheat per hectare, but their costs are so high that the smallest mistake may mean bankruptcy. And the system is not sustainable. The truth is, that most modern farmers produce less on average per acre than the old peasant did.

This is quickly illustrated by a comparison of traditional farmers and large-scale modern farmers in southern Brazil, where big farmers today generally grow soy beans and produce an average of less than 2,000 kilos per hectare in the summer. In our subtropical climate they then grow wheat or other grains on the same field averaging 1,000 kilos per hectare in the winter. But the soy beans are not even used to feed hungry Brazilians, but to feed fat cows in the Common Market, which in turn produce what is called "mountains of butter" and "seas of milk". The European Community spends approximately tens of billions of dollars in tax money every year to support this absurd overproduction and then to destroy a large part of it.

Also, it is said, f.i., that modern farmers produce up to 6,000 kilos per hectare maize while the peasant farmers in Chiapas produce only 2,000. What is not said is that the peasant grows his maize together with beans, squash, potatoes, sweet potatoes, fodder, herbs and a lot more, all on the same field, totalling more food per acre, without chemical fertilizer and with no poisons!

Brazilian peasants also grow food for local consumption. They plant manioc producing up to 15 to 20 tons a hectare. They harvest at least 15 to 20 tons a hectare when they plant Irish and sweet potatoes. They harvest impressive amounts in their orchards and truck gardens. As this indicates, traditional farmers are highly productive.

Feed lots are yet another form of modern farming. In modern feed lots, more food is destroyed than is produced. Feeding cattle soy beans turns them into net consumers of food.

Modern and traditional agriculture are also quite different in terms of their energy costs. The traditional peasant produced his own inputs, including his own energy – using solar energy in the form of photosynthesis which is nothing more than solar energy that has been turned into organic matter. His fertiliser is the dung from his animals. Soil fertility is also restored with crop rotation and green manure, mixed cropping. Using draught animals that are fed from his soil he is also using solar energy.

By contrast, most products used in modern farming – chemical fertilizers, for example – are made from phosphates that come from mines which will be exhausted in the next 30 or 40 years if we continue using and wasting phosphates at our present rate. The same is true of nitrogen fertilizers which are taken from the air using the Haber-Bosch or similar processes that require enormous inputs of energy which, with rare exceptions, also requires the use of fossil fuels. Under these conditions, modern agriculture is simply not sustainable.

Examples of the severe consequences of extensive modern agriculture can be found in Brazil where South America's subtropical rain forest has been and one of the other great biomes – the cerrado (which is the equivalent to the African savannah) – is being systematically destroyed.

Moreover, modern agriculture is destroying not only whole ecosystems and biomes, it is eliminating biodiversity in nature and in our cultivars. Some 600 to 700 varieties of wheat and cereal grains were grown in Germany only 30 to 40 years ago. Today you find only half a dozen kinds being cultivated.

It is estimated that years ago, probably more than ten thousand varieties of rice were cultivated in its countries of origin in the Far East. Today we grow the same half a dozen varieties in all rice growing regions in the world. An enormous and irreversible loss of biodiversity in our cultivars that was caused by what is called one of the greatest advances for mankind in the last 50 years, the Green Revolution. In Southeast Asia, local peasant cultures all had their own locally adapted varieties. These varieties were not, according to our present standards, highly productive but they guaranteed the farmer safe harvests. Those peasant cultures survived for hundreds and sometimes thousands of year, and every one of these varieties was the result of a kind of "local biotechnology" – conscious or unconscious selections by the farmers themselves, after each harvest they kept the best seed for sowing in the next season.

In the early 1950s, the Rockefeller Institute set up the International Rice Research Institute in Manila. The Institute selected and crossed several of the local varieties and produced hybrids that were "apparently" much more productive than the local varieties. In reality, they were never really compared with the local varieties under local conditions. Right from the start, the new hybrids were treated with all kinds of chemical, mechanical and pesticide processes. Something that a peasant farmer never did with his plants. Once these new varieties were out on the market, the governments, ministries of agriculture, banks and parliaments set up special schemes for financing and supplying the necessary inputs. They also passed legislation to force people to use these new strains of rice. This led to a tremendous uprooting of peasants. And this uprooting has taken place all across the globe. If São Paulo has close to 20 million people today, Mexico City 22 million inhabitants and so on, it is due to modern agriculture. Where in the economists' balances do you see an accounting for this horrendous loss and tremendous human misery?

Austria offers another example. When Austria entered the Common Market, the technocrats in Brussels began subsidising the pulling out of old varieties of apples. Although they still had hundreds of different varieties, farmers were supposed to pull them out and replace them with only a half a dozen modern varieties of apples that look better on the supermarket shelf.

The question arises, what is more important to the farmer – profit or maximum yield? Maximum yield is important for the chemical industry and for the machine industry, because they want to sell their products. It is also important to the food processing industry. The farmer must be left with enough money to survive. Today, even in Germany, thousands of farmers go broke and give up every year.

Unless we understand what is going on, we will not be able to help change this situation. Few people realise that most technologies are conceived, developed and imposed in the interest of powerful institutions such as the chemical industry, the machine industry, the food processing industry and the like.

3 Alternatives: the way out

Alternative agricultural methods exist and could be easily put to use. The French biologist Chaboussou developed what is called the Theory of Trophobiosis. Chaboussou studied the question of what makes a plant susceptible to pests, parasites and pathogenic agents such as bacteria and viruses. His conclusion: On a healthy plant, the pest starves. Consequently, the solution to our pest problems lies not in warfare with poisons and other such weapons, but in farming methods that make for healthy plants.

A few examples from Brazil illustrate this theory. Eight years ago, a local businessman in my area was about to pull out the 25,000 guava trees (*Psidium guajava*) he had on his 65-hectare plantation and raise cattle instead because his costs ate away most of his profit. The official extension service had told him to plough his orchard every year – which surely damaged the roots of his trees – harrow every 15 days and apply herbicides regularly. His trees became so sick that he had to apply fungicides and insecticides and later miticides and ultimately nematicides. Nematicides are so expensive that sometimes one application will cost more than the price of the land it is used on.

I advised him not to pull his trees out, to forget the ploughing, the harrowing and the herbicides, the chemical fertilizers, to use only raw phosphate, to let native vegetation come up, to sow soft grasses like rye and oats as well as clover and other legumes. I also told him to get whey from cheese making at a nearby dairy and to apply it to his trees in lieu of insecticides and fungicides. Although he had his doubts at first, he has been doing this for eight years now. At the beginning, he applied the whey at full strength. Today he is down to a two-percent dilution. He pays a little less (only the cost of transportation) for a tankload of 7,000 litres of whey than he used to pay for one pail of insecticide or fungicide. Today he has the most beautiful, healthy guava trees and he has around a hundred cows and a flock of sheep that partly graze in the orchard keeping the green cover short. Of course whey does not kill anything, neither insects, fungi or bacteria. But it does stimulate plants, promoting a balanced metabolism. Of course, the soil has to be managed organically.

Another possibility, when whey is not available, is to work with methanogenically matured slurry: One can take an open 200-litre drum, fill it half full with fresh cattle manure, add one or two kilos of sugar – raw sugar if possible – one or two litres of milk, and then fill it up with water. The biogas can be burned off or allowed to go during fermentation. Once it is totally mature and stops fermenting, you add microelements and then apply it to the trees. Methanogenic fermentation works only at temperatures over twenty degrees centigrade. The above preparation must be made in summer but can be stored indefinitely in the cold season. The easiest way, though, is to use the sludge from a biogas digestor, when available.

Germany wastes some 200 million tons of cattle and pig slurry every year in absurd ways – such as putting 150 to 200 tons of it on one single acre of land just to get rid of it. There have even been suggestions to dry and burn it and for some time it was even dumped it in the ocean.

It seems to me that the organic farming movement that today every year succeeds in converting a few dozen farmers to one hundred percent organic farming practices with certifiction of their products (a process that takes five years) is often more a hurdle than a help for the necessary general reversal towards sustainable farming. If by a stepwise abandonment of poison and aggressive soil management we succeed in getting reduction of 5 or 10% by one hundred or two hundred thousand farmers, that would be much more progress, ecologically and socially, and would really be a significant and irreversible step forward.

In conclusion, it can be said that much is fundamentally wrong with our present economic system, its technology, our present ideology and our anthropocentric world view and ethics. We should take a closer look at the technologies presented us: They are said to be inevitable because they are the result of progress, when in fact most of them are only instruments of power –instruments for creating dependence.

Biodiversity - resource for new products, development and self-reliance

Werner Nader, Nicolas Mateo
Instituto Nacional de Biodiversidad, Santo Domingo, Costa Rica

Abstract. Biodiversity is one important reason for the impressive growth of pharmaceutical, chemical and agro-industry in this century. "Patents" from nature and the ethnological knowledge linked to their use in day-to-day life have inspired generations of inventors to create a vast array of products ranging from Aspirin to *Bacillus thuringensis* and Phosphinothricin to Zocor, some of which generate annual sales of more than US$ 1 billion. Biodiversity's genetic potential for biotechnology might be even greater, and products like tissue plasminogen activator from vampire bats, or high-laurate canola oil from rape seed transformed with a gene from an undomesticated Californian bay will soon be available on the market. Some of these new products might pose a threat to biodiversity or to the economies of tropical and subtropical countries, as in the case of Calgene's high-laurate canola oil which can substitute coconut and palm kernel oil. The research agreement between INBio and Merck, Sharp & Dohme acknowledges that biodiversity can be used for product development in a sustainable manner and thus conserved for future generations. Royalty obligations on net sales of future pharmaceuticals, agrochemicals, fragrances and biotech products to finance conservation measures comprise only a minor fraction of potential benefits which might contribute to maintaining biological diversity and variability. Even more important are the contributions that capacity building and technology transfer can make to the scientific and technological development of biodiversity-rich countries. Fair research partnerships between source countries and industry and academic institutions in industrialised countries are necessary. National scientists must spend more time investigating biodiversity, both at the basic research level and in the development of new products. The use of biological diversity for socio-economic development is an important goal for Costa Rica, and INBio is making a significant contribution to this process.

Biodiversidad - Fuente de nuevos productos, desarrollo y conciencia

Resumen. Desde hace más de cien años la industria farmacéutica y química ha estado aprovechando el valor de la biodiversidad. Se ha desarrollado un espectro amplio de nuevos productos desde Aspirina a Zocor, derivados de compuestos naturales de microorganismos, plantas y animales. Algunos de estos productos tienen un volumen de ventas más de mil millones de dólares por año. El potencial de genes de la biodiversidad para la industria biotecnología podría exceder estos valores. Nuevos productos como un "Tissue Plasminogen Activator" producido a través de la ingeniería genética de un gene del vampiro o un nuevo aceite de colza, alto en ácido laúrico y producido en colza transgénica, van a aparecer pronto en el mercado. El acuerdo entre INBio y Merck, Sharp & Dome fue la primera concesión de un beneficiario hacia la biodiversidad, la que debe conservar el potencial biológico de los trópicos para las generaciones futuras. Regalías de ventas de productos futuros como nuevos fármacos, agroquímicos, fragancias o genes no son los únicos beneficios, que pueden contribuir a la conservación de la biodiversidad. Contribuciones como la transferencia de tecnología a los países en vías de desarollo tienen la misma importancia. Se requieren colaboraciones equitativas entre los países tropicales, la industria y el sector académico en los países industrializados. Los científicos nacionales deben obtener la oportunidad de investigar su riqueza biológica, no sólo en la ciencia fundamental, sino también para el desarollo de nuevos productos que fortalezcan su industria nacional. Este escenario podría conducir a nueva conciencia, motivación y orgullo acerca de "su biodiversidad" pro parte de los países tropicales. Todo esto contribuiría a la conservación de los recursos naturales. La utilización de la biodiversidad como fuente para el desarollo socioeconómico de Costa Rica es la meta del INBio y podría ser considerado como un proyecto piloto.

1 Introduction

Biodiversity refers to the variety and variability of organisms and ecological complexes and comprises species, genetic and ecosystem diversity. Biodiversity is not only the basis for life on earth, it also provides the goods and services essential to support every type of human endeavour. Accordingly, biodiversity enables societies to adapt to different needs and situations (US National Research Council, 1992).

Biodiversity generates economic value in different ways. Agriculture, for example, benefits directly from a functioning ecosystem which enables it to avoid the extensive use of agrochemicals. Biodiversity generates economic value from extractable products obtained from individual species (Wilson 1992) and provides fuels, medicines, materials for shelter, food and energy. The use of compounds, genes and species has been essential for meeting industry needs (Tamayo *et al.*, in press). Markets for pharmaceuticals, pesticides, seeds, cosmetics, toiletries, perfumes and ecotourism are summarised in Table 1. Furthermore, ecosystems contribute to climate regulation, the maintenance of hydrologic cycles and the nitrification of soils. In addition, recreation, science and education figure among the vast array of social, ethical, spiritual, cultural and economic goods and services provided by biodiversity.

The international accounting framework and the models derived from it ignore the economic value of biodiversity, as is the case for other natural resources like wetlands and forests. The general public pays for the damages caused by the loss of these resources through flooding and erosion (Repetto 1992). Except for the genetic variability of cultivated plants, the potential problems which the loss of biodiversity may cause are not yet obvious to the majority of economists.

For many tropical and subtropical countries biodiversity is the most important natural resource available and its potential for socio-economic development must be acknowledged in order to conserve it for future generations. Since this potential lies mostly in the development of innovative products and services, biodiversity-rich countries require equitable partnerships with the academic and industrial sectors of industrialised countries. The UN Convention on Biological Diversity encourages this type of international collaboration and the technology transfer involved. The agreements of the Costa Rican Instituto Nacional de Biodiversidad (INBio) with international industry represent an important pilot undertaking (Reid *et al.*, 1993).

It is often argued during negotiations with industry that royalty payments and patent sharing can be justified only when a significant intellectual contribution has been made. The supply of biological and genetic materials is not considered sufficient grounds because they are viewed as mere raw materials. Indeed, current economic systems value and reward human intellectual achievements and secure this value through patent rights. However, biodiversity also provides "patents" from nature, and ethnological uses of biological resources represent intellectual achievements on the part of previous and current generations. In the context of biodiversity conservation, these contributions need to be economically rewarded and secured through internationally recognised rights.

Table 1. Selected world markets based on biodiversity

1. Pharmaceuticals	
Drug market, 1993	about US $ 235 billion[a]
Pharmaceuticals derived from plants, 1993	about US $ 59 billion[b]
Phytomedicines, 1993	US $ 12.4 billion[c]
2. Agriculture	
Sales of pesticides world-wide	US $ 47 billion[d]
Seeds (estimate for 2000)	US $ 7 billion[a]
Horticulture (UK fresh flowers and plants, 1991)	US $1.6 billion[a]
3. Cosmetics, toiletries and perfumes	
total sales in the USA in 1994	about US $ 20 billion[a]
natural cosmetics and toiletries based on 2.5 %	below US $ 500 million[a]
4. Tourism	
global output, 1995	US $ 3,400 billion[a]
ecotourism in developing countries in 1988	US $ 12 billion[a]

[a] ten Kate, 1995
[b] Based on the estimation that plant derived drugs make up 25 % of the total drug market (Principe 1989)
[c] Grünwald, 1995
[d] Frost and Sullivan, 1992.

Table 2. Examples of potential and existing products from biodiversity

	Drugs	Pesticides	Medical gene technology	Agricultural gene technology
Plants	Aspirin-Xanthotoxin	Pyrethrum, DMDP	hyoscyamine 6-beta-hydroxylase[a]	thioesterases[b]
Fungi	Ampicillin-Zocor	--	Biotrans-formation	--
Bacteria	Adriamycin-Zinostatin	Ivermectin Phosphino-thricin	Streptokinase	Cholesterol Oxidase[c]
Vertebrates	Squalamine[d]	Steroid derivatives	Vampire bat tPA[e]	Oligoadenylate Synthetase[f]
Insects	--	Pheromones	Anticoagulants	Cecropin B[g]
Arachnids	Arylamines[h]	--	Anticoagulants	Neurotoxins[i]

[a] Yun *et al.*, 1992
[b] Calgene Inc., 1994 a-c
[c] Corbin *et al.*, 1994
[d] Moore *et al.*, 1993
[e] Schleuning *et al.*, 1992
[f] Truve *et al.*, 1993
[g] Florack *et al.*, 1995
[h] Jackson & Parks, 1990
[i] FMC Corp., 1993

2 Biodiversity – Resource for new products and socio-economic development

2.1 The products

Table 2 summarises examples of products from wild life, which became or might become important to medicine and agriculture. The majority of organisms could provide resources for the development of new products. In chemistry, numerous pharmaceuticals and pesticides have been derived from secondary metabolites of plants, fungi and bacteria. Outstanding examples include analgesics like aspirin (a derivative of salicylic acid, *Salix*), antibiotics like penicillins (*Penicillium*), cephalosporins (*Cephalosporium*) and macrolides (e.g. erythromycin from *Saccharopolyspora erythrea*), cytostatics like adriamycin (*Streptomyces peucetius*), etoposid (*Podophyllum*) and vinca alkaloids (*Catharanthus roseus*), lipid reducers such as mevacor and zocor (*Aspergillus terreus*), the immunosuppressive cyclosporin (*Tolypocladium inflatum*), the dermatic xanthotoxin (from various species of Leguminosae, Umbelliferae and Rutaceae), the antiglaucoma agent pilocarpine (*Pilocarpus*) and the muscle relaxant tubocurarine (calabash-curare, *Chondodendron tomentosum*); furthermore insecticides such as pyrethrum (*Chrysanthemum*) and its derivatives, the pyrethroides, the herbicide phosphinothricin (*Streptomyces* spp.) and the anti-parasitic agent ivermectin (*Streptomyces avermitilis*), which is used extensively in tropical veterinary medicine.

Secondary metabolites from vertebrates, insects and arachnids like the antibiotic squalamine from the shark (Moore *et al.* 1993) and the arylamines from spider toxins (which are potent antagonists of the neurotransmitter glutamate and might prevent brain damage caused by stroke) are currently undergoing preclinical and clinical trials (Jackson & Parks 1990).

Marine organisms yield new structures with high molecular diversity and important biological activities (König *et al.* 1994). For example, Briostatin and Didemnin B found in molluscs reveal strong anti-tumor activity and have reached preclinical and clinical trials respectively. Tunicate alkaloids are also currently undergoing intensive study (Research Foundation of the State University of New York 1994).

Another important resource for new products are genes from biodiversity. Even prior to the development of gene technology, genetic resources were extensively used in traditional and modern breeding of domesticated plants and animals. New crops and animal breeds with improved characteristics such as disease resistance and better adaptation to extreme environmental conditions are being developed on a continuous basis. Gene technology opened up a new area for genetic prospecting and Table 3 lists selected wildlife genes with potential medical or agricultural applications. A tissue plasminogen activator from the common vampire bat offers an interesting example of a new pharmaceutical product that was developed and produced from wildlife resources by means of genetic engineering. This enzyme also occurs in humans, providing natural protection against thrombosis and heart attacks by dissolving thrombolytic blood clots. Recombinant human tPA has been approved in the USA and Europe as a therapeutic agent against heart attacks and has already saved many lives. Researchers at Schering AG found four similar proteins in the saliva of *Desmodus rotundus*, the common vampire bat, that are more efficient and safer for therapeutic application than their human counterparts (Schleuning *et al.* 1992). These products are presently undergoing preclinical studies. However, numerous new pharmaceutical proteins are expected to be found in animal toxins and the saliva of reptiles, leeches, spiders, tics, scorpions, bees, mosquitoes, flies, etc. which might be suitable as blood coagulation inhibitors or accelerators, therapeutics for high or low blood pressure, muscle relaxants, etc. (Schlee 1992) and become accessible through modern gene technology.

The application of gene technology in agriculture might threaten the environment and economies of developing countries. For this reason, many countries are developing and implementing biosafety strategies and codes of conduct. The UN Convention on Biological Diversity calls for regulations "in the field of the safe transfer, handling and use of any living modified organism resulting from biotechnology that may have adverse effects on the conservation and sustainable use of biological diversity" (Art. 19, Para. 3). The creation of

Table 3. Genes from biodiversity and their applications

Product	Source	Application	Citation
1. Pharmaceuticals			
Calcitonin (Salcitonin)	Salmon	Osteoporosis	a
Hydantoinase	Thermophilic bacteria (Yellowstone)	D-amino acid production	b
Hirudin	Leech	Anticoagulant	c
Tissue Plasminogen Activator	Vampire bat	Anticoagulant	d
Magainin	Frog skin	Antibiotic	e
2. Pesticides			
delta-Endotoxin	*Bacillus thuringensis*	Insecticide	f
Cholesterol Oxidase	*Streptomyces* sp.	Insecticide	g
Neuro-toxic Peptides	Toxins from spiders and scorpions	Insecticide	h
Cecropin B	Silkmoth	Antibacterial	i
3. Engineering of Metabolic Pathways			
12:0-Acyl-Carrier Protein Thioesterase	*Umbellularia californica*	High-laurate Canola oil (soaps)	j
Fatty Acyl-CoA:Fatty Alcohol Acyltransferase	*Simmondsia*	Wax production	k
Fatty Acyl Reductase	*Simmondsia*	Wax production	l
Hyoscyamine 6-beta hydroxylase	*Hyoscyamus niger*	production of scopolamine	m

a) Epand *et al.*, 1986
b) BASF AG, 1987
c) Fortkamp *et al.*, 1986
d) Schleuning *et al.*, 1992
e) Jacob & Zasloff, 1994
f) Koziel *et al.* 1993
g) Corbin *et al.*, 1994
h) FMC Corp., 1993
i) Florack *et al.*, 1995
j) Calgene, Inc., 1994 a-c
k) Calgene, Inc., 1995 a)
l) Calgene, Inc. 1995 b
m) Yun *et al.*, 1992

transgenic plants with neurotoxic (FMC Corp. 1993) and bactericidal (Florack *et al.* 1995) peptides from spider toxins and insects, or enzymes such as the insecticidal cholesterol oxidase (Corbin *et al.* 1994) and the phosphinothricin transferase from actinomycetes (Hoechst AG 1991) for the purpose of developing new resistance against pests and herbicides needs to be evaluated to exclude potential hazards to the environment and consumers.

Although modern biotechnology promises to create new sources of income for farmers, traditional export commodities from developing countries are already threatened by the development of substitutes that can be grown in industrialized countries. It is also quite possible that the transfer may even skip the farms of these countriesand head straight for the bioreactors. In Africa alone, US$ 10 billion in exports are vulnerable to industry-induced changes in raw material prices and requirements (Shand 1993). Most developments in plant biotechnology have been achieved with crops that are cultivated primarily in industrialised countries. This tendency could deny farmers in developing countries the opportunity to benefit from the new agricultural developments offered by gene technology.

Calgene's laurate high canola oil, for example, may displace coconut and palm kernel oil. This would pose a threat to the economic survival of millions of farm families in the tropics (Calgene, Inc. 1994 a - c). Thaumatin, a sweet-tasting basic protein from the tropical plant *Thaumatococcus*, was traditionally used in West Africa as a sweetener. The population in this region earned a large part of its income with the collection of *Thaumatococcus* fruits for the British food industry. Now however, the gene has been cloned and the sweetening protein can be produced at low cost by large-scale fermentation in brewer's yeast (Lee *et al.* 1988). The same applies to natural compounds such as indigo which can be produced by fermenting *Escherichia coli* and engineered with genes from a toluol-degrading subspecies of the soil bacterium *Pseudomonas putida* (Ensley *et al.* 1983). Products like vanilla, pyrethrum and rubber may follow this path. Considering these economic threats, a biodiversity-rich country would be well advised to carefully examine the potential returns of bioprospecting ventures. A technology that has been developed from wildlife genes might come back like a boomerang and negatively affect its economic prospects.

The list of products derived from biodiversity now includes retroviruses and adenoviruses used in numerous clinical trials to test whether gene therapy can be effective in curing cancer, AIDS or inherited diseases like haemophilia. Plant viruses serve as vectors for the development of transgenic plants. Insect viruses like the Baculovirus are used not only as efficient insecticides, but also by biotechnology to produce recombinant proteins.

2.2 Product Development

The development of new products from biodiversity is extremely expensive and risky. Companies like Merck & Co. spend more than 10% of their sales revenue on research and development. The development of a single drug costs US$ 359 million (Mendelsohn & Balick 1995). Only one out of every 10,000 chemical substances that undergo the screening process for new drugs ever reaches the market. Statistics on the screening of natural products are equally discouraging (Table 4). Between 1960 and 1982, the National Cancer Institute tested over 114,000 extracts from some 35,000 plant species for new anticancer drugs (Cragg *et al.* 1994). A number of new natural compounds were discovered during this research, but only one – taxol – is being used clinically today. Camptothecin, a DNA topoisomerase I inhibitor, was also discovered at the Institute and a derivative of this compound will soon be on the market. The National Cancer Institute has screened extracts from over 26,000 plant species for potential new drugs against AIDS. This work yielded several promising new compounds, of which only four are currently undergoing clinical trials: the calanolides from the tree *Calophyllum langerum* from Sarawak on Borneo, prostatin from the

Table 4. Natural compounds from plants with therapeutic potential against Cancer and HIV (Cragg *et al.* 1994)

Compound	Source	Known ethnic use	Status of development
Anti-Cancer			
Etoposide/teniposide	*Podopyhllum*	skin cancer, warts	used clinically
Vinca-alkaloids	*Catharanthus roseus*	diabetes	used clinically
Taxol*	*Taxus brevifolia*	none	used clinically
Camptothecin *	*Camptotheca acuminata*	ornamental	used clinically
Homoharring-tonine*	*Cephalotaxus sp.*	none	under trial
Elliptinium	*Apocynaceae*	anti-cancer	under trial
Anti-HIV			
Prostatin *	*Homalanthus nutans*	yellow fever	under trial
Michellamine B *	*Ancistrocladus korupensis*	none	under trial
Calanoide A *	*Calophyllum sp*	none	under trial
Conocurvone*	*Conospermum sp.*	none	under trial

(compounds labelled with an asterix were found in screening programs at the National Cancer Institute)

medicinal plant *Homalanthus nutans* from Samoa which is traditionally used against yellow fever (caused by a retrovirus), michellamine B from the tropical vine *Ancistrocladus korupensis* from Cameroon, and Conocurvone from a *Conospermum* species from Western Australia. Despite these prospects, all these efforts are worthwhile: A single new successful pharmaceutical can generate market earnings in the range of one billion dollars and such work could possibly lead to a cure for diseases like cancer and AIDS.

Because of the complexity and high cost of modern drug research, numerous academic institutions and contract research companies are teaming up with pharmaceutical companies to carry out joint research projects. This opens up new opportunities for biodiversity-rich countries to enhance their own research and development capacity and to offer either contract research services to international industry or characterised natural compounds with defined biological activities for licensing. The objective here is to make the transition from being a mere supplier of raw biological materials to a research partner.

2.3 Combining biodiversity prospecting with conservation and socio-economic development – The INBio experience

During the past 6 years, INBio entered into collaborative research projects with 8 companies (Table 5) and various academic institutions. A detailed analysis of these collaborations reveals that opportunities to develop new products are numerous and that a biodiversity-rich country can use the benefits drawn from bioprospecting for its scientific, technological and economic development and for biodiversity conservation. The following case studies highlight that this process requires a significant amount of flexibility, creativity and innovation.

Table 5. INBio's collaboration with industry

- **Merck & Co.:** Preparation of extracts from plants and insects for screening for new drugs.
- **Bristol-Myers Squibb:** Preparation of extracts from insects for drug screening within an ICBG project (Universidad de Costa Rica, Cornell University, Guanacaste Conservation Area and INBio); research at INBio and the University of Costa Rica to develop drugs against Malaria, inflammation and microbial infections.
- **Givaudan-Roure:** Prospecting for fragrances and aromatics through gas chromatography.
- **Recombinant BioCatalysis, Inc.:** Prospecting for thermostable enzymes through the direct isolation of nucleic acids from volcanic hot springs.
- **British Technology Group, Inc. and Hacienda La Pacifica:** Development of the pyrrolidine alkaloid 2R,5R-dihydroxy-methyl-3R,4R-dihydroxypyrrolidine (DMDP) as a bio-nematicide.
- **AnalytiCon AG, Berlin:** General collaboration agreement for the isolation and characterisation of natural compounds.
- **INDENA, Milan:** Screening for bioactive compounds at INBio.
- **Intergraph Corp.:** Development of software and hardware for the geographical and biodiversity data management

2.3.1 INBio-Merck

INBio's research agreement with Merck & Co. was signed in November 1991, six months before the Earth Summit was held in Rio. This agreement has been renewed twice and its basic elements include the delivery of a limited amount of samples from plants, insects and micro-organisms to Merck for the development of new drugs in exchange for a research budget, technology and know-how transfer, and the sharing of future benefits through royalty payments. Royalty income is devoted to biodiversity research and conservation in Costa Rica and would be shared equally by INBio and the Ministry for the Environment and Energy. Although the agreement is non-exclusive, INBio does not supply samples which are under study at Merck to any other industrial or academic parties for a limited period of 2 years.

2.3.2 INBio-Recombinant BioCatalysis and INBio-Givaudane Roure

INBio's research agreements with Recombinant BioCatalysis and Givaudane Roure are smaller but share similar elements with the Merck agreement, including the transfer of technology and know-how and the sharing of benefits through royalty payments. As part of its collaboration with Recombinant BioCatalysis, Inc., INBio and the Center for Molecular and Cellular Biology at the University of Costa Rica are preparing DNA from extreme habitats such as hot volcanic springs for the development of new industrial enzymes (Nader & Rojas 1996 a). Thermostable enzymes derived from the volcanic springs at Yellowstone Park are already being used extensively in the Polymerase Chain Reaction (Taq Polymerase) and for chemical catalysis in industrial processes such as in the synthesis of D-amino acids for the production of half-synthetic antibiotics like ampicillin and amoxicillin (BASF AG 1987).

The research agreement with Givaudane-Roure targets the development of new fragrances and aromas from biodiversity.

2.3.3 INBio-ICBG

The International Co-operative Biodiversity Groups were initiated by the United States Agency for International Development (US-AID), the National Science Foundation (NSF) and the National Institute of Health (NIH). To date, five of these groups have been established world-wide with the intention of promoting collaboration between biodiversity-rich countries and US universities and companies (Rosenthal 1996). The agreement between INBio, the University of Costa Rica, the Guanacaste Conservation Area, Cornell University and the pharmaceutical company Bristol-Myers Squibb targets the development of new natural compounds from arthropods. Within the course of this collaboration, samples are delivered to the company under terms and conditions similar to those in the Merck agreement, but are also tested in Costa Rica in screening assays for new anti-inflammatory, anti-malaria and anti-microbial compounds. Through this arrangement, institutions in Costa Rica are active players in researching and developing products from biodiversity.

2.3.4 INBio-INDENA

INBio's collaboration with INDENA, an important Italian manufacturer of extracts and chemicals from plants, targets the development of new phytochemicals and phytomedicines from Costa Rica's plant diversity. Under its agreement with INDENA, INBio is no longer a supplier of extracts but an active partner in the screening process. A large portion of this developmental work is conducted at INBio's laboratories.

2.3.5 INBio-AnalytiCon

The agreement with AnalytiCon AG, a German company for contract drug research, represents another important step in INBio's strategy to acquire research capacity and know-how and consequently increase possible income from biodiversity for the purpose of its conservation. Under this agreement, AnalytiCon will transfer know-how and equipment to INBio and support the institute in characterising natural compounds. In addition, the German company will integrate INBio's research capacity into its on-going contract research with pharmaceutical companies. Both parties have also agreed to share all intellectual property rights which are derived from their joint research.

2.3.6 *INBio-BTG-La Pacifica*

Most of the products developed through INBio's collaboration with international industrial partners will probably be produced and commercialised in industrialised countries. In order to promote the value of biodiversity for the socio-economic development within Costa Rica, it will be crucial for local farmers, co-operatives and industry to be involved in the development and commercialisation of new products.

Legumes like the African mulberry tree (*Derris*) and the Latin-American genus *Lonchocarpus* contain a variety of alkaloids like rotenone which are powerful insecticides. Another natural pesticide, the pyrollidin alkaloid 2R,5R-dihydroxy-methyl-3R,4R-dihydroxypyrrolidin (DMDP), was first described from *Derris elliptica* in 1976 (Welter *et al.* 1976) and was later isolated from *Lonchocarpus* (Janzen *et al.* 1990) and from a bacterium, an actinomycete (Watanabe *et al.* 1995). This pyrollidin alkaloid is an analogue of fructose (Fig. 1) that inhibits hydrolases of polysaccharides and blocks the receptors for fructose in the gustatory nerves of insects (Fellows 1983). The alkaloid is thus a repellent and not a toxin. Its systemic activity against plant parasitic nematodes was first detected in 1993 by the Scottish Crop Research Institute in collaboration with the Royal Botanical Gardens at Kew (Birch *et al.* 1993). The patent on its application as a nematicide is held by the British Technology Group (British Technology Group Ltd. 1994).

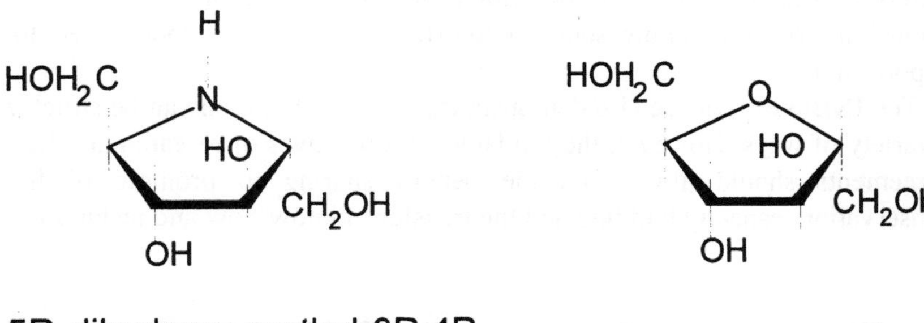

2R,5R-dihydroxy-methyl-3R,4R -dihydroxypyrrolidin (DMDP)

β -D-FRUCTOSE

Fig. 1. The pyrrolidin alkaloid DMDP (2R,5R-dihydroxy-methyl-3R,4R-dihydroxy-pyrrolidin) is an analogue of fructose and a natural nematicide
Ilustración 1. El pirrilidín alcaloide DMDP (2R, 5R-dihydroxy-methyl-3R,4R-dihydroxy-pyrrolidin) es un análogo de fructuosa y un nematicida natural

Benefits for Costa Rica might be significant if this nematicide could be produced in the country itself and used as a natural pesticide in organic agriculture. INBio has started in this sense collaboration with the Costa Rican agricultural company La Pacifica and the British Technology Group to assess the development of DMDP.

INBio's collaboration with national and international companies reveals that biodiversity could provide a broad spectrum of products ranging from pharmaceuticals and biopesticides to fragrances and genes. This spectrum can be extended by new fruits, nuts, ornamental plants, essential oils, preservatives, fibres and gums. New market segments are currently emerging throughout the world because biodegradable and less hazardous natural compounds are being used to an increasing degree to replace chemically synthesised plastics and additives in food, cosmetics and paints. Biodiversity prospecting thus functions in the spirit of the UN Convention on Biological Diversity which encourages the development of environmentally sound uses from genetic resources (Art. 15, Para. 2).

The argument that biological resources become worthless once they undergo industrial research is used quite often against bioprospecting. Looking solely at the development of new drugs however, the biological resources of even a small country like Costa Rica seem inexhaustible. In light of the drug industry's capacity to screen thousands of extracts a month, 13,000 Costa Rican plant species seem to be a rather small number. However, the array of new bioassays and thus, new screening opportunities, are increasing steadily. Furthermore, extracts can be fractionated and new natural compounds isolated and supplied to industry at a higher price. New sources for natural compounds such as insects, molluscs, fungi and bacteria are continuously being discovered. It is therefore important that bioprospecting institutions in source countries pursue only non-exclusive collaboration with industry. If exclusivity for the evaluation of biological materials is granted, it should be limited in terms of time and quantity. Limited exclusivity schemes allow source countries to keep their doors open to new opportunities.

The INBio experience also demonstrates that collaboration can be structured in a variety of ways. However, the fundamental objectives of research and licensing agreements should always include benefit sharing to promote biodiversity conservation, capacity building and the transfer of know-how and technology.

3 Biodiversity – A source of self-reliance

Biodiversity's non-commercial value is just as important as its commercial value. In many countries, biological richness is quite often part of a land's cultural and national identity. INBio was founded in 1989 with the aim of strengthening awareness about the value of biological diversity. Collaboration with international industry developed later.

The conservation of biodiversity requires more than research by a scientific elite. In this regard, INBio targets the integration of the rural population into the

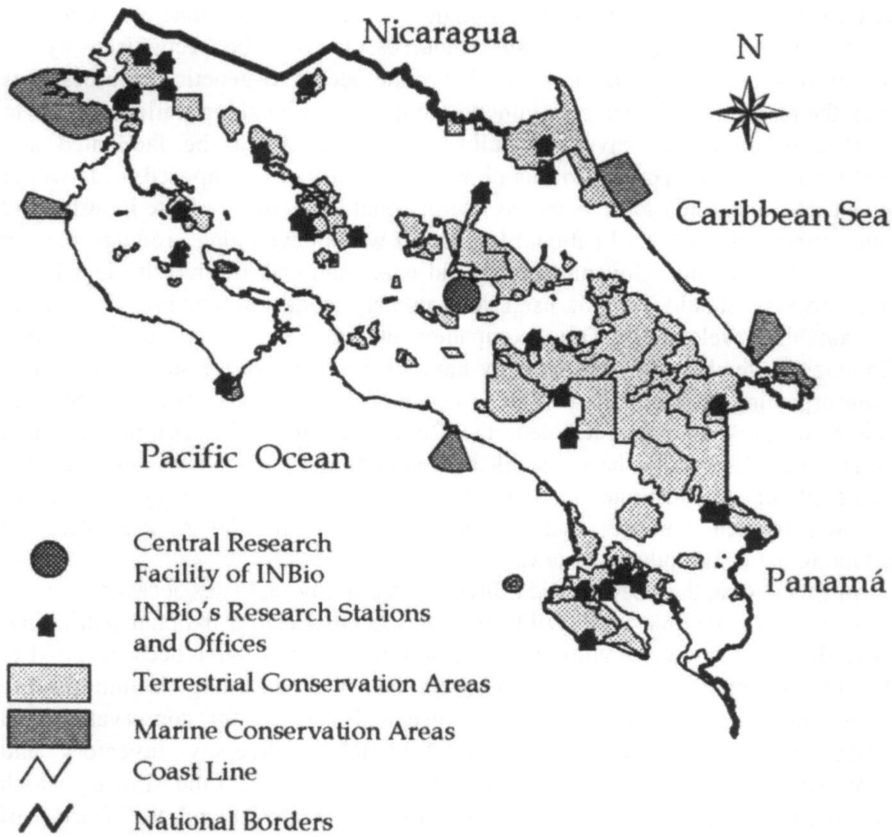

Fig. 2. Costa Rica's Conservation Areas and INBio's biodiversity offices. About 27% of Costa Rica's territory is presently protected. INBio is performing research in these Conservation Areas at its research centre and 28 research offices

Ilustración 2. Áreas de Conservación de Costa Rica y oficinas de biodiversidad del INBio. En la actualidad se encuentra protegido un 27% aproximadamente del territorio costarricense. INBio está llevando a cabo trabajos de investigación en estas Áreas de Conservación en su centro de investigación y en 28 oficinas de investigación

research process. As part of INBio's National Inventory effort, parataxonomists – young people from communities around conservation areas – are trained to collect material in the field and perform initial taxonomic research. Their work helps broaden our fundamental understanding and knowledge of biodiversity. They work within their local communities and promote the idea of biodiversity as a national resource (Fig. 4). Furthermore, an Information Dissemination Division has been created to disseminate information and biodiversity research findings to teachers, students, school children, professionals and decision makers.

4 Biodiversity and legal frameworks

Article 15 of the Convention on Biological Diversity states that the sovereign rights of states over their natural resources have to be recognised by its signatories, and that the authority to determine access to genetic resources rests with the respective national governments, subject to national legislation. Access to genetic resources for environmentally sound uses should be facilitated and restrictions that run counter to this objective should not be imposed. This would give each nation the right – based on international law – to prosecute industry and other users who disregard national legislation while developing products derived from its biodiversity. Governments should enact national biodiversity legislation and industry should commit itself to voluntary codes of conduct to support sustainable development. Drug companies like Glaxo (January 6, 1994) and Sandoz (September 30, 1994) already have decided to observe such codes when bioprospecting. Academic institutions like the National Cancer Institute in the US, the New York Botanical Garden, the Royal Botanical Gardens at Kew and Strathclyde University have established benefit-sharing policies such as the National Cancer Institute's Letter of Collection. An increasing number of biodiversity-rich countries such as the Philippines and the Andean Pact are adopting national biodiversity laws.

In Costa Rica, the "Ley de Vida Silvestre" currently regulates access to genetic resources in a flexible way. Wild fauna has been declared national patrimony; wild flora and its conservation, research and development have been declared to be of public interest. The Ministry of the Environment and Energy is authorised to grant concessions, contracts, rights of use and licenses for conservation and sustainable use. INBio conducts the National Biodiversity Inventory and bioprospecting on the basis of a collaboration agreement with the Ministry which regulates the sustainable collection of biological material and the sharing of benefits from research and future products between the Ministry and INBio.

Clearly, national and international legislation can only provide a framework. It is even more important to reach a general consensus among scientists, managers and politicians regarding the need to contribute to the conservation of biodiversity as an important resource for the future of mankind.

Acknowledgements. The authors wish to thank their colleague Annie Lovejoy for her assistance with the manuscript and her suggestions, the Centre for International Migration and Development in Frankfurt which awarded W. Nader a grant, and Dr. Jasper Köcke for his continuous support.

References

BASF AG (1987) Verfahren zur Herstellung von mesophilen Mikroorganismen, die eine bei höherer Temperatur aktive D-Hydantoinase enthalten. German Patent Nr. 3535987

Birch ANE, Robertson WM, Geoghegan IE, McGavin WJ, Alphey TJW, Phillips MS, Fellows LE, Watson AA, Simmonds MSJ, Porter EA (1993) DMDP - a plant-derived sugar analogue with systemic activity against plant parasitic nematodes. Nematologia 29: 521-535

British Technology Group Ltd. (1994) Control of parasitic nematodes. US-Patent Nr. 5,376,675

Calgene, Inc. (1994a) Plant thioesterase having preferential hydrolase activity toward C12. US Patent Nr. 5,304,481

Calgene, Inc. (1994b) Plant medium-chain-preferring acyl-ACP thioesterases and related methods. US Patent Nr. 5,298,421

Calgene, Inc. (1994c) Plant thioesterases. US-Patent Nr. 5,344,771.

Calgene, Inc. (1995a) Jojoba wax biosynthesis gene. US Patent Nr. 5,445,947

Calgene, Inc. (1995b) Fatty acyl reductase. US Patent Nr. 5,411,879

Corbin DR, Greenplate JT, Wong EY, Purcell JP (1994) Cloning of an insecticidal cholesterol oxidase gene and its expression in bacteria and in plant protoplasts. Applied and Environmental Microbiology 60: 4239-4244

Cragg G, Boyd MR, Cardellina JH, Newman DJ, Snader KM, McCloud TG (1994) Ethnobotany and drug discovery: the experience of the US National Cancer Institute. *Ciba Foundation Symposium 185 (Ethnobotany and the Search for New Drugs)*, Wiley & Sons, Great Britain, pp 178-190, pp discussion 191-196

Ensley BD, Ratzkin BJ, Osslund TD, Simon MJ, Wackett LP, Gibson DT (1983) Expression of naphthalene oxidation genes in *Escherichia coli* results in the biosynthesis of indigo. Science 222: 167-169

Epand RM, Epand RF, Orlowski RC, Seyler JK, Colescott RL (1986) Conformational flexibility and biological activity of salmon calcitonin. Biochemistry 25: 1964-1968

Fellows L. (1983) Sugar shaped bullets from plants. Chemistry in Britain, September, 842-845

Florack D, Allefs S, Bollen R, Bosch D, Visser B, Stiekema W (1995) Expression of giant silkmoth cecropin B genes in tobacco. Transgenic Research 4: 132-134

FMC Corporation (1993) Insecticidally effective peptides. US Patent Nr. 5,441,93.4

Fortkamp E, Rieger M, Heisterberg-Moutses G, Schweitzer S, Sommer R (1986) Cloning and expression in *Escherichia coli* of a synthetic DNA for hirudin, the blood coagulation inhibitor in the leech. DNA 5: 511-517

Frost and Sullivan, Inc. (1992) The U.S. Market for Chemical Pesticides. Frost & Sullivan, Inc., New York, p 8

Grünwald, J (1995) The European phytomedicines market: figures, trends, analyses. HerbalGram (Austin, Tx.) 34: 60-65

Hoechst AG (1991) Phosphinothricin resistance gene. US Patent Nr. 5,077,399

Jackson H, Parks TN (1990) Anticonvulsant action of an acrylamine-containting fraction from *Agelenopsis* spider venom. Brain Research 526: 338-343

Jacob L, Zasloff M (1994) Potential therapeutic applications of magainins and other antimicrobial agents of animal origin. *Ciba Foundation Symposium* 186, Wiley & Sons, Great Britain, pp 197-216, pp discussion 216-223

Janzen DH, Fellows LE, Waterman PG (1990) What protects *Lonchocarpus* (Leguminosae) seeds in a Costa Rican dry forest? Biotropica 22 (3): 272-285

König G, Wright AD, Sticher O, Angerhofer CK, Pezzuto JM (1994) Biological activities of selected marine natural products. Planta Medica 60: 532-537

Koziel MG, Beland GL, Bowman C, Carozzi NB, Crenshaw R, Crossland L, Dawson J, Desai N, Hill M, Kadwell S, Launis K, Lewis K, Maddox D, McPherson K, Meghji MR, Merlin E, Rhodes R, Warren GW, Wright M, Evola SV (1993) Field performance of elite transgenic maize plants expressing an insecticidal protein derived from Bacillus thuringiensis. Bio/Technology 11: 194-200

Lee JH, Weickmann JL, Koduri RK, Gosh-Dastidar P, Saito K, Blair LC, Date T, Lai JS, Hollenberg SM, Kendall RL (1988) Expression of synthetic thaumatin genes in yeast. Biochemistry 27: 5101-5107

Mendelsohn R, Balick MJ (1995) The value of undiscovered pharmaceuticals in tropical forests. Economic Botany 49 (2): 223-228

Moore KS, Wehrli S, Roder H, Rogers M, Forrest JN, McCrimmon D, Zasloff M (1993) Squalamine: an aminosterol antibiotic from the shark. Proc Nat Acad Sci 90: 1354-1358

Nader WF, Rojas M (1996a) Gene Prospecting for sustainable use of the biodiversity in Costa Rica. Genetic Engineering News, New York, April 1, 1996

Nader WF, Rojas M (1996b) New rules for natural compound and biotechnological research after INBio and Rio. BIOForum (Darmstadt), September and October issues

Principe PP (1989) The economic significance of plants and their constituents as drugs. In: Wagner H, Hikino H, Farnsworth NR (eds.) Economic and Medicinal Plant Research, Vol. 3. Academic Press, London, pp 1-17

Reid WV, Laird S, Meyer CA, Gámez R, Sittenfeld A, Janzen DH, Gollin MA, Juma C (1993) Biodiversity Prospecting: Using Genetic Resources for Sustainable Development. World Resources Institute, Washington D.C.

Repetto R (1992) Accounting for Environmental Assets. Scientific American 266: 94-100

Research Foundation of the State University of New York (1994) Biological applications of alkaloids derived from the tunicate *Eudistoma sp*.-US Patent Nr. 5,278,168

Rosenthal JP (1996) Equitable sharing of biodiversity benefits: agreements on genetic resources. OECD International Conference on Biodiversity Incentive Measures, Cairns, Australia, March 25-28, 1996

Schleuning WD, Alagon A, Boidol W, Bringmann P, Petri T, Kratzschmar J, Haendler B, Langer G, Baldus B, Witt W (1992) Plasminogen activators from the saliva of Desmodus rotundus (common vampire bat): unique fibrin specificity. Annals of the New York Academy of Sciences 667: 395-403

Schlee D (1992) Ökologische Biochemie. G Fischer Verlag, Jena, p 394-424

Shand H (1993) Agbio and third world development. Bio/Technology 11: 13

ten Kate Kerry (1995) Biopiracy or Green Petroleum? Expectations & Best Practice in Bioprospecting. Overseas Development Administration, London

Tamayo G, Nader WF, Sittenfeld A (in press, 1997) Biodiversity for the bioindustries. In: Ford-Lloyd BV, Newbury HJ, Callow JA (eds) Biotechnology and Plant Genetic Resources: Conservation and Use. CAB International, Wallingford, Oxon, England

Truve E, Aaspollu A, Honkanen J, Puska R, Mehto M, Hassi A, Teeri TH, Kelve M, Seppänen P, Saarma M (1993) Transgenic potato plants expressing mammalian 2'-5' oligoadenylate synthetase are protected from potato virus X infection under field conditions. Bio/Technology 11: 1048-1052

US National Research Council (Panel of the Board on Science and Technology for International Development) (1992) Conserving Biodiversity: A Research Agenda for Development Agencies. National Academy Press, Washington D.C.

Watanabe S, Kato H, Nagayama K, Abe H (1995) Isolation of 2R,5R-dihydroxy-methyl-3R,4R-dihydroxy-pyrrolidin (DMDP) from the fermentation broth of Streptomyces sp. KSC-5791. Biosc Biotech Biochem 59: 936-937

Welter A, Jadot J, Dardenne G, Marlier M, Casimir J (1976) 2,5-Dihydroxymethyl-3,4-dihydroxypyrrolidone dans les feuilles de Derris elliptica. Phytochemistry 15: 747-749

Wilson EO (1992) Resolution. In: Wilson EO (ed.) The Diversity of Life. Harvard University Press, Cambridge, pp 311-342

Yun DJ, Hashimoto T, Yamada Y (1992) Metabolic engineering of medicinal plants: transgenic Atropa belladonna with an improved alkaloid composition. Proc Nat Acad Sci 89: 11799-11803

Access to genetic resources and benefit sharing

Marc Auer
Federal Ministry for Environment, Nature Conservation, and Nuclear Safety, Bonn, Germany

Abstract. Access to genetic resources and benefit-sharing seems to be a fairly new concept. However, the prospecting of plants and animals dates back to colonial times. In the past few years the challenging issue of how ecology, economics and sociology overlap has witnessed an explosion of publications and talks about the concept of "bioprospecting" and how it can be implemented.

The wealth of millions of fungi, insect and other species that have yet to be discovered constitutes the richest reservoir in this field. Biological diversity, or biodiversity, provides a tremendous array of benefits to humanity. It is a source of food, fuel, fibre and medicines, and provides raw material for industrial products. It also contributes to the maintenance of ecological systems.

The potential and actual economic value of genetic material is thus increasing rapidly. Despite the evident economic significance of these resources, the countries, communities and individuals providing either genetic resources and/or knowledge on their use have received little, if any, direct benefits to date. This eliminates potential incentives and thus undermines efforts to conserve biological diversity and ensure sustainable use of its components.

Against the backdrop of the North-South divide, access to genetic resources has met with increasing criticism in the last few decades. Principles such as "equal development opportunities," "conservation of biological diversity," "sustainable use," "benefit sharing" and "indigenous community rights" have made "access to genetic resources" a highly controversial international issue.

These developments in international agreements are matched by the growing number of countries that are passing legislation to control access to genetic resources. Developing countries opposed the principle that genetic resources are a common heritage of mankind and questioned whether it was fair that gene-poor developing countries can obtain genetic resources free of charge from the gene-rich developing countries. While negotiating the Biodiversity Convention, the developing countries succeeded in having the principle of common heritage replaced by the principle of national sovereignty over genetic resources. International efforts have begun to address existing imbalances.

The right to determine access to genetic resources plays a key role in the development and implementation of the Convention on Biological Diversity (CBD). The CBD is a step forward towards progressive protection of the world's biodiversity because it is based upon a new approach which provides for the protection of biodiversity as a whole rather than focusing on particular species. Moreover, this approach is based on the principle of sustainable development.

Acceso a recursos genéticos y participación en los beneficios

Resumen. El acceso a los recursos genéticos y a la participación en los beneficios parece ser un concepto completamente nuevo. Sin embargo, la prospección de plantas y animales data de la época colonial. En los últimos años, el desafío que representa el tema de cómo se superponen la ecología, la economía y la sociología ha dado lugar a una explosión de publicaciones y conferencias sobre el concepto de "bioprospección" y la manera de implementarlo.

La riqueza de millones de hongos, insectos y otras especies que aún han de ser descubiertas constituye la más rica reserva en este campo. La diversidad biológica o biodiversidad ofrece una enorme gama de aspectos beneficiosos para la humanidad. Significa una fuente de alimento, combustible, fibra y medicinas y proporciona materias primas para los productos industriales. Por otro lado, contribuye también al mantenimiento de los sistemas ecológicos.

El potencial y valor económico actual del material genético está, por consiguiente, experimentando un rápido incremento. A pesar de la evidente importancia económica de estos recursos, los países, comunidades e individuos que proporcionan recursos genéticos y/o bien conocimientos respecto a su uso han recibido hasta el momento pocos beneficios directos, si es que han recibido alguno. Esto elimina posibles incentivos y mina por tanto los esfuerzos para conservar la diversidad biológica y asegurar el uso sostenible de sus componentes.

Ante el telón de fondo de la división Norte-Sur, el tema del acceso a los recursos genéticos se ha tratado de forma cada vez más crítica en las últimas décadas. Principios tales como "igualdad de oportunidades de desarrollo", " conservación de la diversidad biológica", "uso sostenible", "compartimiento de beneficios" y "derechos de la comunidad indígena" han hecho del "acceso a los recursos genéticos" un tema internacional muy controvertido.

Estos avances en los acuerdos internacionales se complementan con el creciente número de países que están aprobando leyes para controlar el acceso a los recursos genéticos. Los países en vías de desarrollo se opusieron al principio de que los recursos genéticos son patrimonio común de la humanidad y cuestionaron el hecho de si era justo que los países desarrollados pobres en recursos genéticos pudieran obtener recursos genéticos gratuitamente de los países en vías de desarrollo ricos en recursos genéticos. Mientras se negociaba el Convenio sobre la Diversidad Biológica, los países en vías de desarrollo lograron sustituir el principio de patrimonio común por el principio de soberanía nacional sobre los recursos genéticos. Se han empezado a realizar esfuerzos a nivel internacional para tratar los desequilibrios existentes.

El derecho a determinar el acceso a los recursos genéticos juega un papel clave en el desarrollo e implementación de la Convención sobre Diversidad Biológica (CDB). La CDB supone un paso adelante hacia la protección progresiva de la biodiversidad mundial, ya que, en vez de orientarse en especies particulares, se fundamenta en un nuevo enfoque que vela por la protección de la biodiversidad como un todo. Además, este enfoque está basado en el principio del desarrollo sostenible.

Preface

Access to genetic resources and benefit-sharing seems to be a fairly new concept. However, the prospecting of plants and animals dates back to colonial times when people began to search for spices, flavourings and other plant products many centuries ago. They found pepper, rubber, coffee, sugar cane, oil palm and hundreds of other plants that laid the foundation for a new world order that exploited the biological resources of the South in order to generate wealth and prosperity in the North. Unfortunately explorers from Columbus onwards did not ask permission before transferring plant and animal genetic resources around the globe.

In the past few years, the challenging issue of how ecology, economy and sociology overlap has witnessed an explosion of publications and talks on the concept of "bioprospecting" and how it can be implemented. I would like to explore some of the fundamental problems underlying this new push to commercialize our planet's biological treasures. Other important inter-related aspects such as biosafety, intellectual property rights, farmers' rights, prior informed consent, legislation, monitoring and enforcement cannot be dealt with in this article due to length constraints.

1 Facts and figures

A number of our present-day medicines are taken directly from nature: antibiotics, atropine, ergot alkaloids – all these drugs are based on natural resources. Examples from the field of cancer therapy are taxol extracted from the Yew tree and alkaloids from the tropical periwinkle. The therapeutical effects of many substances in plant and animal organisms are due to pharmacologically active principles which may be used as a platform for developing new drugs.

The millions of fungi, insect and other species that have yet to be discovered constitute the richest reservoir in this field. To date, the pharmaceutical industry has made large-scale commercial use of only little more than 90 out of the 320,000 plant species on Earth. It is estimated that about 1,400 plant species of the tropical rain forest contain active ingredients that might be effective in fighting cancer. Germany imports approximately 40,000 metric tons of crude vegetable drugs annually and is second only to Hong Kong in this respect. About one third of all over-the-counter medicines sold in Germany are derived from plants.

2 Introduction

Biological diversity, or biodiversity, offers a tremendous array of benefits to humanity. It is a source of food, fuel, fiber, and medicines and provides raw materials for industrial products. It also helps maintain ecological systems. Biodiversity primarily benefits the local communities living in close proximity to biological systems such as forests, farmlands, and coastal habitats. However, vast benefits accrue to urban populations, national economies, and the global community as well. Today, the flow of genetic resources is increasing. New technological developments, particularly in the domain of biotechnology, are making it possible to transfer genes across species and kingdoms. This is revolutionizing agriculture and opening up new opportunities in drug development. A medium-size biotechnological or pharmaceutical company can conduct between 10,000 and 1,000,000 assays a year. However, only one in every 10,000 to 20,000 extracts contains a new drug candidate, and in the case of anti-cancer agents this figure is only one in every 40,000 to 50,000 extracts. However, development costs are high and 10 to 15 years may be invested in drug development before marketing can take place. The cost of new drug development in 1986 was estimated to be $50 million to $100 million per product. Today, a realistic estimate would be closer to $200 million per product. For this reason it is important that collectors, researchers and sponsoring organizations do not give misleading information regarding the possible commercial benefits of research.

The potential and actual economic value of genetic material is thus increasing rapidly, prompting a surge of interest in genetic material and stimulating international trade in genetic resources which is commonly referred to as "biotrade".

Despite the evident economic significance of these resources, the countries, communities and individuals providing genetic resources and/or knowledge regarding their use have received little, if any, direct benefits to date. Local and indigenous communities – in many cases the actual custodians of much of the world's wild and domesticated biodiversity – rarely receive a fair share of the benefits derived from the expanding commercial and industrial use of biodiversity. This eliminates potential incentives and thus undermines efforts to conserve biological diversity and ensure sustainable use of its components. If local custodians of a resource were to receive equitable benefits from its use, they would have a greater incentive to help ensure its conservation.

2.1 Historical background

By the late 1970s, the pharmaceutical industry had largely lost its interest in isolating ingredients of highly developed plant species. The successful isolation of important active ingredients was a rarity – in fact, too much of a rarity. Progress in the field of chemical synthesis and molecular biology promised the development of new tailor-made synthetic drugs. Despite this seemingly unfavourable situation, expeditions were launched in the early 1980s to conduct ethnobotanical research in areas and regions that had received little attention in the past.

Today, plant and insect collectors are roaming tropical rain forests for the pharmaceutical industry, searching for medicinal herbs, venomous snakes and poisonous frogs that might contain ingredients that have potential for pharmaceutical development. For example, the Costa Rican Instituto National de Biodiversidad (INBio) is engaged in taking stock of species of Costa Rica. The rural population has been integrated into this work and trained by experts to become "parataxonomists" who collect plants and insect material and make corresponding inventories. In addition to this stock-taking, awareness raising and educational work, INBio cooperates with pharmaceutical, agroindustrial, cosmetic and biotech companies in order to spur the collection of genes, macro- and microorganisms and other valuable natural products.

Efforts at international level include, for example, the UNCTAD BIOTRADE Initiative, the Joint Biodiversity Groups Program of the US National Institute of Health (NIH) and, of course, the cooperation between Costa Rica's INBio and the US pharmaceutical company Merck, Sharp&Dohme.

Things have changed dramatically since the 1950s when exports of biological material from developing countries were deemed to be of no major significance. The industrialized nations' demand for genetic resources – which is associated primarily with the pharmaceutical industry and the agroindustrial sector – has been rising markedly. Against the backdrop of the North-South divide, access to genetic resources has met with increasing criticism in the last few decades. Principles such as 'equal development opportunities', 'conservation of biological diversity', 'sustainable use', 'benefit sharing' and 'indigenous community rights' have made 'access to genetic resources' a highly controversial international issue.

2.2 Legal background

The principle of 'access to the others' genetic resources without restriction' which is based on the view that genetic resources are the 'common heritage of mankind' was formalized as recently as 1983 when it was recognized in the non-binding FAO Undertaking on Plant Genetic Resources. In a parallel development, intellectual property rights have also been strengthened. The GATT Agreement on Tariffs and Trade and Related Aspects of Intellectual Property Rights – the TRIPS Agreement – obliges signatories to extend patent and other intellectual property rights to certain categories of living organisms.

These developments in international agreements are matched by the growing number of countries that are introducing legislation to control access to genetic resources.

Developing countries opposed the common heritage principle and questioned whether it was fair that gene-poor developed countries can obtain genetic resources free of charge from gene-rich developing countries. While negotiating the Biodiversity Convention, the developing countries succeeded in having that the principle of common heritage replaced by the principle of national sovereignty over genetic resources. International efforts have begun to address existing imbalances.

The Convention on Biological Diversity which entered into force in December 1993 – Germany has been a Contracting Party since December 1993 – has established new fundamental provisions for regulating access to genetic resources for the Convention's (currently) 168 Contracting Parties.

3 The Convention on Biological Diversity and the access issue

The right to control access to genetic resources plays a key role in the development and implementation of the Convention on Biological Diversity (CBD). The CBD is based upon a new approach which provides for the protection of biodiversity as a whole rather than focusing on particular species. It therefore represents a step forward toward the progressive protection of the world's biodiversity. Moreover, this approach is based on the principle of sustainable development.

This new approach is being implemented by acknowledging the sovereign rights of States over their genetic resources and by linking the provision of such resources and their utilization to participation and benefit sharing. The underlying idea is to give genetic resources economic value and thereby create incentives to ensure their sustainable use and conservation. Access to genetic resources is a crucial factor in achieving the Convention's objectives in regard to the sustainable use and protection of biodiversity and is therefore one of the Convention's key issues.

The CBD provides a fairly clear and straightforward basis for the provision and transfer of genetic resources around the world. The Convention does not however provide a complete system of rights for the use of genetic resources. In this regard, it would be advisable to establish uniform standards in order to preclude circumventions which may amount to unfair advantages for certain States. International action aimed at setting some more explicit standards for access and enforcement is to be recommended.

Fair and equitable benefit-sharing between users of genetic resources and countries of origin must be mentioned here. How should we define the fair and equitable sharing of benefits arising from the commercial and other utilization of genetic resources in accordance with the Convention?

Possible ways to achieve this aim include:

- An exchange of information on genetic resources with the countries of origin and the dissemination of relevant information back to the countries of origin, research cooperation, technology transfer, education and training, monetary benefit-sharing, capacity building and mixed models. As already indicated, the Convention provides for a complex system of participation and sharing of benefits based on Arts. 15, 16 and 19. The Convention does <u>not</u> state which kind of transactions or between whom such transactions are to take place. Consequently, the Contracting States need, either jointly or individually, to develop these provisions further. And policy-makers must take on the task of developing a participation and benefit-sharing system. At this stage, legal analysis would not be able to contribute much toward determining the structures of such a system. However, it is possible to discuss the question of if and under what circumstances certain much-discussed transactions would be compatible with the provisions of the Convention.

- One frequently proposed and already practiced model organizes participation and benefit-sharing by private agreement. Such agreements can be concluded between a party interested in genetic resources and the body that is authorized to grant access to such resources. Such an agreement could outline details regarding participation in research and the transfer of research activities to the country of origin, technology transfer and the sharing of findings and benefits.
- The interests of indigenous and local communities have to be taken into account when discussing participation and benefit-sharing. The CBD emphasizes the role that such communities, their traditional lifestyles and traditional knowledge, innovations and practices might play in the conservation and sustainable use of genetic resources.

All in all, the development of a system for participation and benefit-sharing requires urgent consideration at national and international level, particularly in light of the fact that some States and business corporations are already in the process of securing an interest in genetic resources by signing agreements with resource States.

4 Bioprospecting – Risks and chances

This approach has appeared on the international scene under the catchword "bioprospecting." Biodiversity prospecting as defined by the Convention is the exploration and collection of biological material for commercial utilization. There can be no doubt that genetic resources, traditional and indigenous knowledge and experience will be the *gold, silver and diamond mines of the future.* However, this novel approach is aimed at transforming a basically economic activity into a tool that can be used for nature conservation at the same time. There are many indications that biodiversity prospecting will become a major and exciting challenge in the pharmaceutical and agroindustrial sector. Biotechnology in general and genetic engineering in particular are expected to be central technologies in the 21th century. The growth of the biotechnology industry is likely to give rise to a marked increase in the demand for genetic resources.

Economic considerations add value to biological diversity and make nature conservation an economically attractive goal. Profits from prospecting activities could be channelled back to nature conservation.

Instead of always having to ask for technology and financial assistance, developing countries could offer industrialized nations genetic assets and technologies ranging from raw materials to traditional medicines. Provided that the revenues accruing from this development benefit the primary and actual custodians of biological diversity, the incremental income generated in this way would enable marginalized groups to supplement their income and improve their living conditions. Bioprospecting could thus provide an incentive for innovative forces in countries with a high degree of biological diversity.

Technological capacity-building, education and training in these countries will make them more attractive to interested companies and improve their share of revenues from licensing contracts. Cooperation with pharmaceutical companies, for example, offers opportunities to become acquainted with new technologies. The opportunities that this process offers to highly qualified research institutions in the South are already evident in research and development cooperation between corporate groups and specific institutes. Prospecting contracts could also be used to transfer shared benefits back to the South – in the form of products that are developed by corporate groups which are primarily oriented towards markets in industrialized countries, products that would otherwise not have been developed. For example, resource-providing countries from the South might request pharmaceutical companies to engage actively in the search for plant ingredients that could be used to combat tropical diseases. This would open up additional new markets in turn. However, bioprospecting also involves risks for developing countries.

An economy-oriented approach to biodiversity may push other functions of intact ecosystems into the background. Furthermore, we must refrain from viewing traditional, cultural and social uses of genetic resources as merely complementary aspects; they must remain part of the central program for the conservation of biological diversity. The incremental income of developing countries to be expected from genetic resource prospecting remains uncertain. The Convention demands that access to genetic resources be regulated. However, those offering genetic resources will be in a weak negotiating position as long as they depend on financial support and technology transfers from the North. Moreover, most institutions in developing countries are the weaker party when negotiating with large multinational companies. It is evident that competition between developing nations also involves the risk of genetic resource dumping.

Due to their abundant and rapidly proliferating stock of biological resources, some countries may profit more than others from bioprospecting. We will not see raw material monopolies controlling the genetic resources sector in the foreseeable future. Such monopolies might constitute a serious danger in light of South-South disparities. Major technological advances in the synthesis of active substances will be followed by a sharp fall in the price of the respective resource.

Ultimately, there is no guarantee that prospecting will safeguard biological diversity. If species being harvested for their active ingredients are overharvested, reductions in stock may lead to a loss in genetic variability.

5 Research and policy-making requirements – Their implementation in Germany

The German government acts to fulfil its obligations under the Convention by seeking increased research cooperation between developing countries and German industry which is aimed at research and development that meets the specific requirements of the Convention. The Convention stipulates that the states' utilization of genetic resources must comply with specific obligations. These obligations are strongly intertwined with research and development cooperation. These obligations, together with the need for legal security regarding the use of genetic resources from the various countries of origin, have induced the Federal Ministry for the Environment, Nature Conservation and Nuclear Safety to consider suitable German initiatives. At federal level, a special task force has been established to deal with the entire spectrum of issues involved in bioprospecting. This group includes various user associations and other interest groups. It will prepare and provide constructive assistance for developing models for research and development cooperation with German pharmaceutical companies, giving special attention to the specific requirements under the Convention. The German government has also commissioned two studies to clarify the legal requirements that apply to the Federal Republic of Germany in regard to access to genetic resources. Further, the German government is currently holding discussions with pharmaceutical companies interested in developing cooperation models. A two-year cooperation project between users and countries of origin is to be launched in due course. A master plan for cooperation in the bioprospecting field has been drawn up in preparation of this project and an international workshop was held with numerous participants from pharmaceutical companies and countries of origin. The results will be incorporated into these cooperation models.

In addition, the German government is investigating whether international guidelines might help foster compliance with the obligations arising from the Convention. These guidelines would be aimed at avoiding further legislation by inducing voluntary compliance on the part of the users and providers of genetic resources.

6 Outlook for the future

Many countries hope to obtain greater benefits from their biodiversity by developing the technical capability to screen, isolate and develop biologically-derived chemicals. However, in order to earn profits that can be reinvested in further research and biodiversity conservation, suppliers of genetic resources and other biochemical material must have either a cost advantage in some area of biochemical research and development or superior information about where to look for valuable biological material. Countries seeking to use biological resources as a basis for sustainable development must also work toward developing the technical and entrepreneurial skills necessary for biochemical research, development, and marketing.

In conclusion, establishing a biochemicals prospecting market that will promote conservation of biodiversity and create opportunities for sustainable economic development will require coordinated international efforts aimed at stimulating and complementing investments by the governments, private sectors and local communities of developing countries. Competing in this emerging market will require an integrated approach aimed at increasing the capabilities of developing countries. Germany can play an important role in this process by facilitating and enhancing cooperation between the German market and the countries of origin of genetic resources.

Part 5

Bolivia -
A megadiversity country

Part 5.1

The biodiversity of Bolivia

Bolivia is a megadiversity country and a developing country

Pierre L. Ibisch
Institute of Botany, University of Bonn, Germany

Abstract. Bolivia has access to the most important ecological regions of tropical South America. It is one of the few countries in the world to have a major portion of global biodiversity concentrated within its national boundaries. Bolivia is also a centre of crop-genetic resources of global importance. Biodiversity is a main factor in great ethnocultural diversity which implies a broad indigenous knowledge of its utilization for human development. Although Bolivia may be called a 'rich' megadiversity country, it is also one of the world's poorest nations in economic and human development terms. Ironically, poverty has been a key factor in the conservation of Bolivia's biodiversity. Today, Bolivia still has some of the most extensive forests in the world. This is due to a very low human population density, especially in the lowlands, and to a lack of means for accessing and rapidly exploiting the country's natural resources. However, recent years have been marked by dynamic economic development. A surge of economic and development activities such as agro-industry, oil exploitation, timber extraction and road construction is leading to biodiversity degradation not only in Bolivia, but throughout the entire world. On the other hand, poverty – especially in traditionally settled and cultivated regions of the Andes – has set up a vicious circle of famine, migration, agricultural frontier encroachment and deforestation. In Bolivia, this complex mixture of underdevelopment and development is causing biodiversity loss.

The time has come to test and implement concepts for sustainable development which incorporate the non-extractive and sustainable utilization of biodiversity wherever possible, while guaranteeing the conservation of biodiversity. It is necessary to intensify land use in regions that are already disturbed or degraded so that ecosystems that are still intact can be preserved. In some regions, perhaps the only efficient way to guarantee conservation might be to revoke the right to utilize national resources. But this renouncing cannot be taken for granted. Who should foot the bill? In the international game of politics the megadiversity country Bolivia holds the trump in its hand and just needs to play and make it count: It could either demand the international community to reimburse it for conserving its biodiversity – as environmental service to the 'global village' – *or* it could exploit and deplete its natural resources for its own short-term national development! There is no alternative: The world community – and the rich, industrialized countries in particular – have to compensate Bolivia for the environmental services it offers, and which up to now have been seen as being free of charge. This investment, an economic cooperation between partners with a mutual interest, transcends humanitarian aid. In this context, the "Debt-for-nature-swaps" concept is not sufficient; the "Global Environment Facility" seems to be more adequate. However, more compensation instruments and mechanisms should be developed and implemented. A possible goal could be bilateral or multilateral alliances between megadiversity and 'mega-economy' countries which aim at a joint implementation of the internationally accorded conventions and agendas on biodiversity conservation and sustainable development. Development research and policy have a number of tasks to fulfil in developing and assisting this new type of cooperation. Bolivia – a megadiversity country with excellent political conditions – should be an ideal partner.

Bolivia es un país de megadiversidad y un país en vías de desarrollo

Resumen. En Bolivia se encuentran las más importantes regiones ecológicas de la Sudamérica tropical. Pocos países del mundo tienen la suerte de concentrar dentro de sus fronteras nacionales una proporción tan grande de biodiversidad global. Asimismo, Bolivia es un centro de recursos genéticos de cultivos de importancia global. La biodiversidad constituye un factor esencial en la gran diversidad etnocultural, que implica un amplio conocimiento indígena de su empleo para el desarrollo humano.

Aunque Bolivia se pueda calificar como país 'rico' en cuanto a su megadiversidad, es también una de las naciones más pobres del mundo en términos económicos y de desarrollo humano. Irónicamente, la pobreza ha sido un factor clave para la conservación de la biodiversidad de Bolivia, en cuyo territorio se encuentran aún algunos de los bosques más extensos del mundo. Ello se debe a una densidad demográfica muy baja, sobre todo en las tierras bajas, y a la falta de medios para acceder a los recursos naturales del país y explotarlos con rapidez. Sin embargo, los últimos años se han caracterizado por un dinámico desarrollo económico. Una oleada de actividades económicas y de desarrollo - tales como la industria agrícola, explotación petrolífera, explotación maderera y la construcción de carreteras - está llevando a la degradación de la biodiversidad, y no sólo en Bolivia, sino en todo el mundo. Por otro lado, la pobreza, especialmente en las regiones andinas pobladas y cultivadas de modo tradicional, ha impuesto un círculo vicioso de hambre, migración, ampliación de la frontera agrícola y deforestación. Esta compleja mezcla de subdesarrollo y desarrollo está causando pérdidas en la biodiversidad de Bolivia.

Ha llegado el momento de estudiar e implementar conceptos para un desarrollo sostenible que incorporen la utilización no extractable y sostenible de la biodiversidad donde sea posible, pero garantizando la conservación de la biodiversidad. Es necesario intensificar el uso del suelo en aquellas regiones que ya han resultado dañadas o degradadas, de manera que queden preservados los ecosistemas que siguen intactos. En algunas regiones, puede que la única manera eficaz de garantizar la conservación sea revocar el derecho a utilizar los recursos nacionales. Pero esta renuncia requiere de una compensación: ¿Quién correría con los gastos? En el juego de la política internacional, Bolivia, como país de megadiversidad, tiene el triunfo en la mano; todo lo que ha de hacer es jugarlo y hacerlo valer: Podría exigir a la comunidad internacional una compensación económica por la conservación de su biodiversidad - como servicio al medio ambiente para el 'pueblo global' - *o bien* podría explotar y agotar sus recursos naturales para su propio desarrollo nacional a corto plazo. No hay alternativa: La comunidad internacional - y los países ricos e industrializados en particular - tiene que compensar a Bolivia por el servicio al medio ambiente que ofrece y que hasta ahora se ha considerado como un servicio gratuito. Esta inversión - una especie de cooperación económica entre socios con un interés mutuo - va más allá de la ayuda humanitaria. En este contexto, el concepto de 'deuda a cambio de naturaleza' no llega hasta donde debería; el Fondo para el Medio Ambiente Mundial *(Global Enviroment Facility)* parece resultar más adecuado. No obstante, habría que desarrollar e implementar más instrumentos y mecanismos de compensación. Un objetivo posible podrían ser alianzas bilaterales o multilaterales entre países con megadiversidad y países con 'mega-economía', alianzas que tendrían como objeto una implementación conjunta de las convenciones y programas sobre conservación de la biodiversidad y desarrollo sostenible que se han acordado internacionalmente. La investigación y la política de desarrollo tienen una serie de tareas que cumplir para desarrollar y ayudar a este nuevo tipo de cooperación. Bolivia - un país de megadiversidad con excelentes condiciones políticas - sería un socio ideal.

1 Introduction: One of the world's 'richest poorest' countries

Bolivia is one of the richest countries in the world – in terms of biological diversity: a megadiversity country! This wealth is contrasted by enormous economic and human poverty. Bolivia is the poorest country in South America. Its situation poses a major paradox and an enormous dilemma which has been the subject of little attention in past decades. Today, both underdevelopment and development in Bolivia are leading to the irreversible destruction of the country's biodiversity. Now that it has become fashionable to talk about development and the environment, sustainability, biodiversity, genetic resources and bioprospecting, this situation may change: Bolivia could become a model for ways of integrating the conservation and sustainable utilization of biodiversity into development. When we summarize Bolivia's "megadiversity" and analyze current development trends and the effect they have on biodiversity conservation and/or destruction, the following questions arise: Is it possible to develop Bolivia while conserving its unique and special biodiversity? Could biodiversity be used as an instrument to promote development? Why, how and for whom should Bolivia conserve its biodiversity? What would it cost to do so and who will foot the bill? Finding answers to these questions will pose a major challenge for development research and development activities!

2 Bolivia is a megadiversity country

The national 'megadiversity approach' was first developed by Mittermeier (1988; also Mittermeier & Werner 1989) in order to stress the unique importance of those few countries which have a major portion of the world's biological diversity concentrated within their boundaries, and to acknowledge the fact that conservation programs are developed by and with governments of sovereign nations. Of course, plant and animal species are not aware of geopolitical boundaries but the megadiversity country approach acknowledges that conservation is managed at country level (Groombridge 1990). Even recent publications on megadiversity countries neglect Bolivia (e.g., Groombridge 1990, McNeely *et al.* 1990, Olivieri *et al.* 1995).

Moraes & Beck (1992) were the first to propose that Bolivia should be included in the list. Bolivia is still the least investigated country in South America (Solomon 1989, Campbell 1989) and scientists who are unacquainted with it tend to underestimate its diversity. This diversity is manifest at all levels, from its abiotic conditions ('geodiversity', see Barthlott *et al.* 1996) to the hierarchic biological systems which exist within its borders. The term biodiversity as used here refers to the number of biological systems which coexist within defined dimensions of space and time (Solbrig 1991, Ibisch 1996a).

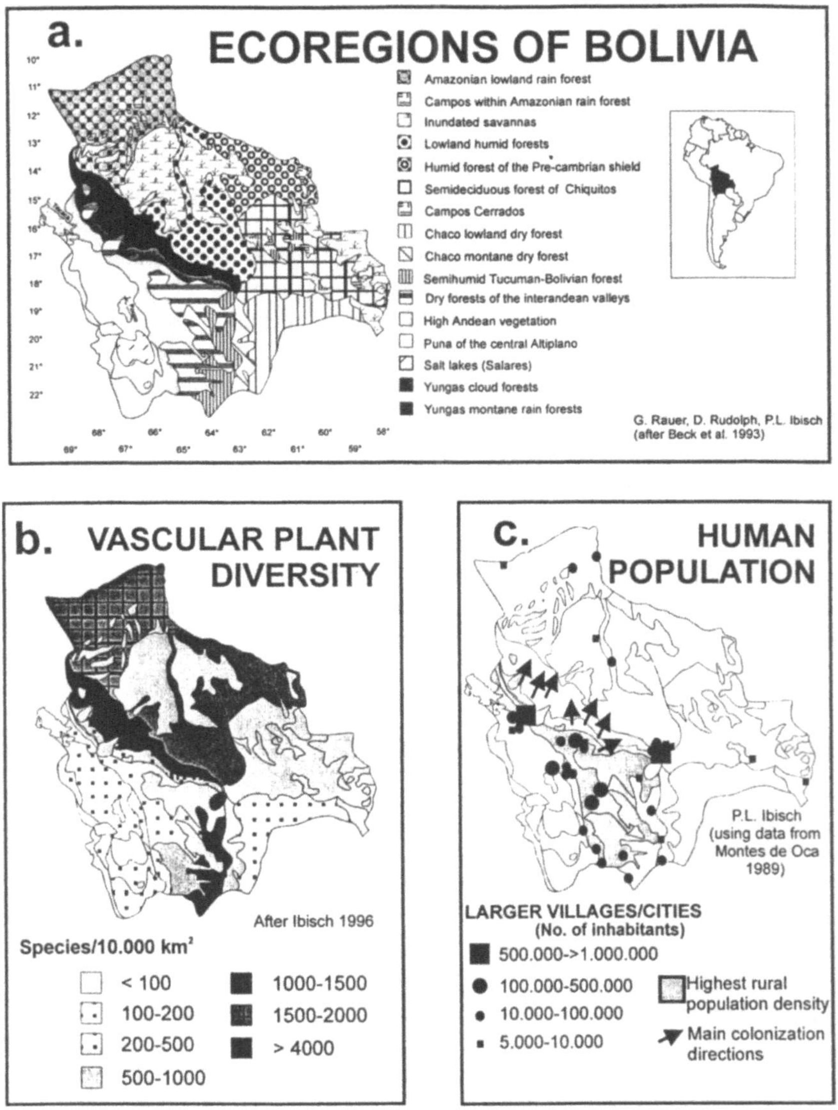

Fig. 1. Maps of Bolivia. a. Ecoregions. b. Plant diversity. c. Human population
Ilustración 1. Mapas de Bolivia. a) Ecoregiones. b) Diversidad de las plantas. c) Densidad demográfica

2.1 Ecosystem diversity

The ecosystem is an important hierarchic level of biological systems. Few countries in the world can match Bolivia's ecosystem diversity. There is probably no other tropical country which has access to as many biogeographical regions and biomes as Bolivia (Fig. 1.a). In his "three dimensional scheme" of tropical vegetation, Lauer classifies (e.g., 1986) zonal vegetation according to hydro-thermical units. Almost all of them – from deserts to rain forests, from hot lowlands to glaciers – can be found in Bolivia. Furthermore, there is a complex edaphical differentiation of zonal and azonal ecosystems[1]. In recent times, a totally new forest type has been discovered in Bolivia and is to be included in the world's plant formations: the Velasco Dry Forest (Parker *et al.* 1993). Of course, forests are of special importance for species diversity and global ecology. Some comparative data on the world's forests illustrate this point: Bolivia has some 500,000 square kilometres of forests, placing it in the ranks of the ten or eleven most forest-rich countries in the world (Fig. 2.b; The International Bank for Reconstruction / World Bank 1994, Ministerio de Desarrollo Sostenible y Medio Ambiente 1995). If we were to take only tropical countries into account, Bolivia would even be ranked fifth or sixth! Of the world's most forest-rich tropical countries, Bolivia is the country with the most extensive per-capita-forest resources (Fig. 2.c).

2.2 Species diversity

Species constitute a second important level of biological systems. One main reason for Bolivia's high level of species diversity is its abiotic diversity in space and time, including all the historical, geological and orographical processes and climatic changes. All factors and mechanisms which stimulate and accelerate speciation or guarantee the maintenance of high species diversity such as habitat heterogeneity, favourableness, extinction-buffering long-term stability, isolation or local medium disturbances are active in Bolivia (Ibisch 1996a). Furthermore, its geographical location facilitates the immigration of very different biogeographical elements. Bolivia is an Amazonian, Andean, Chaco and Cerrado country.

Species diversity cannot be estimated for many groups of organisms in Bolivia (e.g., invertebrates, fungi). Although most other groups are little known, it is possible to give at least estimates. Table 1 reviews some of the known data on Bolivian species diversity and specifies the country's respective rank on 'world-ranking lists'.

[1] The ecosystem diversity encompasses a continuum from very humid to dry tropical forests, savannas, shrublands, swamps, (salt-) lakes, dunes and rocks. Ribera (1992) lists more than 40 ecoregions of Bolivia.

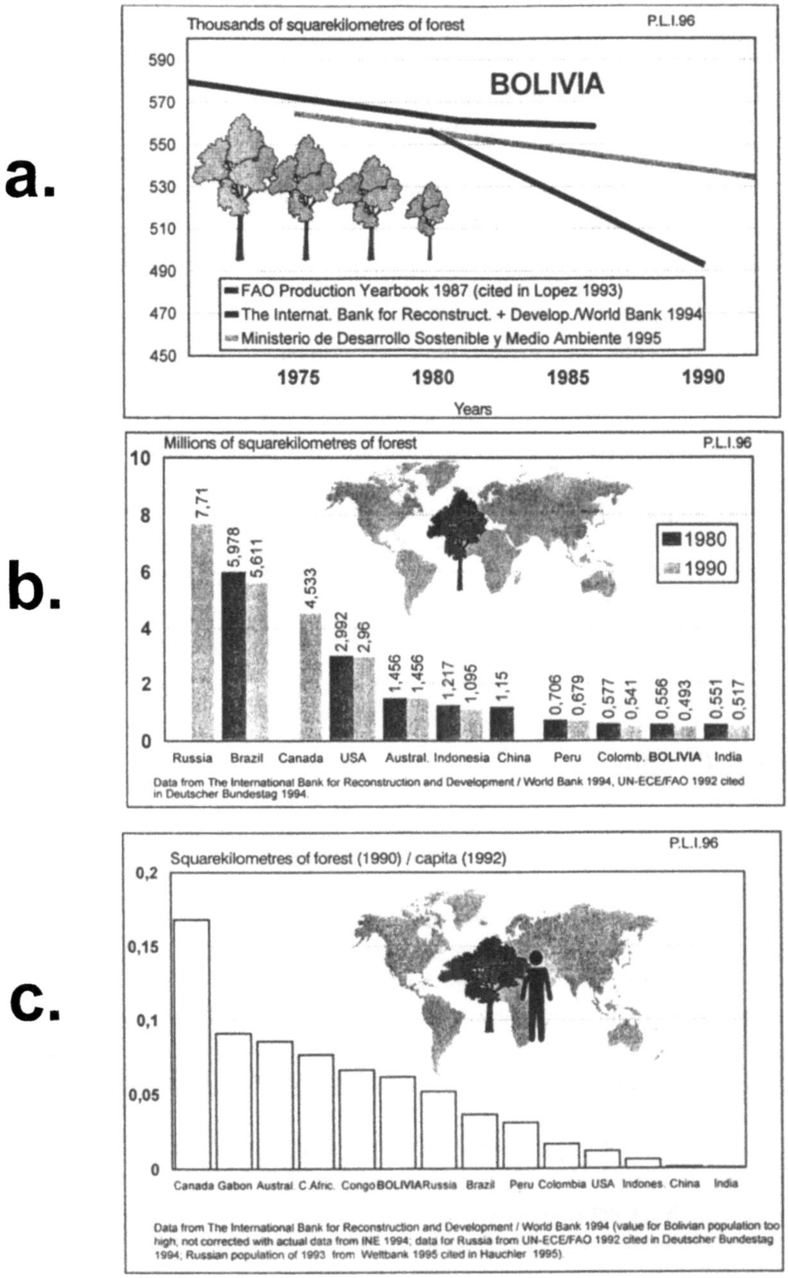

Fig. 2. Forest-richness in Bolivia. a. Bolivia's forest cover during the last decades according to different sources. b. The current world-ranking list of the most forest-rich countries. c. World-ranking list of countries with the largest per-capita forest areas

Ilustración 2. Riqueza forestal de Bolivia. a) Extensión ocupada por los bosques de Bolivia durante las últimas décadas según distintas fuentes. b) Escalafón de los países del mundo que cuentan con una mayor riqueza forestal. c) Escalafón de los países del mundo con las mayores áreas forestales percápita

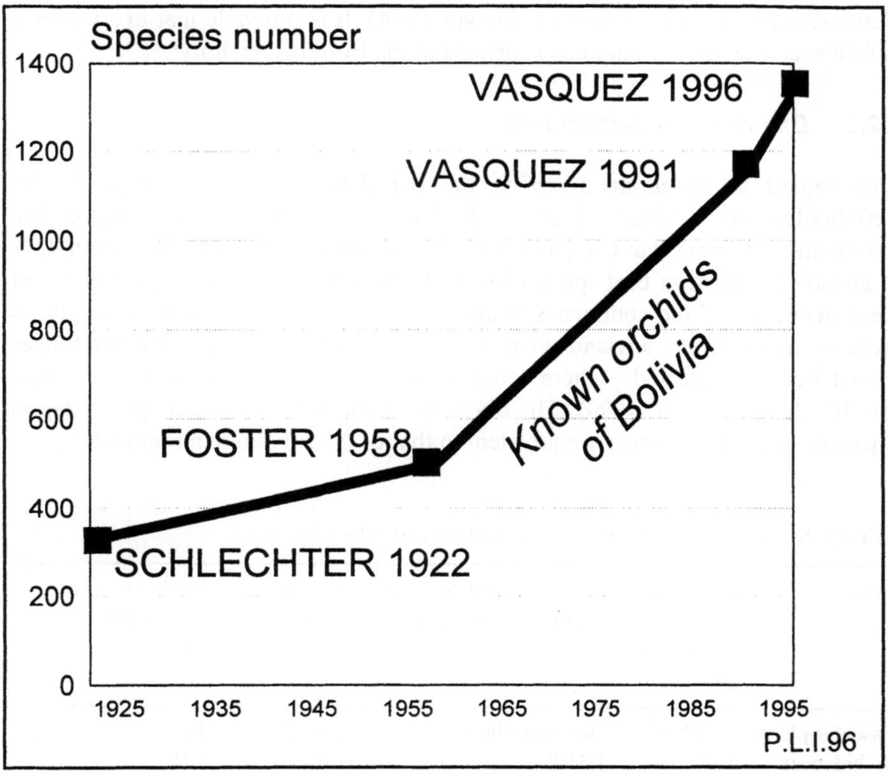

Fig. 3. Known number of Bolivian orchids during the 20th century
Ilustración 3. Número conocido de orquídeas bolivianas durante el siglo XX

2.2.1 Example 1 – Vascular plants

One of the most informative articles on Bolivian flora in recent years was published by Moraes & Beck (1992). It is very probable that at least 18,000 to 20,000 plant species exist on Bolivian territory. The most species-rich families are the Orchidaceae, Asteraceae and Poaceae. The orchids impressively illustrate the evolution of knowledge on Bolivian flora. In 1922, Schlechter listed 322 Bolivian orchids. Foster (1958) counted about 500 species and Vásquez (1996) more than 1,300 (Fig. 3). The real number probably lies around 1,500 to 1,700 (Ibisch & Vásquez in prep.). The lack of knowledge on the number of Bolivian species has led us to underestimate the country's richness and diversity. The centre of plant species diversity probably is located in the montane rain forests of the Yungas (Fig. 1.b; Ibisch 1996a, Barthlott *et al.* 1996). Here, you can find probably up to 1,000 plant species coexisting on one hectare (Ibisch 1996a). Several plant groups have a diversity centre in Bolivia: e.g., Cactaceae (Strecker 1992), Amaranthaceae (pers. com. T. Borsch 1996), *Cleistocactus* (Cactaceae; Strecker 1992), *Puya* or

Fosterella (Bromeliaceae; Smith & Downs 1974). It is probable that every fourth to fifth species will turn out to be restricted to the Bolivian territory[2].

2.2.2 *Example 2 – vertebrates*

With regard to vertebrate diversity, the red data book "Libro Rojo de los Vertebrados de Bolivia" (Ergueta & Morales 1996) provides up-to-date information. It shows that fish and birds in particular are very diverse. About 43% of all South American bird species live in Bolivia (Rocha & Quiroga 1996) and more than 35% of the continent's mammals (Tarifa 1996). Faunistic diversity is probably higher in the lowlands than in the Andes. In this context, the dry forests should not be neglected: "There is more to biodiversity than the tropical rain forests" (Redford *et al.* 1990): The diversity of mammal faunas at some sites in Chaco turned out to be nearly equivalent to that of the Amazonian rain forests.

Table 1. Knowledge of Bolivian species diversity of selected groups

Taxa	Historical records (number of species)	Actual records/estimates (number of species)	Endemics (percentage estimates)	Rank on world "species richness" list [a]
Vascular plants	± 10,000[b]	18,000-19,000[c]	20-25%	10-11
Orchidaceae	323[d]	1,330[e]	20-25%	7-9
Cactaceae	± 200[f]	>320[g]	74%	2
Fish	-	+-500[h]	?	?
Amphibians	112[i]	155[j]	15-20%	< 11
Reptiles	-	229[k]	7-8%	± 15-16
Birds	1,274[l]	1,385[m]	1%	5-6
Mammals	-	319[n]	4-5%	< 10

[a] Own estimates; compare Groombridge (1992). [b] Foster, 1958. [c] Estimate Moraes & Beck (1992). [d] Schlechter (1922). [e] Vásquez (1996; compare Fig. 3). [f] Foster (1958; including many synonyms). [g] Hunt (1992). [h] Estimate, Sarmiento & Barrera (1996). [i] De la Riva (1990). [j] Köhler (1995), pers. com. J. Köhler (1996). [k] Dirksen (1995); Pacheco & Aparicio (1996: 220 spp.). [l] 1989; see Rocha & Quiroga (1996). [m] Rocha & Quiroga (1996). [n] Tarifa (1996).

2.3 Ethno-cultural diversity

As a broad and consistent concept, biological diversity is not limited to the world of unicellular life, plants, fungi or animals but also includes the diversity of human cultures (World Resource Institute *et al.* 1992, cited in Barzetti 1993). It is no coincidence that regions with an elevated level of biodiversity also display

[2] Actual endemism percentage of Peru: 31% (Brako & Zarucchi 1993).

high linguistic richness (Harmon 1993, cited in Barzetti 1993). In Bolivia, most of the semiarid and arid parts of the Andes have historically been occupied by Aymará-speaking indigenous tribes (Cortés 1992) as well as other linguistic groups of minor importance. Most of them were conquered by Quechua-speaking Incas. Today, vast sections of the Andean region are occupied by Aymará and Quechua farmers. Ethnodiversity in the Bolivian lowlands is much greater, especially in those areas that are humid and biologically more diverse. Thirty different cultures that developed special adaptations to their different natural resources can still be found here (about 9 in the Chaco, the rest in humid forests; Libermann 1995). Traditional land use in the Andes, the enormous efforts to achieve sustainable agriculture and the history of ecosystem and biodiversity degradation especially after the Spanish *conquista* are quite well known (e.g., Ruthsatz 1983, Seibert 1993). But human cultures can look back upon thousands of years of land use and ecosystem modification in the Bolivian lowlands as well.[3]

2.4 Genetic resource diversity

Organisms which are actual or potential providers of resources for human life are genetic resources. Indeed, genetic resource diversity in Bolivia is tremendous. Ethnocultural diversity is based primarily on the diversity of natural resources which permit survival in certain types of ecosystems. These resources include staple foods and medicines, construction materials and clothing. Moraes & Beck (1992) cite some publications on economic plant diversity utilized by distinct tribes[4]. Bolivia has dozens of cultivated and wild plant species which are of enormous actual or potential importance for global food security (see Cárdenas 1989). One of the most important staple foods of the world – the potato – originated in the High Andes of Bolivia and Peru. Currently, 31 wild and seven cultivated tuberous species of *Solanum* are known (Ochoa 1990). Small-scale farmers manage hundreds of local varieties. In the department of Santa Cruz alone, Vásquez & Coimbra (1996) identified 130 edible wild botanical species; at least 10 of them have high export potential.

---

[3] Up to 500,000 people may have lived in the Beni (Denevan 1963, 1966 cited in Campos 1992). Here, huge drained fields, numerous canals and small mounded fields are still observed in the savannas (Campos 1992). In many cases a historical agricultural regression to foraging has occurred (Balee 1992), accompanied by a drop in population density.

[4] E.g., Hinojosa 1991: Mosetenes, Sud Yungas, Altobeni: 167 utilized species; Boom 1989: Chacobos, Beni, Alto Ivón, 82% of all trees are utilized, 36% for food, 25% for medicine; Girault 1987: Kallawaya, Andes, 980 medicinal plant species). One of the most recent studies (Birk 1995) shows that the Chiquitanos near Concepción utilize 290 plant species (75% for medicine, 41% for nutrition) of which 77% are not cultivated.

3 Bolivia is an underdeveloped and a developing country

Of course, the terms 'underdeveloped', 'developing' or 'developed' are somewhat debatable, controversial or even misleading and dangerous. However, they are used here because it is obvious that there are more poor people in underdeveloped or developing countries than in developed countries, and that developed or industrialized countries have undergone a dynamic economic development that, in the current world community, is directly associated with political power. Furthermore, most people share a common concept of what 'development' is. This does not automatically mean that 'development' can or should be a model or aim for so-called underdeveloped or developing countries.

3.1 Indicators of Bolivian underdevelopment

According to a recent report on human development issued by the United Nations' development organization (UNDP), Bolivia holds a modest position in global ranking (number 111; report cited in Los Tiempos, 18.7.1996). Despite this, it is the poorest country in South America. The country has made considerable progress in recent years, yet its indicator values in the human development index remain low[5]. It is important to stress that 70% of the Bolivian population is regarded as poor (94% of the rural population). Its population growth curve is typical for a non-industrialized country. The actual growth rate is about 2.1% (Ministerio de Desarrollo Sostenible y Medio Ambiente 1994; see also Fig. 4.a). Looking at the country's economy and infrastructure, we also find a low development level (The International Bank for Reconstruction and Development / World Bank 1994): Bolivia can be found at the lower end of the 'lower-middle-income-countries'[6]. The density of paved roads can also be used as an indicator. Bolivia belongs to those countries having the fewest roads. In comparison, road density in Germany is about 1,000 times greater.

3.2 Indicators of Bolivian development

On the other hand, it can be demonstrated that Bolivia also is a developing country: Road construction – both an indicator of and a motor for development – has increased rapidly (INE 1989, 1994; Fig. 4.c). Especially in the lowland departments, road construction increased by 25 to 65% between 1989 and 1993! This is directly related to the advancing agricultural frontier. In the years

[5] Life expectancy reached 59,4 years, the illiteracy rate sank to 20%, child mortality is 75 per thousand born and 51% of the children are undernourished. Detailed information on poverty in Bolivia is provided by the 'poverty map' of Bolivia (Mapa de pobreza; Ministerio de Desarollo Humano 1993) or by the Ministerio de Desarrollo Sostenible y Medio Ambiente (1994).

[6] Gross national product: US$ 680 per capita.

a.

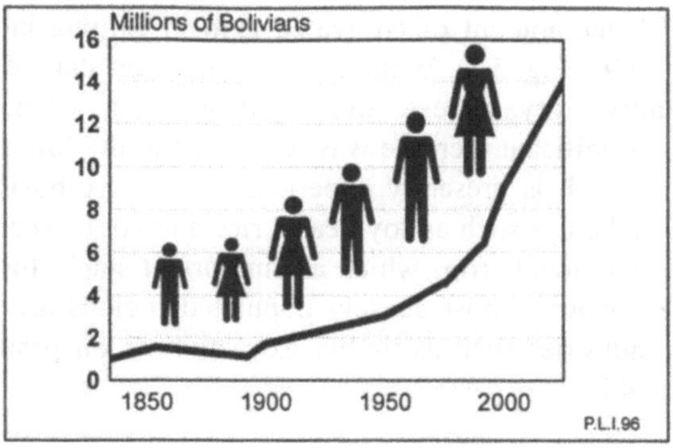

b.

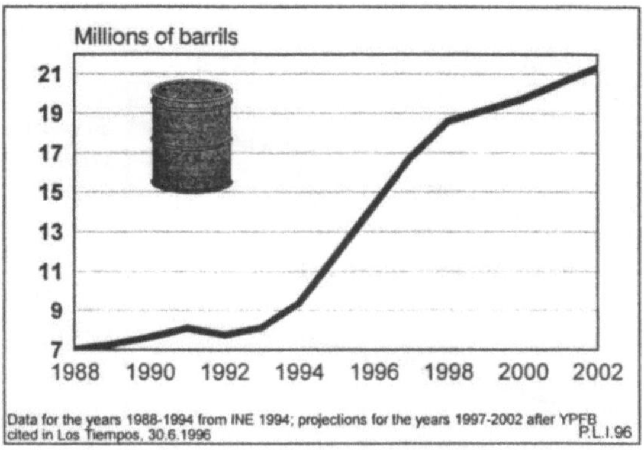

c.

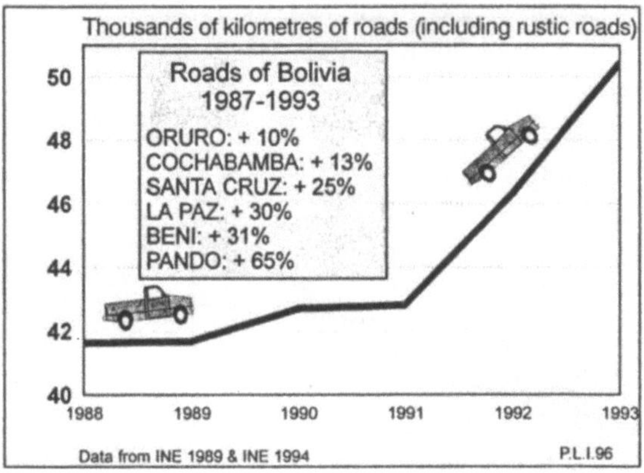

Fig. 4. Development of Bolivia. a. Population growth (according to data from Montes de Oca 1989: 1834-1976; INE 1989/1994: 1988-1994; The International Bank for Reconstruction and Development / World Bank 1994: estimates, 2000-2025; what appears to be a population reduction after 1882 was due to a loss of territory. b. Growth of oil production. c. Growth of the road system

Ilustración 4. Desarrollo de Bolivia. a) Crecimiento demográfico (según datos de Montes de Oca 1989; 1834-1976; INE 1989/1994: 1988-1994; el Banco Internacional para la Reconstrucción y el Desarrollo / Banco Mundial 1994: estimación, 2000-2025); lo que aparece como una reducción de la población después de 1882 es el resultado de una pérdida de territorio. b) Crecimiento de la producción petrolífera. c) Crecimiento de la red de carreteras

between 1987 and 1994, the amount of cultivated land in Bolivia increased by about 46% (INE 1989, 1994; Fig. 5.a). While road construction stagnated or even decreased in the country's dry valleys and highlands (e.g., department of Cochabamba), the most significant increase was registered in the lowlands of the Santa Cruz department which is presently experiencing an agro-business boom that is fuelled by industrial crops such as soy beans, rice and cotton (compare Fig. 5.b). In contrast, some traditional crops which are important staple foods for the rural population in the country's most densely populated regions are stagnating (e.g., potatoes). Other industries such as timber extraction or oil production are also expanding (see Fig. 4.b).

a.

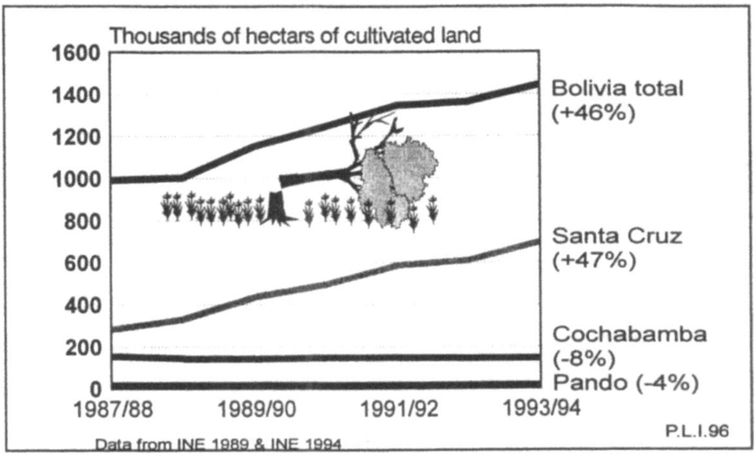

b.

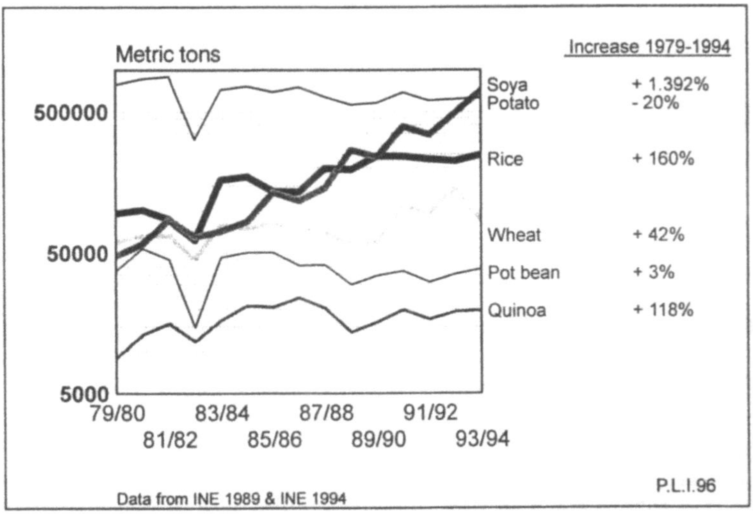

Fig. 5. Agricultural development in Bolivia. a. Growth of areas under cultivation. b. Growth of crop production

Ilustración 5. Desarrollo agrícola de Bolivia. a) Crecimiento de las áreas cultivadas. b) Crecimiento de la producción de cultivos

4 Underdevelopment and development – both present a threat to biodiversity

In Bolivia, a complex combination of underdevelopment and development problems causes biodiversity loss.

4.1 Underdevelopment is dangerous for biodiversity

Ironically, poverty and underdevelopment have contributed much to the conservation of biodiversity. In Bolivia one can find pristine areas with rain forests – and with dry forests which are especially rare throughout the world (see above; compare Parker *et al.* 1993). This is due to Bolivia's very low population density, especially in the lowlands, and to a lack of physical and technological means to access and exploit the country's natural resources rapidly. Bolivia's non-agricultural and semi-agricultural indigenous tribes, which – according to 'modern' technical definitions – are generally poor, probably practice the most sustainable utilization of natural resources. But indigenous people in a changing world are not automatically good conservationists: If they have the opportunity to sell meat to mestizo-colonizers, they increase their hunting activities (Maclean 1992, G. Vargas 1996, pers. com.) and if markets for ornamental plants or for plant drugs develop in nearly every village, some people will integrate themselves into a monetary economy.

Poverty can cause local and regional biodiversity destruction. This has been observed in various areas, such as the poorer areas in Bolivia's dry valleys and altiplano regions (especially the Southwestern Cochabamba and Northern Potosí departments). Here, for several historical reasons, the critical resource base that is necessary for sustainable management had been lost over the years, and the short-term exploitation of soil and biotic resources (firewood, for example) has led to biodiversity destruction and poverty. This triggers increased migration to Andean and lowland rain forests where uprooted highland farmers destroy valuable ecosystems which they never learned to manage. In this way, poverty causes the destruction of biodiversity, which in turn causes poverty, which in turn causes more biodiversity destruction, and so on (compare Harborth 1992). Experience shows that migration and colonization cannot solve the poverty and biodiversity problems of traditionally settled regions. Some highland *campesinos* who seasonally migrate to the lowland rain forest region of the Río Chapare said that they have got to deforest one hectare of their approximately 15 hectares of land every year (I. Condori 1996, pers. com.). For a *campesino*, 15 years of supplementary income is quite a long time... It is clear that one of the biggest threats to biodiversity is lack of development. And it is equally clear that development must be directed at all environments and all communities in order to protect the remaining intact ecosystems (Myers 1994). Of course, "no one-sided view, which takes hunger and famine as enemy number one of the tropical forest and the result of the population explosion, is a good basis for working out a strategy to counter destruction" (Burger 1992) because ...

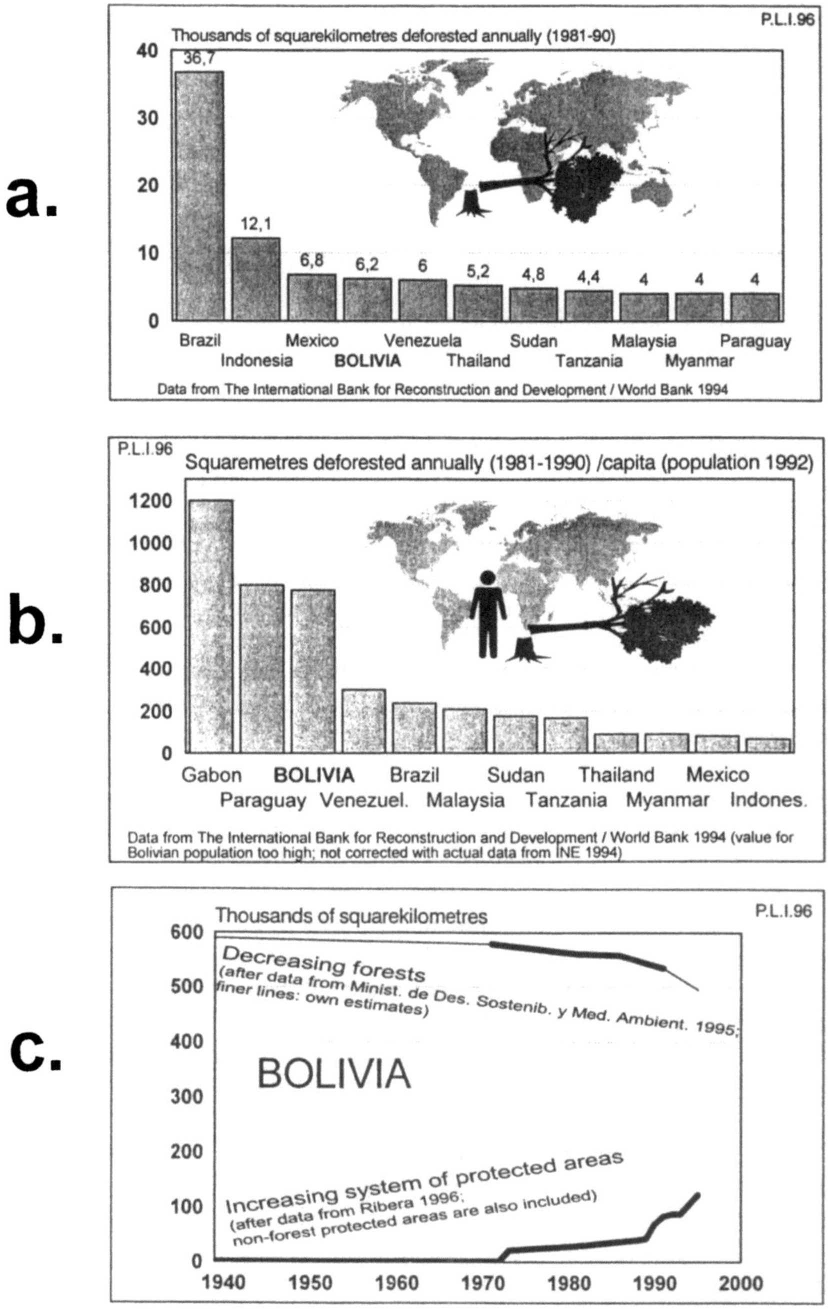

Fig. 6. Deforestation in Bolivia. a. Current world-ranking list for deforestation. b. Current world-ranking list for per-capita deforestation. c. Deforestation and growth of the national protected areas' system

Ilustración 6. Deforestación en Bolivia. a) Escalafón actual de la deforestación mundial. b) Escalafón actual de la deforestación mundial percápita. c) Deforestación y crecimiento del sistema nacional de áreas protegidas

4.2 ... development often tends to make matters worse

"Why do forests disappear? Not because there is an evil conspiracy aimed at clearing the world of tropical forests. There are two main reasons: People and nations need new land for food production and they need wood for a variety of different purposes" (Lundgren 1985), and deforestation is a part of millions of people's struggle for survival (Repetto 1990). Like all species and all human beings, today's "*Homo oeconomicus*" is fighting to guarantee and maximize the availability of resources. The problem is that he and his technology are more efficient than the poor small-scale farmer. At the same time, a country's 'development level' seems to be defined by consumption requirements for food, natural resources and energy. In Bolivia, as in many developing countries, a part of the population has already entered the industrialized world. Development – up to now and throughout the world – has had only one effect on the world's living creatures: less diversity at all hierarchical levels. Natural ecosystems tend to be transformed into agro-ecosystems or even urbano-ecosystems. Man's colonization of the earth is accompanied by the irreversible extinction of species. And the expansion of 'developed', more aggressive and powerful cultures throughout human history has always meant – without exception – a reduction of cultural and ethnic diversity. Even agricultural technologies aimed at 'improving' the utilization of natural resources have led to a severe diminishment of genetic resource diversity. It is therefore not surprising that all economic and development activities such as agro-industry, oil exploitation, timber extraction and road construction are leading to biodiversity degradation not only in Bolivia, but worldwide. Today, direct utilization and exploitation of the land pays off and conservation does not.

4.3 The dimensions and consequences of biodiversity destruction

Deforestation is the primary and most general reason for the loss of biodiversity in Bolivia. Although the dimension and rate of deforestation vary, the tendency is clear (Fig. 2.a, 6.c). Every year Bolivia loses between 1,780 km^2 (Ministerio de Desarrollo Sostenible y Medio Ambiente 1995) and 6,200 km^2 of forest (The International Bank for Reconstruction / World Bank 1994). Bolivia is currently ranked number four on the annual deforestation list (Fig. 6.a)! It even holds third place on the list for annual per-capita deforestation (Fig. 6.b). And when only the most forest-rich countries are considered, it is ranked number one! Koopowitz *et al.* (1994) analyzed data on annual deforestation, plant species numbers and endemism and concluded that about 300 endemic plant species have become extinct in Bolivia since 1950. Especially in montane rain forests, even small-scale deforestation can endanger species existence because in such areas many endemic herbs, shrubs and trees may be restricted to a very small territory. Local and regional extinction has been a very common process in the degraded areas of the Andes in the last several centuries: Populations of dozens of trees and forests species which may have very special adaptations and qualities are vanishing in a growing number of regions (such as the Province of Arque, Dept. Cochabamba; Ibisch & Rojas 1994, Ibisch 1994). Ecosystem transformation and biodiversity loss in the semiarid, subarid, and arid highlands of the Andes is already a historical process; this area is covered by anthropozoogenically-changed plant communities – substitutes for unknown climax communities (Ruthsatz 1983). But anthropogenic vegetation change has occurred even in more humid regions (e.g. compare anthropogenic montane savannas; Beck 1993).

The extraction of ornamental plants such as orchids or bromeliads along roads in montane rain forests is a danger of increasing importance. Bolivia has maintained a great portion of its natural ecosystems and therefore is not yet a hot spot of faunal extinction[7].

When identifying future problems of biodiversity degradation, it is very useful to look closely at the past (Kangas 1997) and the present. Kessler & Driesch (1994) showed that in the high Andes, 99% of the forests have been lost in historical times. Here the consequences of biodiversity loss for human development and natural potentials can be studied on a local and regional scale (Ibisch 1994, 1996b). Information regarding the consequences of biodiversity destruction tends to be more speculative when it uses larger scales. But some authors are certain that Amazon deforestation, for example, would cause severe climate change ranging from local to global level – which means that it could be irreversible (Shukla *et al.* 1990).

5 Development for biodiversity and biodiversity for development

In light of the fact that the enormous diversity of tropical countries has often been viewed as an obstacle to development, is a "true partnership between conservation and development" even possible (Gentry 1991)? Agenda 21 and the Rio Declaration turned away from sectoral environmental and development policy. Their message: Conservation should be an integral element of the development process (BMZ 1992). The Bolivian government has seemingly adopted this theory. It has elaborated an "Agenda Bolivia 21" which defines four main pillars for sustainable development of the country: (1) economic growth, (2) human development, (3) conservation and management of natural resources, and (4) good governance (see also Jímenez 1996). What about these pillars? Will they resolve the contradictions between development and biodiversity conservation?

[7] The recently published vertebrate red data book (Ergueta & Morales 1996) lists the following numbers on endangered species or species of uncertain status: more than 25 fish species (*Orestias cuvieri* from the Titicaca lake extinct; Sarmiento & Barrera 1996), 2 amphibians (Ergueta & Harvey 1996), 12 reptiles (Pacheco & Aparicio 1996), 139 birds (Rocha & Quiroga 1996), 62 mammals (1 probably extinct in wildlife: *Chinchilla brevicaudata*; Tarifa 1996).

5.1 Biodiversity conservation: Development of national policy, public awareness and popular participation

Development is indicated not only by economic growth or an ongoing improvement in living conditions. In Bolivia the development process is also reflected by the maturing of the country's biodiversity conservation policies and activities. Bolivia has had laws to protect biodiversity since the very first years of the republic – such as a law that prohibited the hunting of chinchillas in 1832 (Marconi 1991). The first protected area was established in 1939; the system of protected areas has been steadily expanded ever since (Fig. 6.c). Today, more than 10% of Bolivia's national territory is legally protected. In recent years, the management of several protected areas has been improved and the Ministry of Sustainable Development and Environment was established, marking an important political step towards an efficient conservation policy. Bolivia ratified CITES (Convention on International Trade in Endangered Species) in 1991 and the Convention on Biological Diversity in 1994. Recently, a forestry law was passed; the government has been working on a biodiversity law since 1992 (Ribera 1996).

Real political development permits and leads to broad popular participation in relevant issues. With regard to biodiversity conservation, Bolivia's civil society is increasingly interested and involved. Many "ecologists" from industrial countries that are concerned about tropical rain forests and nature in developing countries may have the impression that people in developing countries do not care about environmental problems. This is not true. For example, most Bolivian newspapers publish articles on threatened biodiversity nearly every week (and sometimes every day)[8].

There has also been an increase in non-journalistic publications on the subject, such as "El desafío ambiental en Bolivia" (The environmental challenge in Bolivia; FOBOMADE 1992), "Medio Ambiente vs. Desarrollo" (Environment versus Development; Cosio & Farfán 1994).

---

[8] The titles of some recent articles illustrate that the subject of biodiversity is quite present in public discussion: "Bolivia is a wildlife reserve", "Plants and animals are becoming extinct", "FOBOMADE claims Law for Biodiversity", "The Amboró National Park conserves a great diversity of plant species", "The ruthless extraction of the natural richness of the *oriente* is continuing", "The forests are vanishing while the state is still discussing a forest law", "Bolivia is heading for ecological disaster and nobody is doing anything to stop it", "Bolivia on the list of countries which attack their forests", "Deforestation reaches 20 thousand hectares in Cochabamba", "Deforestation could modify Boliva's climate" Original titles and newspapers: "Bolivia es una reserva de vida" (El Diario, 23.1.96), "Plantas y animales están en extinción" (Los Tiempos, 27.1.96), "FOBOMADE reclama Ley de Diversidad Biológica" (Primera Plana, 3.2.1996), "Parque Amboró conserva enorme diversidad de especies vegetales" (Los Tiempos, 5.2.96), "Continua el saqueo despiadado de riquezas naturales del Oriente" (Presencia, 20.2.96), "Los bosques se agotan mientras el país aún discute una ley forestal" (Presencia, 1.3.96), "Bolivia camina hacia el desastre ecológico y nadie hace nada por evitarlo" (Hoy, 2.3.96), "Bolivia en la lista de países que atentan contra bosques" (Opinión, 1.4.96), "Deforestación alcanza a 20 mil Has. en Cochabamba" (Opinión, 26.4.96), "Deforestación puede modificar escenario climático en Bolivia" (Los Tiempos, 26.4.96). All these articles and many more can be found in an article collection on the 'Environment and development - Amazonia and Biodiversity' which is published monthly by a local NGO ("Medio Ambiente y Desarrollo - Amazonía y Biodiversidad", CEDIB - Centro de Documentación e Información - Bolivia).

Increasingly, the participation of indigenous and rural people who often are the real victims of both biodiversity destruction and wrongly implemented conservation activities, is being sought. Looking at the history of biodiversity conservation, it originally concentrated on species, then on ecosystems and only recently started to incorporate people and their needs (Campos 1992, Glick & Wright 1992).

The political and social base to integrate biodiversity conservation into Bolivian development is preparing!

5.2 Differentiated development and integrated management of national territory

Bolivia's human population is distributed very unevenly throughout the country (Fig. 1.c). Most of the country's cities and its most densely populated rural areas are found in the Andes. Here, many traditionally settled regions are losing their carrying capacity as a result of anthropogenic degradation while large intact forest ecosystems are maintained in the eastern lowlands. For many people, Bolivia's central development problem is its "bad population distribution" ("mala distribución de la población en el territorio nacional"; Donoso de Baixeras 1992). The notion that a major population shift to the lowland forests could end poverty and underdevelopment is dangerous and illusory. The lowland ecosystems have an extremely low agricultural carrying capacity; it is well known that deforested lands in the lowlands have poor soils and tend to lose their agricultural potential after a few years' utilization.

Differentiated land-use planning which takes the biodiversity risks and the most adequate utilization potentials of different regions into account is necessary to guarantee long-term development and to maintain Bolivia's natural resources – including its biodiversity. This kind of planning can be put to use at national, regional and local level. Intensifying the utilization of land already in use is certainly more sustainable than transforming and degrading a further number of natural ecosystems. Intensifying land use does not mean industrializing land use but rather increasing its carrying capacity via an appropriate mix of traditional and innovative technologies such as agroforestry with mechanical soil conservation and micro-irrigation systems. Standardized development and biodiversity assessment procedures should be developed in order to evaluate the correct strategy for differentiated land use: Criteria should include quantitative and qualitative biodiversity, intactness, actual land use, history, human population density, poverty and potential for indirect and direct utilization (watershed management, tourism, agriculture, etc.). Biodiversity conservation is much more than the establishment of protected areas – its instruments range from conservation that is aimed at increased and better utilization of natural resources to conservation that renounces their utilization (Ibisch & Karlowski 1996). On the one hand, developing poor regions can help slow down migration and colonization and should therefore be regarded as a major contribution to biodiversity conservation (see also Friedrichsen 1995, FOBOMADE 1992). On the other hand, nature must be provided absolute protection in many regions, not only in order to conserve rare and unique species, but to maintain environmental services on which human communities depend (such as soil and climate stabilization, watershed protection).

5.3 What Bolivia can achieve through economic growth

It costs money to promote development and conserve biodiversity by recovering degraded lands and renouncing direct land use. Can Bolivia achieve sustainable development and biodiversity conservation through economic growth – one of the pillars of the government's "Agenda Bolivia 21"?

Which is more true: "The conservation of an area may sometimes be best secured by minimizing contact of the area with the international economy" (Tisdell 1994) *or* "Free trade is green, protectionism is not" (Yu 1994)? Economists like Yu (1994) say that the liberalization of markets is associated with urbanization and industrialization which reduce population growth rates and the pressures of colonization. But: "Unfortunately a number of environmental changes occurring during the earlier stages (...) are irreversible at the later stages. (...) As economies develop further and reach a mature stage, political support for protection of natural areas increases but by this time much biodiversity is usually lost" (Tisdell 1994).

Many authors think that the magic word 'sustainable development' is an illusion, or even worse, a new, consciously misleading name for the old concept of economic growth and expansion (Willers 1994, Robinson 1993, Oates 1995). Only a few industries promote truly sustainable and ecological development. One hope for the commercialization of biodiversity is the 'eco-tourism' concept (Giannecchini 1993, Arbeitsgruppe Ökotourismus 1995, Boo 1992)[9]. The exportation of green and/or fair products (such as the brazil nut, coffee, quinoa and ornamental plants) could be another. But public awareness and global markets must change considerably before it is possible to generate substantial sales volume and make an effective contribution to biodiversity conservation. And developing countries have to make it clear that the cost for obtaining genetic resources within their boundaries is higher than just the cost of travelling around and collecting them (Khoshoo 1996).

The pharmaceutical utilization of biodiversity and benefit sharing with poor biodiversity-rich countries is currently the focus of a lively discussion (see for example, Fowler 1992, Janzen 1992) which neglects the fact that we have entered the age of 'rational drug design' and that modern chemistry soon will no longer require imperfect and expensive natural drugs (see for example, Barnickel 1996, Skerra 1996). It is quite dangerous to fund biodiversity conservation arguments that are based on pharmaceutical potentials that do not exist. Certainly, a more

[9] Eco-tourism should probably change its name into 'fair tourism' to be more honest and better express its desired aims: Having tourists enjoy nature and culture while also paying for nature protection and local rural development (and not only for the growth of a few travel agencies).

"aggressive commercialization" (McNeely 1993) of biodiversity is inalienable, but we should think less in terms of consuming products and more in terms of services as analyzed in the following chapters!

5.4 Global ecological services, debts and swaps

The Bolivians have initiated a process that came to an end in Europe a long time ago: They have begun to transform their ecosystems into agroecosystems. So what? In this article it is not possible to discuss all the reasons why biodiversity should be conserved. The value and benefits of preserving biodiversity are of an economic, evolutionary, aesthetic, ethical and ecological nature (Pitelka 1993) and are still quite abstract to many people. They are however very clear to the poor *campesinos* in the degraded regions of the Bolivian Andes who have lost them, especially the critical ecosystem services like water, soil and microclimate protection. On a global scale, warnings about potential threats to food security and life – resulting from a collapse of ecosystem services such as climate stabilization, or from the genetic erosion of our food crops, for example – are also clear. One thing is for sure: The destruction of Bolivia's forests has local, regional and global-scale effects. If the world's industrial countries are increasingly concerned about the tropical environment, it is not – and should not be – for sentimental reasons. The protection of our tropical ecosystems is a factor in global security and countries like Bolivia could be "purveyors of environmental services" (Commission for Development and Environment for Amazonia 1992) for the world, purveyors of services which up to now have been considered to be free of charge. The most valued services could include the provision of genetic models and resources for technological innovations and food security, as well as for the stabilization of the world´s climate or even for recreation and pleasure. However, "When those benefits may never accrue to those who conserve, it is unreasonable to expect people to engage in conservation management" (Kaplan & Kopischke 1992). If the world wants the above mentioned services, it has got to pay for them! But how? The need for industrialized countries to participate financially in tropical biodiversity conservation has already been acknowledged. 'Debt for nature swaps' were one of the first compensation instruments to be developed and Bolivia was the very first country to test it (1987; see for example, Kloss 1994). Debt for nature swaps have played an important historical role by initiating the compensation discussion. But they are insufficient as a compensation mechanism (Kloss 1994)[10].

---

[10] Much criticism has risen: the main problem is that the concept regards biodiversity conservation as part of humanitarian help; furthermore, the sovereignty of the developing countries is affected (Kloss 1994), the effectiveness is low (Amelung 1992) and the real discharge of debts often is very limited as Bolivia had to learn in 1988 (Bolivia paid 34 millions US$ but the real debts discharge finally was 400,000 US$; Pearce et al. 1995).

5.5 What the others should pay: economic compensation for ecological services and bilateral partnerships

Developing countries are not the only ones with sizeable debts. Demands have already been made for compensation for the ecological debts of former colonialists. In Bolivia's case, the Spanish *conquista* undoubtedly caused great ecological damage (FOBOMADE 1992). The world's industrialized countries have achieved economic prosperity at least in part by externalizing their environmental costs (Brock 1994). They should accept their historical responsibility by internalizing these costs. However, the externalization mechanisms tend to diversify – even the idea of "joint implementation" of nature and climate conservation by industrialized and developing countries could turn out to be one of these mechanisms (Brock 1994). Global funds are needed. These funds must pay for biodiversity conservation where it is necessary, especially in poor megadiversity countries whose huge reserves of biodiversity are vanishing, but they must be provided primarily by economically powerful nations that have caused most of the global environmental damage. Theoretically, the Global Environment Facility (GEF) – which has been the preliminary financial instrument for the Convention on Biological Diversity since 1995 – is one such fund, but many developing countries and NGOs still do not trust it because of its connection to the World Bank (French 1995). However, there is no doubt (and many economists are aware of this) that a financial transfer from industrialized to developing countries is the only way to conserve our global biodiversity heritage (Kloss 1994). Amelung (1992) says that such compensation is adequate to internalize external costs of internationally important environmental goods. He argues that compensation must not replace traditional development aid but should rather be payment for the exportation of global environmental services (compare Lachmann 1992)[11]. Of course, representatives of industrialized countries may fear that developing countries might use biodiversity to practice extortion (see e.g., Wöhlcke 1991). Conservationists may tend to say, so what? If pressure is the only thing that works.... However, biodiversity conservation would be an ideal task for fair international co-operation. "Every nation – small or big, rich or poor, mega diverse or scantily diverse has something to offer and can gain from cooperation. (...) By co-operating with each other, the developed and developing countries could turn their advantages into a 'win-win' situation" (Chakraborty 1995). We need new models for North-South alliances between megadiversity countries and 'mega-economy countries' (such as Bolivia and Germany) which are willing to start working on the joint implementation of internationally agreed conventions.

Bolivia and its people are not necessarily forced to utilize or consume their natural resources directly. Environmental services for the 'global village' could develop into Bolivia's most important 'industry'. Using funds generated by environmental services, Bolivia's government could finance the recuperation and stabilization of degraded lands in poorer regions which would stop the growing

[11] Costa Rica is a pioneer in trying new approaches of biodiversity conservation and is already earning money by selling eco-services (Wulf 1996).

migration of 'eco-refugees' to virgin forests. These funds would also make it possible to guarantee the protected area system.

6 The challenges: biodiversity conservation, development research and development cooperation in and with Bolivia

Many people will have to change their opinions and learn: Politicians and development professionals must acknowledge the vital role that biodiversity plays for development and for mankind. Furthermore they will have to learn that biodiversity conservation is not necessarily confined to the creation and management of protected areas (GTZ 1992, *inter alia*). It is a multidisciplinary task which includes intensifying land use and rural development of already settled regions. The lists of projects that are relevant to biodiversity conservation are longer than generally assumed (e.g. by BMZ 1995, BML 1995) which means that all projects should be assessed in terms of how they could make a more active contribution toward biodiversity conservation. On the one hand, it is important to be aware of the fact that our knowledge of biodiversity is still insufficient and that we need much more research on the mechanisms, distribution and acceptable utilization of biodiversity. On the other hand, conservationists must accept that conservation biology is not the solution to some of the primary problems of biodiversity conservation which actually lie "in social, economic and political arenas" (McNeely 1992). But this means that ecologists must, among other things, learn to talk to economists and decision makers (Goodland 1992). They have to educate and recruit people from other disciplines and work effectively with governments and rural people to optimize the use of biological resources (Deshmukh 1989). As advocates of biodiversity they have to explain to the world of politics and economics that nature and species conservation is anything but an expensive and unnecessary luxury.

The lack of 'global markets' for the benefits of biodiversity and ecological services still constitutes a central problem. "As long as those global values cannot be captured by host countries, biodiversity will be a risky investment in many contexts. (...) It is essential that we secure an improved idea of what these global values are in terms of economic quantities" (Pearce & Moran 1994). Many scientists think that politics and economics are dirty business. However, the knowledge that the sum of benefits to be gained from flood control, tourism and climate preservation for one hectare of intact forest outweighs the benefits obtained by timber extraction could be more effective in convincing a government to protect biodiversity than a bunch of numbers on endemism or alpha-diversity. "Appropriate use of prices, discount rates, public-goods theory, and resource accounting could have forestalled much of the irreversible destruction of the earth's endowment of tropical forests over the past quarter century" (Gillis 1991). Cooperation between conservationists and business "should be encouraged rather than focusing on negative aspects" (Kangas 1997). Our greatest challenge is to induce all the governments and civil societies of the world to participate in biodiversity conservation as part of sustainable development[12]. The challenges are there and we urgently need partners for taking them up. Prof. Dr. Bohnet, representative of the German Federal Ministry for Economic Cooperation and Development, said in 1992, "It is important for the developing countries not to sit back and wait for solutions from outside (recipient mentality) but to take the necessary environment policy decisions themselves ...." In this context, it should be added that rather than sitting back, Bolivia is trying to develop the national structures required for sustainable development that includes biodiversity conservation. So, besides being a megadiversity country and a developing country, Bolivia should also be an ideal partner!

Acknowledgements. I especially want to thank Dr. Michael Kessler, Binzen, Dr. Jürgen Nieder, Bonn, and my wife Claudia for their critical revisions of the first drafts of this paper and Laura Siklossy, Bonn, for correcting the abstract. Thomas Borsch, Bonn, Ismael Condori, Cochabamba, Roberto Vásquez, Santa Cruz and Galia Vargas, Cochabamba, helped with valuable information. Eunice Vedia de Heins, La Paz, the former Cultural Attaché of the Bolivian Embassy, Bonn, prompted and encouraged all the work leading to the success of the symposium's 'Bolivian day' and to this paper.

References

Amelung T (1992) Kompensationszahlungen für Entwicklungsländer beim Nutzungsverzicht auf natürliche Ressourcen. In: Sautter H. (ed) Entwicklung und Umwelt. Schrift d Vereins f Socialpolitik, Gesellschaft f Wirtschafts- und Sozialwissenschaften 215: 139-162

Arbeitsgruppe Ökotourismus (1995) Ökotourismus als Instrument des Naturschutzes? Möglichkeiten zur Erhöhung der Attraktivität von Naturschutzvorhaben. Forschungsberichte des Bundesministeriums für wirtschaftliche Zusammenarbeit und Entwicklung 116, BMZ, Bonn

Barnickel G (1996) Modellieren von Molekülen. Spektr d Wissenschaft 6/96:98-106

Barthlott W, Lauer W, Placke A (1996) Global distribution of species diversity in vascular plants: towards a world map of phytodiversity. Erdkunde 50: 317-327

Barzetti V. (ed) Parques y progreso. IUCN, Interamerican Development Bank, Washington

Beck SG (1993) Bergsavannen am feuchten Ostabhang der bolivianischen Anden - anthropogene Ersatzgesellschaften? Scripta Geobot. 20: 11-20

Beck SG, Killeen TJ & García E (1993) Vegetación de Bolivia. In: Killeen TJ, Garcia E & Beck SG (eds) Guía de árboles de Bolivia.- Herbario Nacional de Bolivia, Miss. Bot. Gard., La Paz, pp 6-24

Birk G (1995) Plantas útiles en bosques y pampas chiquitanas. Un estudio etnobotánico con perspectiva de género. APCOB, Santa Cruz

[12] We must all learn that the preservation of global biodiversity – for its own sake and for the sake of future generations – is not free of charge. Yes, future generations will pay for our biodiversity destruction: by suffering and by living poorer lives. We still have the chance to pay this debt with money.

BML (1995) Tropenwaldbericht der Bundesregierung, Bonn

BMZ (1992) Umwelt und Entwicklung. Bericht der Bundesregierung über die Konferenz der Vereinten Nationen für Umwelt und Entwicklung im Juni 1992 in Rio de Janeiro. Entwicklungspolitik Materialien 84, Bonn

BMZ (1995) Tropenwalderhaltung und Entwicklungszusammenarbeit. BMZ Aktuell 051, Bonn

Bohnet M (1992) Umweltschutz in Entwicklungsländern als Aufgabe der Entwicklungszusammenarbeit. In: Sautter H (ed) Entwicklung und Umwelt. Schrift.d.Vereins f.Socialpolitik, Gesellschaft f.Wirtschafts- und Sozialwissenschaften 215: 253-274

Boo E (1992) The ecotourism boom. WHN Technical Paper Series 2, WWF

Boom B (1989) Use of plant resources by the Chácobo. Adv.Econ.Botany 7: 78-96

Brako L, Zarucchi JL (eds) (1993) Catalogue of the flowering plants and gymnosperms of Peru. Miss Bot Garden

Brock L (1994) Borniertheit auf Gegenseitigkeit. Nord-Süd-Kontroversen in der Umweltpolitik. E+Z 35(7): 160-161

Burger D (1992) Der Hunger als wahrer Feind des Tropenwaldes? Entwickl u ländl Raum 1/92: 3-6

Campbell DG (1989) The importance of floristic inventory in the tropics. In: Campbell DG & Hammond HD (eds) Floristic inventory of tropical countries. The New York Bot.Garden, New York, pp 6-29

Campos Dudley LC (1992) The Chimane conservation program in Beni, Bolivia: an effort in local participation. In: Redford, KH, Padoch C (eds) Conservation of neotropical forests. Working from traditional resource use. Columbia Univ.Press, New York, pp 228-244

Cárdenas M (1989) Manual de plantas económicas de Bolivia. Editorial Los Amigos del Libro, 2nd edition, La Paz, Cochabamba

CDF 1989, 1991 cited in Lopez 1993), 8

Chakraborty M (1995) Conserving biodiversity: a task for cooperation. In GTZ (ed) Biologische Vielfalt erhalten! Eine Aufgabe für die Entwicklungszusammenarbeit. Publikationsreihe 402/95 - 15d - Biodiv, Eschborn

Commission on Development and Environment for Amazonia (1992) Amazonia without myths. Inter-American Development Bank, UNDP, Amazon Cooperation Treaty, New York

Cortés J (1992) Uso de los recursos naturales en Bolivia: una aproximación histórica. In: Marconi M. (ed) Conservación de la diversidad biológica en Bolivia. Centro de Datos para la Conservación (CDC-Bolivia), La Paz, pp 165-181

Cosio O & Farfán JH (eds) (1994) Medio ambiente vs. desarrollo?. Estudios de caso: Bolivia, Ecuador, Perú. Programa de Asesoramiento Ambiental para la Región Andina, LIDEMA, La Paz

De la Riva I (1990) Lista preliminar de los anfibios de Bolivia con datos sobre su distribución. Estratto del Bollettino del Museo Regionale di Scienze Naturali, Torino 8: 261-319

Denevan WM (1963) Additional comments on the earthworks of Mojos in northeastern Bolivia. Am.Antiquity 28: 540-544

Denevan WM (1966) The aboriginal cultural geography of the Llanos de Mojos of Bolivia. Ibero Americana No. 48, Univ California, Berkeley

Deshmukh I (1989) On the limited role of biologists in biological conservation. Conserv Biology 3: 321

Deutscher Bundestag (Enquete-Kommission "Schutz der Erdatmosphäre des Deutschen Bundestages)(1994) Schutz der grünen Erde. Economica-Verl, Bonn

Dirksen L (1995) Zur Reptilienfauna Boliviens unter spezieller Berücksichtigung taxonomischer und zoogeographischer Aspekte. Thesis, Fac Math Nat Scienc , Univ Bonn, Bonn (unpubl)

Donoso de Baixeras S (1992) La población rural en Bolivia. In: Marconi M (ed) Conservación de la diversidad biológica en Bolivia. Centro de Datos para la Conservación (CDC-Bolivia), La Paz, pp 181-196

Ergueta P, Harvey MB (1996) Anfibios. In: Ergueta P, Morales C (eds) Libro rojo de los vertebrados de Bolivia. CDC-Bolivia, La Paz, pp 67-72

Ergueta P, Morales C (eds) (1996) Libro rojo de los vertebrados de Bolivia. CDC-Bolivia, La Paz

FOBOMADE (Foro Boliviano de la Sociedad Civil sobre Medio Ambiente y Desarrollo) (1992) El desafío ambiental en Bolivia. Documento de trabajo No. 1

Foster RC (1958) A catalogue of the ferns and flowering plants of Bolivia. Contrib Gray Herb Harvard Univ, No. CLXXXIV

Fowler C (1992) Biotechnology, patents and the Third World. In: Sandlund OT, Hindar K & Brown AHD (eds) Conservation of biodiversity for sustainable development. Scandinav.Univ.Press, Oslo. 270-279

French H (1995) Wirksame Gestaltung von Umweltabkommen. Spektr.d.Wissenschaft, Febr.95: 62-66

Friedrichsen J (1995) Grandiose Vergeudung. Akzente - Aus der Arbeit der GTZ 2/95: 29

Gentry AH (1991) Tropical forest diversity vs. development: obstacle or opportunity? Diversity (A News J Int Pl Genet Resourc Community) 7: 89-90

Giannecchini J (1993) Ecotourism: new partners, new relationships. Conservation Biology 7: 429-432

Gillis M (1991) Economics, ecology, and ethics: mending the broken circle for tropical forests. In: Bormann FH, Kellert SR (eds) Ecology, economics, ethics: the broken circle. Yale Univ, Yale, pp 155-179

Girault L (1987) Kallawaya. Curanderos itinerantes de los Andes. Impresora Quipus, La Paz

Glick D, Wright, M (1992) The wildlands and human needs program: putting rural development to work for conservation. In: Redford KH, Padoch C (eds) Conservation of neotropical forests. Working from traditional resource use. Columbia Univ Press, New York, pp 259-275

Goodland RJ (1992) Neotropical moist forests: priorities for the next two decades. In: Redford KH, Padoch C (eds) Conservation of neotropical forests. Working from traditional resource use. Columbia Univ Press, New York, pp 416-433

Groombridge B (ed) (1990) Global biodiversity. Status of the earth's living resources. A report compiled by the World Conservation Monitoring Centre. Chapman & Hall, London

GTZ (1992) Handlungsfelder der Technischen Zusammenarbeit im Naturschutz. Deutsche Ges.f.Techn.Zusammenarbeit, GTZ, Eschborn

Harborth H-J (1992) Armut und Umweltzerstörung in Entwicklungsländern. In: Sautter H (ed) Entwicklung und Umwelt. Schrift d Vereins f Socialpolitik, Gesellschaft f Wirtschafts- und Sozialwissenschaften 215: 41-72

Harmon D. (1993) Indicators of the world's cultural diversity.- Cited contribution to the IV. Congreso Mundial de Parques y Areas protegidas, Caracas, Venezuela. In: Barzetti V (ed) Parques y progreso. IUCN, Interamerican Development Bank, Washington

Hauchler I (ed) (1995) Globale Trends 1996. Stiftung Entwicklung und Frieden Bonn

Hinojosa I (1991) Plantas utilizadas por los Mosetenes de Santa Ana (Alto Beni, Depto. La Paz). Thesis, Instituto de Ecología, La Paz

Hunt D (1992) CITES Cactaceae checklist. Roy.Bot.Gard., Kew & Int Org Succul Pl Study, Kent

Ibisch PL, Karlowski U (1996) Biodiversitätserhaltung durch nachhaltige Nutzung - eine Sackgasse? Entwickl u ländl Raum 2/96: 29-30

Ibisch PL (1994) Flora y vegetación de la Provincia Arque, Departamento Cochabamba, Bolivia - III. Vegetación. Ecol en Bolivia 22: 53-92

Ibisch PL (1996a) Neotropische Epiphytendiversität - das Beispiel Bolivien. Martina-Galunder-Verlag, Wiehl

Ibisch PL (1996b) "Reparieren" von degradierten Agrar-Ökosystemen in den Anden zwischen Theorie und Praxis - Beispiel Provinz Arque, Bolivien. Krit Ökologie 14(2): 9-15

Ibisch PL, Rojas P (1994) Flora y vegetación de la Provincia Arque, Departamento Cochabamba, Bolivia - I. Flora. Ecol en Bolivia 22: 1-14 (anexo 15-41)

Ibisch PL, Vásquez R (in prep.) Diversidad de las orquídeas epifíticas de Bolivia. Revista de la Sociedad de Estudios Botánicos, Santa Cruz

INE (Instituto Nacional de Estadística) (1989) Bolivia en cifras. La Paz

INE (Instituto Nacional de Estadística) (1994) Anuario estadístico 1994. La Paz

The International Bank for Reconstruction and Development / World Bank (1994) World development report 1994. Infrastructure for development. Oxford Univ Press, Oxford

Janzen DH (1992) A south-north perspective on science in the management, use, and economic development of biodiversity. In: Sandlund OT, Hindar K, Brown AHD (eds) Conservation of biodiversity for sustainable development. Scandinav Univ Press, Oslo. pp 27-52

Jímenez JI (1996) Agenda Bolivia for the XXI century. Bolivian Times May 16: 7

Kangas P (1997) Tropical sustainable development and biodiversity. In: Reaka-Kudla ML, Wilson DE, Wilson EO (eds) Biodiversity II. Joseph Henry Press, Washington. pp 389-409

Kaplan H, Kopischke K (1992) Resource use, traditional technology, and change among native peoples of lowland South America. In: Redford KH, Padoch C (eds) Conservation of neotropical forests. Working from traditional resource use. Columbia Univ Press, New York. pp 83-107

Kessler M, Driesch, P (1994) Causas e historia de la destrucción de bosques altoandinos en Bolivia. Ecol en Bolivia 21: 1-18

Khoshoo TN (1996) Biodiversity in the developing countries. In: di Castri F, Younès T (eds) Biodiversity, science and development: towards a new partnership. CAB International, Wallingford, Oxon

Killeen TJ, García E, Beck SG (eds) (1993) Guía de árboles de Bolivia. Herbario Nacional de Bolivia, La Paz, Miss Bot Garden, St. Louis

Kloss D (1994) Umweltschutz und Schuldentausch. neue Wege der Umweltschutz-finanzierung am Beispiel lateinamerikanischer Tropenwälder. Vervuert-Verl, Frankfurt/M

Köhler J (1995) Untersuchungen zur Taxonomie, Ökologie und Zoogeographie bolivianischer Froschlurche (Amphibia: Anura). Thesis, Fac Math Nat Scienc , Univ Bonn, Bonn (unpubl)

Koopowitz H, Thornhill AD, Andersen M (1994) A general stochastic model for the prediction of biodiversity losses based on habitat conversion. Conserv Biol. 8: 425-438

Lachmann W (1992) Kompensationszahlungen für Entwicklungsländer beim Nutzungsverzicht auf natürliche Ressourcen. Korreferat zum Referat von Torsten Amelung. In:

Sautter H (ed) Entwicklung und Umwelt. Schrift d Vereins f Socialpolitik, Gesellschaft f Wirtschafts- und Sozialwissenschaften 215: 163-170

Lauer W (1986) Die Vegetationszonierung der Neotropis und ihr Wandel seit der Eiszeit. Ber Deutsch Bot Ges 99: 211-235

Libermann M (ed) (1995) Mapa étnico territorial y arqueológico de Bolivia. CIMAR, UAGRM, La Paz

López J (1993) Recursos forestales de Bolivia y su aprovechamiento. Artes Gráficas Latina, La Paz

Lundgren B (1985) Global deforestation, its causes and suggested remedies. Agroforestry Systems 3: 91-95

Maclean Stearman A (1992) Neotropical indiginous hunters and their neighbors: Sirionó, Chimane, and Yuquí hunting on the Bolivian frontier. In: Redford KH, Padoch C (eds) Conservation of neotropical forests. Working from traditional resource use. Columb Univ Press, New York, pp 108-128

Marconi M (1991) Catálogo de legislación ambiental en Bolivia. Centro de Datos para la Conservación (CDC-Bolivia), La Paz

McNeely JA (1992) The biodiversity crisis: challenges for research and mangement. In: Sandlund OT, Hindar K, Brown AHD (eds) Conservation of biodiversity for sustainable development. Scandinav Univ Press, Oslo, pp 15-26

McNeely JA (1993) Prologo. In: Barzetti V (ed) Parques y progreso. IUCN, Interamerican Development Bank, Washington

McNeely JA, Miller KR, Reid WV, Mittermeier RA, Werner TB (1990) Conserving the world's biological diversity. IUCN, Gland, WRI, CI, WWF-US, World Bank, Washington

Ministerio de Desarollo Humano (1993) Mapa de pobreza. Una guía para la acción social. UDAPSO, INE, UPP, UDAPE, La Paz

Ministerio de Desarrollo Sostenible y medio Ambiente (1994) Encuesta nacional de demografía y salud. INE, La Paz, Macro International Inc., Calverton, Maryland

Ministerio de Desarrollo Sostenible y Medio Ambiente (1995) Mapa forestal de Bolivia. Memoría explicativa. La Paz

Mittermeier RA (1988) Primate diversity and the tropical forest: Case studies from Brazil and Madagascar and the importance of of the megadiversity countries. In: Wilson EO, Peter FM (eds) Biodiversity. Nat Academ Press, Washington, pp 145-154

Mittermeier RA, Werner TB (1989) Wealth of plants and animals unites "megadiversity" countries. Tropicus 4(1), 1: 4-5

Montes de Oca I (1988) Geografía y recursos naturales de Bolivia. Edit Educacional del Minist.d.Educación y Cultura, La Paz

Moraes M, Beck SG (1992): Diversidad florística de Bolivia.- In: Marconi M. (ed) Conservación de la diversidad biológica en Bolivia. Centro de Datos para la Conservación (CDC-Bolivia), La Paz, pp 73-111

Myers N (1994) Protected areas - protected from a greater 'what'? Biodiversity and Conservation 3: 411-418

Oates JF (1995) The dangers of conservation by rural development - a case-study from the forests of Nigeria. Oryx 29: 115-122

Ochoa C (1990) The potatoes of South America: Bolivia. Cambridge Univ Press

Olivieri S, Bowles IA, Calvacanti RB, da Fonseca GAB, Mittermeier RA, Rodstrom CB (1995) A participatory approach to biodiversity conservation: the regional priority setting workshop. Conserv Internat Policy Papers, revised discussion draft

Pacheco LF, Aparicio J (1996) Reptiles. In: Ergueta P, Morales C (eds): Libro rojo de los vertebrados de Bolivia. CDC-Bolivia, La Paz, pp 73-93

Parker III TA, Gentry AH, Foster RB, Emmons LH, Remsen Jr. JV (1993) The lowland dry forests of Santa Cruz, Bolivia: a global conservation priority. Conserv International, Fund Amigos de la Naturaleza. RAP Working Papers 4

Pearce, D, Moran D (1994) The economic value of biodiversity. IUCN, Earthscan Publications Ltd., London

Pearce D, Adger N, Maddison D, Moran D (1995) Verschuldung und Raubbau in der Dritten Welt. Spektr d Wissenschaft August 1995: 32-36

Pitelka LF (1993) Biodiversity and policy decisions.- In: Schulze ED, Mooney HA (eds) Biodiversity and ecosystem function. Springer, Berlin, pp 481-493

Redford, KH, A. Taber, Simonetti, JA (1990) There is more to biodiversity than the tropical rain forests".- Conserv Biol 4: 328-330

Repetto R (1990) Incentives for sustainable forest management. In: Woodwell GM (ed) The earth in transition: patterns and processes of biotic impoverishment. Press Synd Univ Cambridge, pp 239-255

Ribera MO (1992) Regiones ecológicas.- In: Marconi, M. (ed.) Conservación de la diversidad biológica en Bolivia.- Centro de Datos para la Conservación (CDC-Bolivia), La Paz, pp 9-71

Ribera, MO (1996) Guía para la categorización de vertebrados amenazados.- Centro de Datos para la Conservación (CDC-Bolivia), La Paz

Robinson JG (1993) "Believing what you ain't so": response to Holdgate and Munro. Cons Biol 7: 941-942

Rocha, O, Quiroga C (1996) Aves. In: Ergueta P, Morales C (eds): Libro rojo de los vertebrados de Bolivia. CDC-Bolivia, La Paz, pp 95-164

Ruthsatz B (1983) Der Einfluß des Menschen auf die Vegetation semiarider bis arider tropischer Hochgebirge am Beispiel der Hochanden. Ber Deutsch Bot Ges 23: 535-576

Sarmiento J, Barrera S (1996) Peces. In: Ergueta P, Morales C (eds) Libro rojo de los vertebrados de Bolivia. CDC-Bolivia, La Paz, pp 33-65

Schlechter R (1922) Die Orchideenfloren der südamerikanischen Kordillerenstaaten. V. Bolivia. Rep spec nov reg veg (Beih.) 10: 1-80

Seibert P (1993) Vegetation und Mensch in Südamerika aus historischer Sicht. Phytocoenologia 23: 457-498

Shukla J, Nobre C, Sellers P (1990) Amazon deforestation and climate change. Science 247: 1322-1325

Skerra A (1996) Design künstlicher Proteine mit neuen Eigenschaften.- Spektr d Wissenschaft 6/96: 110-113

Smith LB, Downs RJ (1974) Pitcairnioideae (Bromeliaceae). Flora Neotropica, Monographs No. 14, Part 1. Hafner Press, N.Y.

Solbrig OT (1991) From genes to ecosystems: a research agenda for biodiversity. IUBS-SCOPE-UNESCO, Cambrige, Mass

Solomon JC (1989) Bolivia. In: Campbell DG, Hammond HD (eds) Floristic inventory of tropical communities. The New York Bot Garden/Miss Bot Garden/WWF, New York, pp 456-463

Strecker S (1993) Arealgeographie der Familie der Kakteen. Thesis, Fac Math Nat Scienc, Univ Bonn, Bonn (unpubl)

Subsecretaría de Asuntos Etnicas (1992) Sistema Nacional de Información sobre pueblos indígenas - Censo Indígena de 1992. Cited in "Bolivia: Un mar de étnias, identidades y lenguas distintas". HOY (La Paz), 28.2.1996

Tarifa T (1996) Mamíferos. In: Ergueta P, Morales C (eds): Libro rojo de los vertebrados de Bolivia. CDC-Bolivia, La Paz, pp 165-263

Tisdell C (1994) Conservation, protected areas and the global economic system: how debt, trade, exchange rates, inflation and macroeconomic policy affect biological diversity. Conserv Biol 3: 419-436

Vásquez R (1991) Lista preliminar de las orquídeas del Parque Nacional Amboró y sus alrededores.- 27. reunión anual de la Organización de Flora Neotropica del 18 al 20 de mayo 1991, Santa Cruz, Bolivia (unpubl.)

Vásquez R (1996) Listado de las orquídeas de Bolivia. Santa Cruz (unpubl.)

Vásquez R, Coimbra G (1996): Frutas silvestres comestibles de Santa Cruz. Santa Cruz

Willers B (1994) Sustainable Development: a new world deception. Conserv Biol 8: 1146-1148

Wöhlcke M (1991) Umweltorientierte Entwicklungspolitik: Schwierigkeiten, Widersprüche, Illusionen. In: Hein W (ed) Umweltorientierte Entwicklungspolitik. Dt Übersee-Inst, Hamburg, pp 109-126

Wulf, K. (1996) Abschied vom Mythos Wald. Akzente - Aus der Arbeit der gtz 4/96: 16-21

Yu D (1994) Free trade is green, protectionism is not. Conserv Biol 8: 989-996

Floristic inventory of Bolivia – An indispensable contribution to sustainable development

Stephan G. Beck
National Herbarium of Bolivia, La Paz, Bolivia

Abstract. The first botanical collections and publications dealing with Bolivia date back to the beginning of the nineteenth century. Collectors of the historical material and the institutions housing them are listed. The floristic inventory was given important impetus during the last 15 years. The three major Bolivian botanical institutions built up plant collections with foreign aid and incorporated over 120,000 specimens. A map shows the area where specimens were collected. The flora is estimated at 17,000 flowering plants, 1,500 ferns and 1,200 bryophytes; several new records of families are given. A comparison with data contained in the Flora Neotropica monographs confirms the growing number of new records and new species for the flora. The commercial uses of floristic inventories are illustrated using examples and lists of little-known species of timber trees, fruits, various crops and tubers, forage species, medicinal plants. These examples and lists provide evidence that Bolivia shares a large amount of neotropical genetic resources and proof of their strong potential for country's development. A 'Flora of Bolivia' project is proposed for the purpose of reinforcing and integrating various activities aimed at sustainable development.

Inventario de la flora de Bolivia: Una contribución indispensable al desarrollo sostenible

Resumen. Las primeras colecciones y publicaciones botánicas sobre Bolivia se remontan a los inicios del siglo XIX. Se presenta una lista parcial de los coleccionistas del material histórico y de las instituciones que los acogen. Al inventario de la flora se le dio un importante impulso durante los últimos 15 años. Las tres instituciones botánicas más relevantes de Bolivia establecieron colecciones de plantas con ayuda extranjera incorporando más de 120,000 especímenes. Un mapa muestra el área en la que se recolectaron los especímenes. La flora se estima en 17,000 plantas angiospermas, 1,200 briofitas y 1,500 helechos; se presentan varios registros nuevos de familias. Una comparación con datos contenidos en las monografías de la Flora Neotrópica confirma que se está produciendo un incremento en la cantidad de nuevos registros y nuevas especies en la flora. Los usos económicos de los inventarios de la flora se ilustran mediante ejemplos y listas de especies poco conocidas de árboles, frutas, diversos productos cultivables y tubérculos, especies forrajeras y plantas medicinales. Estos ejemplos y listas son una muestra de que Bolivia dispone de una gran cantidad de recursos genéticos neotropicales y constituyen la prueba de su enorme potencial para el desarrollo del país. Se propone llevar a cabo un proyecto sobre "La Flora de Bolivia" con el propósito de reforzar e integrar una serie de actividades que tienen como objetivo el desarrollo sostenible.

1 Introduction

Bolivia's flora is one of the richest in the world and still one of the least altered with about 20,000 species over a territory of nearly 1,100,000 km^2.

Inventories are conducted by use of systematic and taxonomic studies. Their objective is to identify all species of plants in a given geographical area (Campbell 1989) in order to obtain valuable information for of land-use planning and natural-resource management. Inventories are usually based on vouchers of collected specimens that have been pressed and dried, and are kept in herbaria. However, they can in addition be based on conserved wood samples, pollen, fern spores and living germ plasm. Botanical inventories in Bolivia have received little attention in the past. There is a lack of adequate data since the value of natural ecosystems and biodiversity conservation was only recently established by the 1992 Convention on Biological Diversity in Rio de Janeiro.

2 Brief history of botanical exploration

The first known botanical collections of Bolivian plants that are still kept in museums date from the late eighteenth century and were the work of the Bohemian naturalist Thaddaeus Haenke (1761-1816). His collection was not very large. It included several common species such as *Festuca dolichophylla*, a widespread grass from the more humid parts of the Altiplano. In the nineteenth century, botanical explorations in Bolivia were generally conducted by European naturalists who included the Frenchmen d'Orbigny (1802-1857), Weddell (1819-1877) and Mandon (1799-1866), the British naturalists Bridges (1807-1865), Pearce (1835-1868) and Pentland (1797-1873), the Germans Kuntze (1843-1907) and Meyen (1804-1870). The Danish horticulturist Bang (1853-1895 ?) collected substantial material (with little data) most of which was named by Rusby (1855-1940). The botanists Rusby and Britton from New York were representative of a new era of rather intensive floristic studies in Bolivia during the early twentieth century. They published a few thousand new species for the flora of Bolivia. Asplund (1888-1974), Buchtien (1859-1946), Fiebrig (1869-1951), Fries (1876-1966), Herzog (1880-1961), Troll (1899-1975), José Steinbach (no data) and others also contributed effectively to our knowledge of Bolivia's flora prior to World War II. The well-known Bolivian botanist Cárdenas (1899-1973), who participated with Rusby on the famous 'Mulford Biological Exploration of the Amazon Valley' (1921-1922), collected numerous specimens that were of considerable economic importance.

Even though the 'convenio' did not support pure floristic studies, floristic inventories were given new impetus by the Instituto de Ecología which was created in 1978 by the partnership between the universities of La Paz and Göttingen. The Herbario Nacional de Bolivia was also created at that time. It received its official status in 1983 through a 'convenio' that was supported by UNESCO's Organization for Flora Neotropica.

Table 1. Foreign herbaria with important collections of Bolivian plants

Herbarium	Specimens (No.)	Collectors	Region
NY Botan. Garden New York (NY)[a]	60,000	Bang, Rusby, Kuntze, Tate, Krukoff R.S.Williams, Nee, Boom, Daly	Yungas, Amazonian, Chaco, Santa Cruz
MO Botan. Garden St. Louis (MO)	40,000	Bang, Solomon, D.Smith, Killeen Gentry, Croat, D'Arcy	Yungas, Amazonian, Chiquitania
US National Herbarium Washington (US)	30,000	Bang, Buchtien, Hitchcock, King, Funk, Wasshausen	Yungas, Altiplano
Göttingen (GOET)	30,000	Haenke, Hauthal, Fiebrig, Ellenberg, Beck (?), Kessler	Altiplano, Yungas, Dep. La Paz, Beni
Mus. Nat. Hist. Nat. Paris (P)	10,000	d'Orbigny, Weddell Mandon, Fournet	Altiplano, Sorata area Yungas
Hamburg (HBG)	10,000	Buchtien, Ule, Feuerer (?)	Yungas, Altiplano, Amazonian
Royal Botanic Gardens Kew (K)	10,000	Pentland, Pearce, Bridges, Renvoize	Altiplano, Yungas
Instituto Darwinion San Isidro (SI)	10,000	Steinbach, Cabrera, Kiesling Rugolo, Zuloaga	Altiplano, Yungas Santa Cruz, dry valleys
Inst. Bot. Nordeste Corrientes (CTES)	10,000	Krapovickas, Toledo	Santa Cruz, Beni, S-Bolivia
Instituto Lillo Tucuman (LIL)	7,000	Cárdenas, T. Meyer	whole country
Rijksherbarium Leiden (L)	3,000	Herzog	Andes, Santa Cruz
Rijksherbarium Stockholm (S)	3,000	Asplund, Fries, Werdermann	S. Bolivia, lowland

Other herbaria with important collections: Buenos Aires (BAF), La Plata (LP), Castelar (BAB); Cambridge (GH/A), Chicago (F), Berkeley (UC); Berlin (B), Jena (JE), München (M), Zürich (Z)

[a] Herbaria abbreviations based on Holmgren *et al.* (1981).

In the 1980s, the Missouri Botanical Garden and the Herbario Nacional became involved in conducting floristic inventories along the humid eastern Andean slopes. The New York Botanical Garden also became involved in the Beni lowlands.

In 1989, Funk & Mori published a bibliography of nearly 100 plant collectors in Bolivia which provides an excellent survey of plant collecting in Bolivia. Collecting activities and the number of collectors have increased noticeably since then. The Herbario Nacional de Bolivia's database was started in 1989 and presently (March 1997) includes 284 collectors. However, most of the historical collectors are not mentioned due to the fact that many vouchers are not kept in Bolivia. Approximately 150 more collectors are registered in the herbaria of Cochabamba and Santa Cruz.

3 Where are the collections deposited?

Nearly all herbarium specimens from explorations as recent as 1979 are deposited in Europe, the United States of America and Argentina.

The following table shows a partial list of foreign herbaria with important Bolivian plant collections and collectors. This table includes only primary data and estimates of the number of specimens. There might be about 250,000 specimens of Bolivian vascular plants abroad. This number far exceeds Solomon's estimate (1989) of 90,000 specimens (= 8 specimens/100 km^2). Compared to other tropical countries, Bolivia is 'undercollected'. In contrast, Central America has more than 10 times as many collections as Bolivia.

The location of the collections of the Bolivian naturalists Peña Flores (1822-1901), Montalvo (1816-?) and several others who studied the country's flora is unknown. Even the important Cárdenas specimens are not in Bolivia. Cárdenas feared that a Bolivian institution would not take adequate care of his material and therefore bequeathed his personal herbarium to the Instituto Lillo in Argentina. The situation in Bolivia has changed since then. Table 2 below shows the rapid development of the country's three main herbaria on the basis of the number of specimens they house:

Table 2. Number of specimens in Bolivian herbaria

YEAR	COCHABAMBA (BOLV) Herbario Forestal Martín Cárdenas	LA PAZ (LPB) Herbario Nacional de Bolivia	SANTA CRUZ (USZ) Herbario del Oriente Boliviano
1970	[5,000][a]	---	---
1980	1,000	3,000	---
1990	10,000	40,000	8,000
1996	20,000	90,000	30,000

[a] Cárdenas' herbarium at the agriculture school of the University of San Simon (Coch), discontinued.

Small herbaria have also been established in: TARIJA, SUCRE, POTOSI, ORURO, RIBERALTA and TRINIDAD.

4 The floristic inventory of Bolivia: where do we stand?

The enclosed map (Fig. 1) shows known botanical collections in Bolivia. The rather dense net of points overrates the current inventory status because the map takes poorly collected areas also into account. Areas with more than 1,000

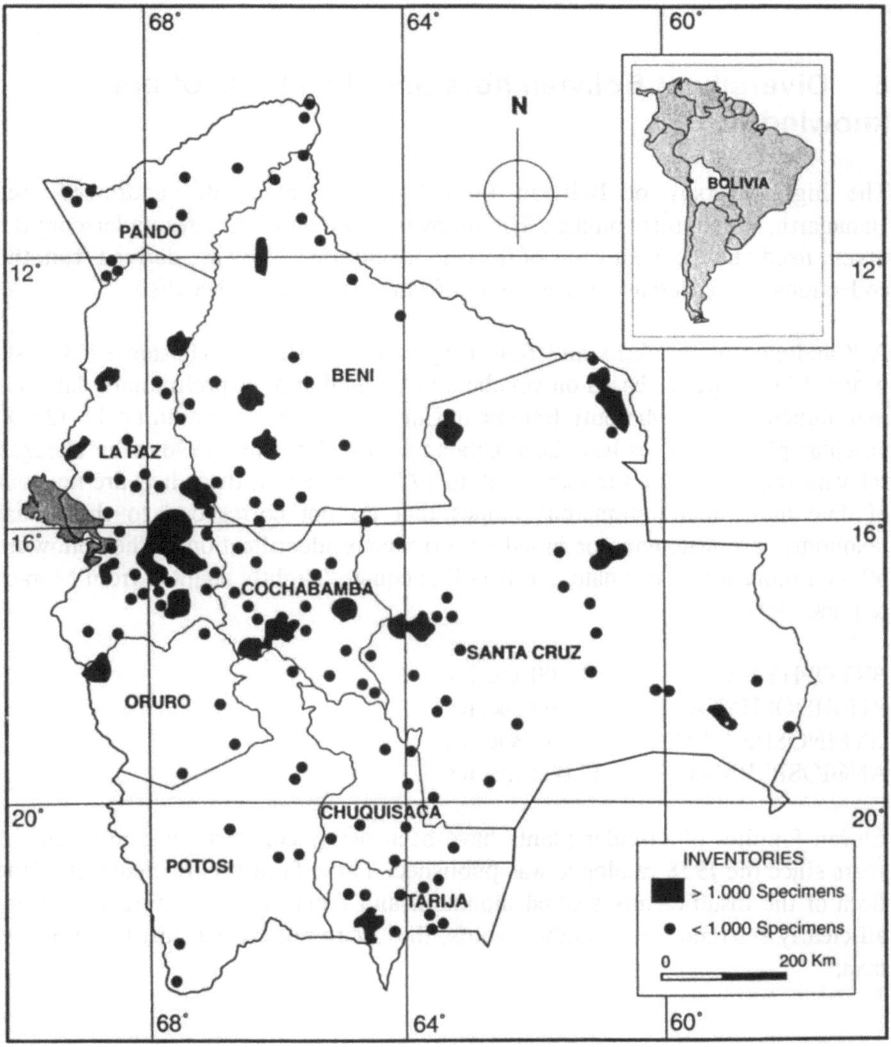

Fig. 1. Botanical collections of Bolivia
Ilustración 1. Colecciones botánicas de Bolivia

collected specimens are usually located near the towns of La Paz, Cochabamba and Santa Cruz, along main roads and at study areas such as Estación Biológica Beni and Espiritu in the Department of Beni, and the National Parks of Amboró and Noel Kempff Mercado in the Santa Cruz Department. Moraes and Beck (1992) published a list of commonly collection localities up to 1992. The Departments of Pando, Oruro, Potosi and Chuquisaca are completely undercollected, and the remaining departments have extensive areas where there has been no collection at all. Although its nearly 25,000 km^2 contains immense stretches of savanna, the province of Mamoré in the Department of Beni, for example, is not included in any grass collection because it has not been explored and because it is difficult to get there.

5 Diversity of Bolivian flora and the status of our knowledge

The high diversity of Bolivian flora has been repeatedly addressed (see Hanagarth, Ibisch, this volume). The following data and estimates underscore the urgent need and the great value of floristic inventories. They are derived from the collections of the Herbario Nacional and from various local checklists.

A 'Catalogue of the ferns and flowering plants of Bolivia' (Foster 1958) lists nearly 10,000 species based on vouchers in US herbaria. A preliminary database maintained by the Missouri Botanical Garden includes approximately 12,000 vascular plants. Mosses have been catalogued by Hermann (1976) who checked relevant literature from the years 1869 to 1974. Obviously, these lists are now out of date and contain numerous names that do not correspond to the actual taxonomy, are synonyms or based on erroneous identifications. The following offers a more recent estimate that is still adequate (slightly adapted from Moraes & Beck 1992):

BRYOPHYTA	1,200 species
PTERIDOPHYTA	1,500 species
GYMNOSPERMAE	16 species
ANGIOSPERMAE	17,000 species

Eleven families of vascular plants have been newly collected in Bolivia in the years since the 1958 catalogue was published. These families are elements of the flora of the insufficiently studied mountain and Amazonian ecosystems and are all clearly separate taxa. In other words, they were not created just by 'splitting' taxa.

List of vascular plant families recorded after 1958:

- CORSIACEAE[a]
- DIALYPETALANTHACEAE
- DICHAPETALACEAE
- GNETACEAE[a] (s.str.)
- HUMIRIACEAE
- HYDNORACEAE
- MAGNOLIACEAE
- RAPATEACEAE[a]
- RHIZOPHORACEAE
- SABIACEAE
- THYMELAEACEAE

[a] not included in Moraes & Beck (1992)

Table 3. Status of knowledge of selected plant families from 1958 to 1996 according to the monographs of Flora Neotropica (1970-1990) and current records in the Herbario Nacional de Bolivia

Plant family in Flora Neotropica Monograph	Species in 1958 (No. of 'good' species)	No. of species (year of Monogr.)	New records + new species	'Good' species in 1996
Brunelliaceae	4 (3)	3 (1970)+2 (1985)	1	6
Caryocaraceae	2 (2)	3 (1973)	7	10
Chloranthaceae	4 (2)	7 (1988)	1 (?)	8 (?)
Chrysobalanaceae	12 (4)[a]	13 (1972)+7 (1989)	10	30
Cochlospermaceae	4 (1)	4 (1981)	--	4
Connaraceae	7 (4)	6 (1983)	4 (?)	10
Flacourtiaceae	27 (9)	39 (1980)[b]	3	42 (?)
Meliaceae	33 (8)	28 (1981)	5	33
Olacaceae	8 (3)	12 (1984)	1	13
Sapotaceae	15 (7)	41 (1990)	1	42

[a] included in the Rosaceae
[b] including Lacistemataceae.

One habitat of the recently discovered Corsiaceae (*Arachnites uniflora*) in the central Andes of Bolivia was disturbed and the small population destroyed during archaeological excavations (Ibisch *et al.* 1996). Although there are certainly more *Arachnites* in the area, the small plant could easily be overlooked.

Other examples of important contributions to our knowledge of Bolivia's plant resources are the documents on the natural history of palms (see Moraes in this volume) and the monographs in the Flora Neotropica[1] series which started in the sixties. At that time, little modern floristic data was available for monographers. Table 3 above provides a survey of the number of species of selected groups in

[1] Flora Neotropica is the official publication of the Organization for Flora Neotropica. The monographs have been published irregularly since 1967 and are available from the New York Botanical Garden.

Bolivia. The taxonomic knowledge in the catalogue of ferns and flowering plants in 1958 reveals conspicuous deficits (Foster 1958): More than 50% of the names are wrong, an extreme case being Meliaceae, the mahogany family where only 8 clearly separate species out of 33 are listed. At that time, insufficient material was available on Amazonian groups such as Caryocaraceae, Chrysobalanaceae and Sapotaceae which have been collected more intensively in recent years.

Extraordinarily high species diversity is found in several groups of plants in small patches of mountain forest: Fourteen of the 15 known Bolivian species of Symplocaceae grow in the mountain forests of the Yungas of the La Paz Department which Ståhl (1994, 1995) calls the richest area of the Andes, along with southern Ecuador. Similar data is available on the genus *Freziera* of the Theaceae, where 11 species occur in a single degree-square of the cloud forest of La Paz (Weitzman 1995). These groups are usually part of undisturbed forests. Both are located in the Parque Nacional y Area de Manejo Integrado Cotapata in the Nor Yungas Province, just an hour's drive from La Paz towards Coroico.

Based on collections in *the Smithsonian Institution of the US National Herbarium*, Funk (1995) recorded 601 species of the Asteraceae, a family frequently associated with open, altered vegetation, at altitudes of more than 1,000 m in the Department of La Paz. This is the largest number recorded for similar areas in the Andes.

Bolivia is a centre of high diversity for the Asteraceae and the following groups: Amaranthaceae, Basellaceae, Bromeliaceae, Cactaceae, Erythroxylaceae, Fabaceae, Malvaceae, Myrtaceae (?), Orchidaceae, Oxalidaceae, Passifloraceae, Piperaceae, Rubiaceae, Rutaceae, Sapindaceae, Solanaceae, Tropaeolaceae, Zygophyllaceae.

Unfortunately, the growing 'development' activities revolving around colonization and land use in the last 15 years have not taken the conservation of natural ecosystems into account. At least 150,000 hectares of natural forest are destroyed in Bolivia every year. Consequently, species and ecosystem diversity is being reduced dramatically – and without ever having been thoroughly researched.

6 Commercial use of botanical inventories

Besides ideological and ethical reasons (see Ibisch, this volume), strong economic interests play a role in the conservation and sustainable use of biodiversity. Any rational conservation and utilization plan should be developed on the basis of an inventory of natural resources. An accurate plant inventory is an essential instrument for obtaining knowledge of a region's ecology and biogeography, and offers much useful data on soil fertility, hydrology, climate and the potential use of plant resources. Questions concerning the distribution and conservation status of species and ecosystems cannot be addressed without basic inventories. How are we to know if a species is rare, endemic or extinct without an inventory?

In a country with a multitude of diverse natural resources, a floristic inventory should be an obligatory part of land-use planning activities. A knowledge of vegetation types and composition enables plant ecologists and project planners to recognize the natural fertility and potential risks for land-use conversion in areas being considered for agriculture, silviculture, and construction projects for roads, bridges, industries and houses.

Converting areas with natural vegetation into agricultural land ('opening the agricultural frontier'!) no longer generates enough profit for small farmers. The financial return on products such as citrus fruits, bananas and even rice and maize is generally miserable in Bolivia. In addition, *campesinos* are converting more and more land into degraded 'chumi'[2] and 'badlands' in their struggle to survive. *Campesinos* have to grow their own food of course, but natural, sustainably managed ecosystems should be used for earning any additional income. Producers of agricultural and forest goods must seek new types of natural products that do not require the colonization of pristine areas or result in increased migration.

There is a growing consensus in Bolivia that calls for using biodiversity to earn money that can be employed for the country's development. Profit generated by accessing botanical genetic resources should benefit local people in particular.

Bolivia's broad cultural diversity includes approximately 30 different indigenous cultures with at least seven language families. Combined with the country's great species diversity, this would indicate an enormous reservoir and potential for prospecting genetic resources. A multitude of scarcely known plants are used to feed people and cattle as well as to cure physical and mental illnesses, and are employed for textiles and construction purposes. Intellectual property rights must be established in order to make it possible to compensate indigenous people for their knowledge about the uses of biological biodiversity. These rights should be based on fairness and equity. Besides the long-term benefits arising from royalties that are paid for commercialized products – royalties which very often entail a five-to-ten-year waiting period and might not ever be paid – local residents must receive immediate benefits that are commensurate with their most urgent needs. King *et al.* (1996) provide examples for a reciprocity strategy.

The following examples help illustrate the importance of floristic inventories:

6.1 Timber trees

Three species make up some 97% of all rough-sawn lumber exports! Mara or mahogany (*Swietenia macrophylla*) is virtually the most important in economic terms. A CITES[3] regulation restricts commercial use of mara wood and will

[2] Local name for secondary vegetation that grows after primary forest has been cut and the land is not cultivated.

[3] The Convention on International Trade in Endangered Species (Washington 1973) to control trade in flora and fauna as a measure to prevent extinction, which Bolivia also signed.

consequently help valorize other species of trees. The recently published guide to the tree species of Bolivia (Killeen *et al.* 1993) refers to ± 2,700 species, based on herbarium specimens and includes many potential timber tree species.

Table 4. Little known timber species of Bolivia with economic potential

Botanical name	Common name	Use
Fabaceae (Leguminosae)		
Apuleia leiocarpa (Vogel) Macbr.	Almendrillo amarillo	Hardwood, railroad ties, parquet
Cedrelinga catenaeformis (Ducke) Ducke	Mara macho	Construction
Centrolobium ochroxylum (Rose) Rudd	Tarara	Construction, agroforestry
Diplotropis purpurea (Rich.) Amshoff	Sucupira	Hardwood, construction, furniture
Dipteryx micrantha Harms	Coumarou, almendrillo	Hardwood, parquet
Enterolobium schomburgkii Benth.	Jevío	Plywood, furniture
Hymenaea oblongifolia Huber	Paquillo ?	Hardwood with resin
Lecointea amazonica Ducke		Hardwood
Macrolobium acaciaefolium (Benth.) Benth.	Arapari	Plywood, furniture
Myrocarpus frondosus Allemâo	Bálsamo ?	Hardwood
Myroxylon peruiferum L.f.	Quina quina	Construction, parquet
Schizolobium parahybum (Vell.) Blake	Cerebó	Agroforestry, plywood
Stryphnodendron purpureum Ducke	Palo yugo	Agroforestry, construction
Meliaceae		
Cabralea canjerana (Vell.) C.Mart.	Trompillo macho	Construction, furniture
Guarea gomma Pulle	Trompillo colorado	Construction
Guarea purusana C.DC.	Trompillo	
Ruagea glabra Triana & Planch.		
Trichilia inaequilatera Pennington	Pitón, sama colorada	
Trichilia quadrijuga H.B.K.		
Trichilia rubra C.DC.		
Trichilia septentrionalis C.DC.		
Podocarpaceae		
Podocarpus celatus de Laubenf.	Pino	Construction, furniture
Podocarpus ingensis de Laubenf.	Pino	Construction, furniture
Prumnopitys exigua de Laubenf.	Pino colorado, pino negro, pino de Castilla	Construction, furniture

Since there are no reliable forest inventories available due to a lack of taxonomic knowledge, quantifications on occurrence do not exist for most species.

A recently published handbook of 100 uncommon tree species of the humid Andean forests between Venezuela and Bolivia was supposed to show 20 species with a potential economic value for Bolivia. But a brief review reveals more than 50 species that grow in Bolivia. Approximately 30 of these species are unfamiliar and not usually included in reference books on timber trees of the New World (INIA/OIMT 1996).

The enclosed list (Table 4) of three timber-producing plant families shows species that had been unknown or little known for Bolivia until recently and are now of increasing importance. The accurate classification of tropical products will be a priority for international trade activities. Local names do not correspond to scientific species names, and frequently cause confusion and create conflicts. For example, the list contains three different species of 'trompillo'. Two other common species (*Guarea guidonia, G. kunthiana*) have the same name even though their physical and chemical properties do not match. Even Appendix I to the general regulations of Bolivia's new forest law which contains the cutting rights for 1997 is based only on the popular names of 69 'species' which consist of about 160 scientific species(!).

Without proper taxonomic identification and standardization it will not be possible to apply the 'green seal' or 'eco-certificate' to timber which indicates that certain environmental requirements have been met.

6.2 Fruits

Tropical countries are known for their multitude of fruits for human and animal consumption, all of which have different economic value. In Bolivia, small peasant farmers cultivate a great number of common species such as pineapple, mangos and citrus fruits which generally generate little income. The list (Table 5) of approximately 60 not very common species was developed on the basis of floristic inventories conducted in several Bolivian lowland and mountain areas and Cárdenas' important book (1969) 'Manual de plantas economicas de Bolivia'.

Many of these species are new for Bolivia. They are not mentioned in the Foster catalogue (1958). Some are new to science. The list contains mostly insufficiently studied and used species. The international fruit market is becoming interested in some of these such as the ocoró, achachairú[4] and guapurú[5] (Liebster 1995). Most of them are of some local importance and are used for making refreshing drinks and ice cream or are just eaten in the field. There is still a large number of little-known or unknown species of typical tropical plant

[4] The various species of *Rheedia* (Clusiaceae) were recently included in the pantropical genus *Garcinia*, such as the mangosteen (G.mangostana) from the Malayen Islands, and are currently being considered for export to Germany.

[5] This species of fruit has an exceptionally high concentration of vitamin C which makes it and its juice interesting to consumers.

families such as Melastomataceae and Chrysobalanaceae which could be incorporated in plant breeding programmes.

Vasquez and Coimbra (1996) recently published for the Department of Santa Cruz an account listing 130 edible fruit species accompanied by valuable information on ecology, distribution and use.

Table 5. Little known edible fruits of Bolivia

Botanical name	Common name	Distribution	Econ. value	Comments
Anacardiaceae				
Spondias mombin L.	Cedrillo	Amazonian	+	Juice; medicinal
Annonaceae				
Annona muricata L.	Guanábano, sinini	Amazonian	++	Juice, icecream; medicinal
Annona squamosa L.	Chirimoya crespa	Amazonian, savanna	+	
Rollinia herzogii	Chirimoya del monte	Amazonian	+	
Apocynaceae				
Couma sp.		Amazonian, cerrado	+	Rare
Hancornia speciosa Gomes	Mangaba	Amazonian, cerrado	++	Rare
Arecaceae (=Palmae) see Moraes, this volume				
Berberidaceae				
Berberis agapatensis Lechl.		Páramo yungueño	+	Rare, ornamental
Berberis spp.		Andean	+	Ornamental
Cactaceae				Cardenas (1969)
Cleistocactus spp.	Sitikira	Chaco, valleys	+	
Hylocereus setaceus?.	Pitahaya	Near Sta. Cruz	++	
Trichocereus spp.	Pasakana	Andean, altiplano	+	
Caprifoliaceae				
Sambucus peruviana H.B.K.	Sauco, uvilla	Andean	+	Jam, juice; medicinal
Chrysobalanaceae				
Hirtella lightioides Rusby		Amazonian	+	Endemic
Hirtella racemosa Lam.	Coloradillo	Amazonian	+	Widespread medicinal

Licania octandra (Hoffm. ex R. & S.) Kuntze		Amazonian	+	
Clusiaceae				
Rheedia macrophylla (Mart.) P. & T.	Achachairú	Amazonian	+++	Various species
Rheedia acuminata (R. & P.) P. & T.	Ocoró	Sta. Cruz	+++	Insufficiently known
Elaeocarpaceae				
Muntingia calabura L.	Uvillo	Andean, amazonian	+	Widespread
Ericaceae				
Gaultheria erecta Vent.	Machamacha	Andean	+	Widespread; various spp.
Fabaceae-Mimos.				
Acacia feddeana Harms	Palqui	Andean, dry valleys	+++	Agroforestry medicinal
Inga spp.	Pacay	Andean, amazonian	++	Agroforestry
Prosopis spp.	Algarrobo, cupsí	Andean	++	Agroforestry
Fabaceae-Papil.				
Dipteryx sp.	Almendrillo	Amazonian	+	Medicinal
Flacourtiaceae				
Casearia combaymensis Tul.		Amazonian	+	
Hippocrateaceae				
Salacia elliptica (Mart.) Don	Guapomó	Amazonian	++	Various species
Malpighiaceae				
Bunchosia sp.	Cereza	Amazonian	+	
Byrsonima sp.	Cereza	Amazonian	+++	Rare
Melastomataceae				
Bellucia beckii Renner		Amazonian, savanna	+	Endemic
Bellucia grossularioides (L.) Triana	Huicama??	Amazonian, savanna	++	
Miconia spp.		Amazonian, andean	+	
Moraceae incl.Cecropiaceae				
Brosimum alicastrum Swartz	Quechu	Amazonian	+	
Pourouma cecropiifolia Mart.	Ambaibillo	Amazonian	+	Common
Pseudolmedia laevis (R. & P.) Macbr.	Nuí	Amazonian	++	Widespread
Myrtaceae				
Campomanesia lineatifolia R. & P.		Amazonian, subandean	+	Rare
Hexachlamys boliviana Legrand	Ocoró, ocorocillo	Sta. Cruz	+	
Hexachlamys edulis (Berg) Kausel & Legr.	Ocorocillo	Sta. Cruz	+	

Myrcianthes callicoma McVaugh	Guapurucillo, sahuinto	Andean, S-Bolivia	++	
Myrciaria cauliflora Berg	Guapurú	Sta. Cruz	+++	
Psidium acutangulum DC.	Guayaba blanca		+	Rare
Passifloraceae (see Vasquez & Coimbra 1996)				
Passiflora mollisima (H.B.K.) Bailey	Tumbo	Andean	+++	Widespread
Passiflora quadrangularis L.	Granadilla real	Subandean	++	Seeds, medicinal
Rhamnaceae				
Condalia weberbaueri Perk.	Yana yana	Chaco, dry valleys	+	Rare
Ziziphus joazeiro C. Mart. (=*Z. guaranitica* Malme)	Mistol	Chaco, dry valleys	+	Common
Ziziphus mistol Griseb.	Mistol	Chaco, dry valleys	+	Common
Rosaceae				
Rubus roseus Poir.	Zarzamora	Andean	++	
Rubus spp.	Mora	Andean	+	
Rubiaceae				
Alibertia edulis (L. Rich.) A. Rich.	Tutumillo	Amazonian	+	
Borojo asorbilis (Huber) Cuatrec.	Bicillo	Amazonian	+	Rare
Genipa americana L.	Bi	Amazonian	+	Dye, medicinal, insecticide
Sapindaceae				
Melicoccus lepidopetalus Radlk.	Motoyoe	Sta.Cruz	+	Rare
Talisia esculenta Radlk.	Pitón	Amazonian	+	Rare
Sapotaceae				
Pouteria caimito (R. & P.) Radlk.	Aguaycillo	Amazonian	++	Widespread
Pouteria lúcuma (R. & P.) Kuntze		Subandean	++	
Pouteria nemorosa Baehni	Coquino	Amazonian	+++	Restricted known
Solanaceae				
Cyphomandra uniloba Rusby	Tomate de monte	Subandean	++	Endemic
Jaltomata herrerae (Morton) T. Mione	Chilto	Andean	++	
Lycianthes asarifolia (K. & B.) Bitter	Motojobobo	Amazonian	+	Jam
Physalis peruviana L.	Tomate del monte	Andean., amazonian	++	Widespread
Salpichroa glandulosa (Hook.) Miers	Pepino	Andean	+	

Solanum muricatum Ait.	Pepino	Subandean	+++	Common in Peru
Sterculiaceae				
Theobroma speciosum Willd. ex Sprengel	Chocolatillo	Amazonian	++	
Verbenaceae				
Vitex pseudolea Rusby	Tarumá, aceituna	Amazonian	+	

+ = low, only locally traded; ++ = medium, locally sometimes of high value; +++ = high, important for export

6.3 Crops – Tubers, bulbs and roots

Besides typical Andean tubers like potatoes (*Solanum tuberosum* subsp. *andigenum*), oca (*Oxalis tuberosus*), papalisa (*Ullucus tuberosus*) mashua or isaño (*Tropaeolum tuberosum*) and pseudocereals like quinoa (*Chenopodium quinoa*), cañahua (*Chenopodium pallidicaule*) and millmi (*Amaranthus caudatus*), there are some lesser known species (Table 6) such as the roots and tubers of the legumes *Pachyrhizus erosus* and *P. ahipa*, from Asteraceae such as *Polymnia* (incl. *Smallanthus*), of Brassicaceae such as *Lepidium meyenii* and other species for which other uses other than as food may be discovered in the near future.

Table 6 contains a list that includes some rare species which recently became famous as the 'Ginseng of Peru' because of their invigorating properties. Demand will hopefully fuel their cultivation and protect them from extinction.

The well known Andean crops quinoa (*Chenopodium quinoa*) and amarant (*Amaranthus caudatus* and other species) are currently cultivated in the US. There is a growing market for 'alternative' food all over the world and in Europe and the US in particular. These pseudocereals are becoming more and more important (Liebster 1995). The daily Bolivian newspaper 'Presencia' mentioned in two brief reports in its financial section on March 18 and April 13, 1997 that a North American company had patented Bolivian quinoa cultivar from Lake Titicaca in 1994. This is obviously a reference to androsteril quinoa. Bolivian and Peruvian products in particular will lose markets if they have to compete with US-based agrotechnical food companies.

Other Andean crops such as oca (*Oxalis tuberosus)* and papalisa (*Ullucus tuberosus*) are already being cultivated in New Zealand. These and more highly perishable fruits such as chirimoyas (*Annona cherimola*) are flown to the US on direct weekly flights.

6.4 Forage species

The inventory of the nearly finished grass flora of Bolivia to be published by Renvoire helped discover many new species and revealed that about 50 species of the 170 most important Poaceae for agriculture are native to or naturalized (adventive, self-propagating) in Bolivia (Whyte *et al.* 1959). Perhaps some of the most striking pasture grasses are the 'cañuelas' and 'arrocillos' of the temporarily flooded Beni savannas which have received little attention from plant breeders to date. These are various grass species of *Eriochloa, Hymenachne, Leersia, Luziola* and *Paspalum* that are important for cattle raising. Large areas in Venezuela with a similar habitat are now apparently being artificially flooded to produce these grasses all year round.

Pasture legumes are of particular interest. Numerous species are able to bind atmospheric nitrogen through endosymbiosis with bacteria in their root nodules. The cattle in the natural grasslands of the altiplano and the savannas feed on many little-known species. Areas with extreme climatic, edaphic and/or hydrological conditions which can only be used for grazing are of special interest. One example is the flooded savannas in the Beni Department which have a high diversity of the *Aeschynomene* fodder shrub species. An inventory conducted in the Yacuma savannas in the Beni revealed six different species which were previously unknown for Bolivia (Beck 1984).

Table 6. Little known "tubers" of Bolivia that are used for food

Botanical name	Common name	Distribution	Econ. value	Comments
Araceae				
Xanthosoma spp.	Gualuza	Amazonian, Andean	++	Widespread
Asteraceae				
Smallanthus sonchifolius (P. & E.) H. Rob.	Yacón, aricoma	Andean, valleys	++	Medicinal
Balanophoraceae				
Ombrophytum subterraneum (Asplund) Hansen	Ñoke	Altiplano	+	Endemic; medicinal
Brassicaceae				
Lepidium meyenii Walp.	Maca	Altiplano	++	Medicinal
Cactaceae				
Neowerdermannia vorwerckii Friç	Achacana	High andean	+	Endemic
Cannaceae				
Canna edulis Ker.	Achira	Amazonian, andean	+	Widespread
Dioscoreaceae				
Dioscorea trifida L.f.	Ñame	Amazonian, andean	+	Rare
Fabaceae				
Pachyrhizus ahipa (Wedd.) Parodi	Ajipa	Andean, valleys	+	Medicinal?

Different annual and perennial species of edible Chenopodiaceae, especially of the *Suaeda* and *Atriplex* groups that are adapted to arid areas and semi-deserts, grow in the Bolivian Altiplano. These species together with the salt-resistant grasses *Muhlenbergia ligularis* and *Distichlis humilis* provide high-quality grazing and are important for the sheep industry in the Oruro Department.

6.5 Medicinal plants

Bolivia's indigenous cultures have a profound knowledge of their plant resources and how they can be used to prevent and heal sickness. Lucca and Zalles' extensive book on Bolivian medicinal flora (1992) well reflects the multicultural background and extraordinary diversity of plants that are useful to humans. Difficulties arise because of the uncertain taxonomic status of many species due to the lack of taxonomic and floristic knowledge and gaps in herbarium collections that are the result of inadequate inventory work.

The majority of the booklets on medicinal plants refer to common species of open and altered ground, as well as to cultivated and semicultivated plants in Bolivia and other countries. Only few people recognize unfamiliar, locally occuring species. These plants usually grow in natural ecosystems and are insufficiently known to science. Some examples are included in the table of records of uncommon medicinal plants of Bolivia (Table 7).

The species used by the Chimane Indians to heal leishmaniasis and which have proved effective in the laboratory (Fournet *et al.* 1994) provide a good example. Two of them (*Pera benensis* and *Galipea longiflora*) are limited to a rather small area of the pre-Andean Beni lowland, and the third (*Ampelocera edentula*) is more widespread. A special leishmaniasis program developed by ORSTOM/IBBA[6] discovered these plants after several years of collecting and the screening of many species. Another example is the small, herbaceous Gentianaceae *Schultesia guianensis* which was botanically and pharmaceutically unknown in Bolivia until the floristic inventory of the Yacuma savannas in the Beni Department was conducted. According to studies in Brazil, it is an effective cardiotonic (Souza & Souza 1996).

[6] The Institut Français de Recherche pour le Développement en Coopération and the Instituto Boliviano de Biología de Altura in La Paz did research in the endemic areas of the cutaneous leishmaniasis.

Table 7. Records of uncommon medicinal plants of Bolivia

Botanical name	Common name	Distribution	Use	Notes
Anacardiaceae				
Loxopterygium grisebachii Hieron. & Lorentz	Soto mara	Dry montane forest	Antiparasitic, antiofidic	
Thyrsodium schomburgkii Benth.		Amazonian forest	Anti-inflammatory ??	
Toxicodendron striatum (R.& P.)Kuntze		Humid montane forest	Allergenic, poison!!	
Annonaceae				
Cymbopetalum sp.		Amazonian forest	Antirheumatic	Boom 1987
Guatteria discolor R.E. Fries	Xahuisi (Chacobo)	Amazonian forest	Antidiarrhea	Boom 1987
Apocynaceae				
Geissospermum laeve (Vell.) Miers	Quina quina ?	Amazonian forest	?	
Woytkowskia spermatochorda Woodson		Amazonian forest	Anti-inflammatory	Boom 1987
Bignoniaceae				
Arrabidaea chica (H. & B.) Verlot		Amazonian forest	Cicatrisant, desinfecting	Estrella 1995
Tabebuia impetiginosa (Mart.) Standley	Tajibo	Savanna, forest	Antitumor	
Caryocaraceae				
Caryocar brasiliense Camb.		Savanna	Antitumor	Souza & Souza 1996
Euphorbiaceae				
Croton lechleri Müll. Arg.	Sangre de drago	Amazonian forest	Antiviral, antiseptic	Ubillas *et al.* 1994
Fabaceae				
Entada polyphylla Benth.	Ñaña ?	Amazonian forest	Anti-inflammatory	Estrella 1995
Gentianaceae				
Schultesia guianensis (Aubl.) Malme		Savanna	Cardiotonic	Souza & Souza 1996
Lauraceae				
Aniba canelilla H. B. K.	Canelón	Amazonian forest	Antidiarrhea, antipyretic	Fournet *et al.* 1994
Menispermaceae				
Abuta pahni Mart. (Mart.) Krukoff & Barneby		Amazonian forest	Antiprotozoal	Fournet *et al.* 1994

Myristicaceae				
Virola surinamensis (Rol.) Warb.	Sangre de toro	Amazonian forest	Antitumor	Souza & Souza 1996
Phytolaccaceae				
Gallesia integrifolia (Spreng.) Harms	Ajo ajo	Humid lowland, montane forest	Antidiarrhea, antypiretic, antibronchitis	
Piperaceae				
Piper callosum R. & P.	Matico	Amazonian forest	Antipyretic	Boom 1987
Rubiaceae				
Uncaria tomentosa (Willd. ex R. & S.) DC.	Uña de gato	Amazonian forest	Antiinflammatory	Estrella 1995
Rutaceae				
Galipea longiflora Krause	Evanta	Amazonian forest	Leishmaniacid	Fournet *et al.* 1994
Solanaceae				
Brunfelsia grandiflora Don ssp. *schultesii*		Amazonian forest	Antithreumatic, leishmaniacid	Mejia & Rengifo 1995
Physalis angulata L.	Embolsado, chilto	Ruderal	Diabetes, hepatitis, diarrhea	Mejia & Rengifo 1995
Ulmaceae				
Ampelocera edentula Kuhlm.	Sou'sou	Amazonian forest	Leishmaniacid	Fournet *et al.* 1994
Zingiberaceae				
Renealmia thyrsoidea (R. & P.) P. & E.	Achira del monte	Amazonian, montane forest	Antipiretic	

7 Plant genetic resources

Bolivia is an important country in terms of the diversity of its genetic resources of not only wild but cultivated species as well.

Several of the worldwide most valuable cultivated species and their relatives originated in Central and South America, which includes Bolivia:

- Potatoes (genus *Solanum*): According to different taxonomic concepts (Hawkes & Hjerting 1989, Ochoa 1990, Rea & Vacher 1992) there are approximately 115-150 wild and cultivated tuber-bearing botanical taxa in Bolivia
- Peanuts (genus *Arachis*): 7 of the 69 known species are native to Bolivia (Krapovickas & Gregory 1994)
- Manioc or yuca (genus *Manihot*): eastern Bolivia is perhaps an important area of origin (Nassar 1978)
- Common bean (genus *Phaseolus*): 25 acquisitions from Bolivia went to the core collections of CIAT. The Centro Internacional de Agricultura Tropical in Cali, Colombia, holds the largest collection of all food legumes in the world

and recently worked on a limited number of core collections to conserve and use crop germ plasm more effectively (Tohme *et al.* 1995)
- Chilies (genus *Capsicum*): Eshbaugh *et al.* (1984) quoted central Bolivia as the place of origin

In addition to the already mentioned timber trees, fruits, pseudocereals, 'tubers' and forages, the following groups constitute important germ plasm in Bolivia: Pineapples (*Ananas, Pseudananas, Bromelia* species), Chirimoyas (*Annona* and *Rollinia* species), papayas (*Carica* spp.), avocados (*Persea* spp.), lucumas (*Pouteria* and *Manilkara* spp.), guavas (*Psidium guajava, P.guineense*), Brazil nuts (*Bertholletia excelsa*), cacao trees (*Theobroma* spp.), tobacco (*Nicotiana* spp.), rubber trees (*Hevea* spp.) and many more.

There are also poorly studied species within the Sapotaceae, Clusiaceae, Apocynaceae and Moraceae with potential value as sources of rubber and chicle, and plants that are used for oil, aromatic substances and as ornamentals.

Only a few groups such as potatoes and peanuts had drawn major germ plasm collectors' attention to Bolivia in the past. Apparently collectors had been concentrating their explorations on neighbouring countries. The reason for this might be the negligible number of commercial plants collected during inventories, which might also explain the lack of data in reference books. Who ever mentions rubber trees, Brazil nuts or cacao plantations after a trip to Bolivia? Some information on the use of natural resources is given in historical descriptions, such as in Eder's account (~1772) of the Mojos culture in the Beni lowlands.

Genetic resources became important in the political discussion after the development and conservation summit held in Rio de Janeiro in 1992. Last year (1996), a 'Global Plan of Action' was adopted at the 4th International Technical Conference on Plant Genetic Resources to ensure the conservation and better use of plant genetic resources for food and agriculture purposes. The 150 delegates attending the conference from around the world (including Bolivia) were concerned with strengthening the conservation of wild plants which are important as a source of food, with on-farm 'in situ' management, and with cooperating with developing countries in particular for the purpose of building up national capabilities. How can this be accomplished without an accurate knowledge of plant resources, their distribution or their taxonomy?

8 Problems conducting in getting an inventory

Work that deals with natural resources such as flora should be conducted by experienced botanists. They know how to collect and preserve plant specimens correctly and should be able to recognize major taxa. In most tropical ecosystems and in tropical rain forests in particular, the great diversity of forms, colours and textures of plant organism makes this a rather complicated task. Few botanists are able to identify plants in their natural habitat. As a rule, visitors, like botanists, will see a green mixture of trunks, stems, leaves – and only very seldomly notice a flower or fruit. What does one do with this green mass: How does one inventory it? What part belongs to which plant? A trained botanist knows that he should collect 'fertile' specimens, otherwise correct identification will be virtually impossible in a region with an unknown flora. The collection and preparation of herbarium specimens of large trees, epiphytes of the canopy, palms and other large-leaved species, spiny or succulent plants is difficult and very time consuming. Furthermore, bulky herbarium specimens are not easy to handle. As a result, these specimens are often inadequately represented in general collections.

An area's native inhabitants generally know its biota quite well because they depend on it. They attribute special properties to different species or groups of plants, and name them according to their traditions. For examples see Boom (1987) for the Chacobos, and Vargas (1996) for the Mosetenes.

These local people certainly make ideal partners for a botanical inventory – but what should be done in areas where there are no native people? Unfortunately, there are extensive regions where the original population has vanished and 'new' settlers have taken over the land, as happened in the lower slopes of the Yungas which are now managed mostly by Quechua highland Indians and mestizos. As a rule, their knowledge of local plants is rather limited.

9 Conclusion – Towards a Flora of Bolivia

There is no doubt that a comprehensive floristic inventory would be of tremendous importance for the development and conservation of natural resources in Bolivia. The country needs a basis for the qualification and quantification of its vegetative resources. The multitude of examples evidencing the importance of a floristic inventory should induce the political, financial and scientific authorities to initiate an integrated 'Flora of Bolivia' project.

Rather than being conducted like most classical floras of the tropics – as a purely academic activity by a selected group of scientists – this project must be directed toward applicable objectives involving experienced professionals, political and financial decision-makers, and the local inhabitants of the respective areas. A computerized database should be established to make this flora very user-friendly and easy to access.

Objectives of the Flora project:

- Establish a reference collection of species from different areas
- Document classified species
- Develop a database with information on the:
    - taxonomy of species
    - biogeography and distribution of species
    - ecology – habitat conditions, optimal cultivation
    - ethnobotanical knowledge

- actual and potential commercial use – importance for people, valuation
- conservation status – extinct, endangered, etc.
- special aspects – indicator species, bioprospecting...

The flora project should also have a strong educational function which includes research in selected areas and the joint development of specific small projects such as textbooks or guides for natural resources, particularly on subjects such as common trees of ..., endangered plants of ..., useful herbs of ..., edible fruits and nuts of ..., medicinal plants of ..., forage plants of ..., timber trees of ..., etc.

Raising the local population's awareness of their resources should lead to greater respect for nature, which could also be expressed as 'What I know I do not destroy'.

In order to make this a real national movement, it will be important to involve as many people as possible.

A flora project also poses a unique academic challenge: Bolivia is the only neotropical country without a written flora. This offers Bolivia and the international community the opportunity to create a model for a new type of flora.

What fundamental conditions must exist before the 'Proyecto Flora de Bolivia' can start?

A political decision to promote the flora project is an indispensable prerequisite and requires the support of the Secretary of Biodiversity in the Ministry of Sustainable Development and Environment. National and international organizations should provide professional support in managing the project, and clearly delineate their functions and responsibilities. Financial commitments and appropriations will be sought from funding bodies worldwide. The information provided by approximately 150,000 national collections and 250,000 collections abroad suffices to start the project with selected groups of plants.

Botanical basis:

There are three major Bolivian herbaria. In addition, every Department has some basic botanical support organization and/or interested people from 'alcaldias', 'municipios', 'prefecturas', NGOs, universities, etc. Some 20 trained Bolivian botanists and a growing community of students constitute up a national task force. The 'People and Plants' project (WWF/UNESCO/KEW) which reported excellent experience with local people in the Beni Biosphere Reserve can serve as a model. Local people can be trained to be parataxonomist, as Costa Rica's InBio does in its projects. Obviously, the project must involve foresters, agronomists, plant breeders and amateurs. Interested people should find their 'niche' in the project.

The international botanical community is becoming more and more interested in the flora of Bolivia. The Organization for Flora Neotropica held its annual meeting in Bolivia in 1984 and 1991, and conducted local excursions for participants. Some monographs were donated on that occasion to the Herbario Nacional de Bolivia in La Paz and the Herbario del Oriente in Santa Cruz. Most of the monographers would be delighted to work with Bolivians on updating their neotropical accounts for a flora. The 'Red Latinoamericana de Botánica' (latin american botanical network) supported some scientific courses in Bolivia to train young botanists and funded a few scholarships for studies abroad.

In the 'New World', several conservation organizations and the herbaria of the Field Museum in Chicago, the Missouri Botanical Garden in St. Louis, the New York Botanical Garden, the Selby Botanical Garden in Sarasota, Florida, and the US National Herbarium in Washington are interested in collaborating with Bolivians on floristic studies. Unfortunately, 'Latinos' provide only little support for floristic work in Bolivia, with the exception of traditional explorations conducted by Argentinean botanists. Several 'Old World' botanists and institutions also have a long-term interest in Bolivian flora and vegetation. The partnership between the University of Göttingen and the University of La Paz has concentrated on training activities in basic biology, ecology and management. The Instituto de Ecología's recently launched postgraduate program in ecology and conservation places its emphasis on training (usually) non-biologists to prepare them to administer and manage government and NGO agencies. A new 'convenio' with Göttingen will help implement the Convention of Biodiversity. It aims at developing research collaboration with several research institutes, with special focus on floristic work. The following institutions and their scientists have a long history of commitment to the botanical exploration and conservation of Bolivia's ecosystems: The British Royal Botanic Gardens, Kew – a grass flora of Bolivia; the French ORSTOM (Institut français de recherche scientifique pour le dévéloppment en coopération) – medicinal plants; the Spanish Department of Plant Biology of the University Complutense, Madrid - vegetation studies; the Danish Aarhus Botanisk Institute – training botanists; the Swiss Geographisches Institut, Zurich – palynological studies; and the German Botanisches Institut, Bonn – ecological land-use studies.

Acknowledgements. I wish first and foremost to express my profound gratitude to and admiration of the people I met in the forests and grasslands who contributed to the idea of a floristic inventory, and to whom I wish to give back some of the information I obtained during a 'Proyecto Flora de Bolivia'.

I am grateful for the valuable comments and discussions with my colleagues in La Paz, Monica Moraes, Emilia Garcia, Rossy de Michel, Esther Valenzuela, Lourdes Vargas and their students. I am also grateful to the directors of the Cochabamba and Santa Cruz herbaria Susana Arrázola and to Teresa Centurion and Mario Saldias respectively. I wish to thank my colleagues at the Organization Flora Neotropica and the Missouri Botanical Garden staff formerly stationed at La Paz, Jim Solomon, the late David Smith and Tim Killeen for productive discussions on Bolivian flora. And last but not least, I am grateful to my wife Carola for reading the text and making valuable corrections.

References

Beck SG (1984) Comunidades vegetales de las sabanas inundadizas en el NE de Bolivia. Phytocoenologia 12 (2/3): 321-350

Boom BM (1987) Ethnobotany of the Chácobo Indians, Beni, Bolivia. Advances in Economic Botany 4: 1-68

Campbell, DG (1989) The importance of floristic inventory in the tropics. In: Campbell DG, Hammond HD (eds) Floristic inventory of tropical countries. New York Botanical Garden, pp 5-29

Cardenas M (1969) Manual de plantas economicas de Bolivia. Icthus, Cochabamba

Eder FJ (~1772) Breve descripciones de las reducciones de Mojos. Traducción y edición Josep M Barnadas 1985. Historia Boliviana, Cochabamba

Eshbaugh WH, Guttman SI, McLeod MJ (1984) The origin and evolution of domesticated Capsicum species. Journal of Ethnobiology 3(1): 49-54

Estrella E (1995) Plantas medicinales amazonicas: Realidad y perspectivas. Tratado de Cooperación Amazonica, Lima

Foster RC (1958) Catalogue of the ferns and flowering plants of Bolivia. Contributions of the Gray Herbarium of Harvard University 184: 1-223

Fournet A Barrios AA, Muñoz V (1994) Leishmanicidal and trypanocidal activities of Bolivian medicinal plants. J. Ethnopharmacol. 41: 19-37

Funk VA, Mori SA (1989) A bibliography of plant collectores in Bolivia. Smithsonian Contributions to Botany 70: 1-20

Funk VA, Robinson H, McKee GS, Pruski JF (1995) Neotropical montane Compositae with an emphasis on the Andes. In: Churchill SP, Balslev H, Forero E, Luteyn JL (eds) Biodiversity and conservation of neotropical montane forest. Proc. of the Symposium of June 1993., New York Botanical Garden, pp 451-472

Hawkes J, Hjerting P (1989) The potatoes of Bolivia. Clarendon, Oxford

Hermann FJ (1976) Recopilación de los musgos de Bolivia. Bryologist 79: 125-171

Holmgren PK, Keuken W, Schofield EK (1981) Index Herbariorum, part 1: The herbaria of the world, ed 7. Junk, The Hague, Boston

Ibisch PL, Neinhuis C, Rojas P (1996) On the biology, biogeography, and taxonomy of Arachnites Phil. nom.cons. (Corsiaceae) in respect to a new record in Bolivia. Willdenowia 26: 321-332

INIA/OIMT (1996) Manual de identificación de especies forestales de la subregión andina. INIA, Lima - Perú , (supported by the International Trade and Timber Organization, ITTO=OIMT)

Killeen TJ, García E, Beck SG.(eds) (1993) Guía de árboles de Bolivia. Herbario Nacional de Bolivia, Missouri Botanical Garden, La Paz

King SR, Carlson TJ, Moran K (1996) Biological diversity, indigenous knowledge, drug discovery, and intellectual property rights. In: Brush S, Stabinsky D (eds) Valuing local knowledge: Indigenous people and intellectual property rights. Island Press, pp 167-185

Krapovickas A, Gregory WC (1994) Taxonomía del género Arachis (Leguminosae). Bonplandia 7: 1-186

Liebster G (1995) Warenkunde. Obst und Gemüse. Bd. 1 Obst, Bd.2 Gemüse, Morion, Düsseldorf

Lucca de MD, Zalles AJ (1992) Flora medicinal boliviana. Los Amigos del Libro, La Paz, Cochabamba

Mejia K, Rengifo E (1995) Plantas medicinales de uso popular en la amazonía peruana. 249 pag. AECI-GRL -IIAP (Iquitos), Lima

Moraes M, Beck SG (1992) Diversidad florística de Bolivia. In: Marconi M (ed) Conservación de la diversidad biológica en Bolivia. Centro de Datos de Conservación - Bolivia, La Paz, pp 73-111

Nassar NMA (1978) Conservation of genetic resources of cassava (Manihot esculenta): Determination of wild species location with emphasis on probable origin. Economic Botany 32: 311-320

Ochoa CM (1990) The potatoes of South America: Bolivia. Cambridge University Press, Cambridge

Rea J, Vacher JJ (eds) (1992) La papa amarga. 1. Mesa redonda: Peru-Bolivia, La Paz, 7 y 8 de Mayo 1991. ORSTOM, La Paz

Solomon, JC (1989) Bolivia. In: Campbell DG, Hammond HD (eds) Floristic inventory of tropical countries. New York Botanical Garden, pp 456-463

Souza Brito ARM, Souza Brito AA (1996) Medicinal plant research in Brazil: Data from regional and national meetings. In: Balick MJ, Elisabetsky E, Laird SA (eds) Medicinal ressources of the tropical forest. Biodiversity and its importance to human health. Columbia University Press, New York, pp 386-401

Ståhl B (1994) The genus Symplocos (Symplocaceae) in Bolivia. Candollea 49: 369-388

Ståhl B (1995) Diversity and distribution of Andean Symplocaceae. In: Churchill SP *et al.* (eds) Biodiversity and conservation of neotropical montane forest. NY Botanical Garden, pp 397-405

Tohme J, Jones P, Beebe S, Iwanaga M (1995) The combined use of agroecological and characterisation data to establish the CIAT Phaseolus vulgaris core collection. In: Hodgkin T *et al.* (eds) Core collections of plant genetic resources. International Plant Genetic Resources Institute. Wiley & Sons, Chichester, UK, pp 95-107

Ubillas R et al. (1994) SP-303, an antiviral oligomeric proanthocyanidin from the latex of Croton lechleri (sangre de drago). Phytomedicine 1(2): 77-106

Vargas Ramirez VL (1996) Etnobotánica de las plantas medicinales de los Mosetenes que viven en la comunidad de Muchanes. Tesis Carrera de Biología, Universidad Mayor de San Andrés, La Paz

Vasquez ChR, Coimbra G (1996) Frutas silvestres comestibles de Santa Cruz. Gobierno Municipl, Santa Cruz, Bolivia

Weitzman AL (1995) Diversity of Theaceae and Bonnetiaceae in the montane neotropics. In: Churchill et al. (eds) Biodiversity and conservation of neotropical montane forest. NY Botanical Garden, pp 365-376

Whyte RO, Moir TRG, Cooper JP (1959) Las gramineas en la agricultura. Estudios agropecuarios N° 42. FAO, Roma

Richness and utilization of palms in Bolivia – some essential criteria for their management

Mónica Moraes R.
National Herbarium of Bolivia, La Paz, Bolivia

Abstract. Information about diversity, distribution and utilization of native palms in Bolivia is provided along with basic data on production of and human impact on four palm species: *Attalea phalerata* (motacú), *Euterpe precatoria* (asaí), *Geonoma deversa* (jatata) and *Parajubaea sunkha* (sunkha). Relevant management criteria are presented in the Bolivia context.

Riqueza y utilización de las palmas en Bolivia: Algunos criterios esenciales para su manejo

Resumen. Se presentan los datos de diversidad, distribución y uso de las palmas de Bolivia. Se considera la información básica de producción e impactos humanos entre los criterios relevantes de manejo en base a cuatro especies como estudio de caso: *Attalea phalerata* (motacú), *Euterpe precatoria* (asaí), *Geonoma deversa* (jatata) y *Parajubaea sunkha* (sunkha).

1 Introduction

There are about 200 genera and 1,500 palm species in the world; approximately 20% of this total species diversity is present in the Neotropics. Bolivia is home to 40% of the genera and 15% of the species present within the neotropical region (Moraes 1996a, see Table 1). A preliminary approach to the biogeography of native palms shows the influence of four main phytogeographic provinces: Amazonia, Andes, Gran Chaco, and Cerrado (Moraes 1996a). Four species are endemic: The *Parajubaea sunkha* in the inter-Andean valleys of southwestern Santa Cruz, the *Parajubaea torallyi* in the inter-Andean valleys of northeastern and central Chuquisaca, the *Syagrus cardenasii* in the alluvial plain and subandean piedmont of Santa Cruz and Chuquisaca, and the *Syagrus yungasensis* in the dry inter-Andean valleys of La Paz.

The distribution of palms covers a wide range of altitudinal strata (140-3,400 m) and vegetation types; and this plant family occupies about 70% of Bolivia's total territory (Moraes 1992, 1996a). In the humid forests of the Andean foothills and lowlands, palms meet distinct soils. The most species-rich altitudinal range is between 140 and 500 m with 22% of the genera and 66% of the species

(Moraes 1996a). The montane humid forests of the Andes (> 500 m) have a diversified palm flora adapted to steep slopes, while in inter-Andean valleys, the *Parajubaea torallyi* sets an altitudinal record by growing at elevations of up to 3,400 m where precipitation is less than 500 mm per year (Moraes 1996c). The savanna lowlands have a distinct component of palm species. The open areas of the Gran Chaco are represented by two palmate-leafed species – *Trithrinax campestris* and *Copernicia alba* (see Table 2). The former is highly tolerant to less than 300 mm per year. Amazonia and the Cerrado have wetland formations with water levels of up to 1-3 m; palm stands are frequent and represented by pinnate-leafed species such as *Astrocaryum huaimi*, *Mauritiella martiana*, *Acrocomia aculeata*, and *Copernicia alba*.

Table 1. Comparative numbers of Bolivian palms in the Neotropics

Taxa	Neotropical palms	Bolivian palms	Percentage
Subfamilies	5	5	100
Genera	67	27	40
Species	550	84	15

Many palms are conspicuous components of tropical ecosystems and some are considered to be indicators for typifying certain habitats or formations (Moraes 1996a). Palms are characterized by a high diversity of life forms, sizes and life strategies, particularly where competition for light and nutrients supports a diverse palm flora. Most species are arborescent and occupy the forest canopy. Shrub or acaulescent species are found in the understory level, while scandent species predominate in disturbed areas and along ecotones between rivers and forests.

2 Traditional use of palms

Information on species richness or biogeographic distribution does not reveal the entire dynamic scope of palms, especially with regard to the productivity and stability of palm communities in time and space. There is a historical relationship with human populations that is reflected in the present distribution of this plant group. After grasses and legumes, palms have one of the highest utility values for human civilization (Moore 1973). There is a long historic interaction between ethnic groups and palms; several species of palms have been continuously cultivated by indigenous people.

This plant family is a frequent indicator of archaeological sites in the Amazon; the boundaries of palm distribution coincide with migratory routes of some indigenous tribes (Balée 1989).

Palms are used for a variety of purposes, ranging from fibres to food (see Table 3). Different parts of palms are used for the construction of houses; stems are used for support or are split for walls, and leaves are widely used as thatch. Many species produce edible fruits that are rich in fats and gathered for oil extraction. A few palms have sweet meristems which are exploited as palmhearts for human consumption. Fibres are extracted from leaves, fruits and spathes for a number of uses which also include medicinal, artesanal, ceremonial, and forage uses.

About 90% of the species offer multiple resources and only a few species are exploited for a single product (Moraes 1992, see Table 3).

Table 2. Characteristic palm species of selected ecosystems in Bolivia

Ecosystems of Bolivia	Palm species
ANDES	
Rain forests in Andean mountains	*Ceroxylon parvifrons, C. parvum, C. vogelianum, Dictyocaryum lamarckianum, Geonoma* spp., *Iriartea deltoidea, Oenocarpus bataua*
Dry forests in Andean mountains	*Aiphanes aculeata, Syagrus yungasensis*
Evergreen forests in Andean foothills	*Astrocaryum murumuru, Attalea phalerata, Euterpe precatoria, Socratea exorrhiza*
AMAZONIA	
Seasonally flooded savannas in alluvial plain	*Acrocomia aculeata, Attalea phalerata, Copernicia alba, Mauritiella martiana*
Rain forests in the Amazonian plain	*Attalea butyracea, Astrocaryum* spp., *Chelyocarpus chuco, Euterpe precatoria, Oenocarpus* spp., *Phytelephas macrocarpa, Socratea exorrhiza*
Riparian lowland forests	*Astrocaryum jauari, Bactris riparia, Syagrus sancona*
Well-drained sandy soils in alluvial forests	*Chamaedorea* spp., *Geonoma deversa, Hyospathe elegans*
CERRADO	
Well-drained rocky soils in low hills, savanna-like	*Allagoptera leucocalyx, Astrocaryum campestre, Syagrus* spp.
Semideciduous forests	*Attalea speciosa*
Poorly drained swamps with black water	*Mauritia flexuosa*
CHACO	
Dry thorny forests from the SE	*Copernicia alba, Trithrinax campestris*
Inter-Andean valleys with dry forests	*Parajubaea sunkha, P. torallyi*

Table 3. Main uses' categories of Bolivian palms

	Construction		Food			Medicinal	Oil	Artesanal
	Roofs	Walls	Fruits	Palmito	Beverages			
Acrocomia aculeata	-	+	+	-	+	+	(+)	-
Allagoptera leucocalyx	-	-	+	-	-	-	-	-
Astrocaryum murumuru	-	-	(+)	-	-	-	-	+
Attalea phalerata	+	+	+	+	-	+	+	+
Attalea speciosa	+	+	+	+	-	+	+	+
Bactris gasipaes	-	-	+	+	+	-	-	-
Bactris major	-	-	+	-	-	-	-	-
Ceroxylon parvifrons	(+)	-	-	-	-	-	-	+
Chamaedorea angustisecta	-	-	-	-	-	+	-	-
Chelyocarpus chuco	(+)	-	+	-	-	-	-	+
Copernicia alba	-	+	-	-	-	+	-	-
Dictyocaryum lamarckianum	-	-	-	+	-	+	-	-
Euterpe precatoria	(+)	-	(+)	+	(-)	+	-	+
Geonoma deversa	+	-	-	-	-	-	-	-
Mauritia flexuosa	-	-	-	-	+	-	-	-
Oenocarpus bataua	(+)	(+)	+	-	+	+	+	-
Parajubaea sunkha	-	-	+	+	-	-	-	+
Parajubaea torallyi	-	-	+	+	+	-	-	-
Phytelephas macrocarpa	+	-	+	-	-	-	-	-
Socratea exorrhiza	-	+	-	-	-	+	-	+
Syagrus cardenasii	-	-	+	-	-	-	-	-
Trithrinax campestris	-	-	-	-	-	-	-	+

(+) = Less frequent.

3 Management of palms

Basic research carried out during the last ten years has identified some species of palms as having management and development potential (Moraes 1996b). Due to their economic value and a potential regional market, selected species have been studied in greater detail. The following section presents lay aspects of distribution, production, harvesting practices, and potential negative impacts that may affect the conservation of existing populations.

Attalea phalerata (motacú palm)

The motacú palm is economically the most important palm species in Bolivia. It has a solitary stem and grows up to 18 m in height. Each plant bears two infructescences per year and each infructescence produces 350-500 fruits with 2-5 seeds per fruit (Moraes *et al.* 1996). It flowers throughout the year, but has two peaks a year; fruit is produced in the dry season and at the onset of the rainy season. The motacú reaches reproductive maturity after 7-10 years when it reaches a height of one meter.

Attalea phalerata is found in the lowlands and in the subandean forest belt, at elevations of 150-1,000 m. It occurs in evergreen montane and lowland forests, as well as in gallery forests, forest islands and in secondary forests. In dense stands, growth of as many as 115-236 individuals per hectare have been documented (Moraes *et al.* 1996). This species is used for multiple purposes (Table 3): Roofing made from leaves lasts for 5-7 years; the fruits are edible - and dispersed by rodents, wild pigs, cattle and monkeys - and also can be used for oil extraction (Moraes *et al.* 1996). The oil of the motacú palm is used for a variety of home remedies and in the production of cosmetics. Local communities gather mature and immature fruits for oil extraction. The kernel fat content reaches 60-70% and potential oil production from natural stands is estimated to be 1.1 to 2.4 tons/ha/year (Moraes *et al.* 1996).

Negative human impacts include burning following clearing for agriculture, particularly in the montane forest region. In savanna regions, extensive cattle management also reduces the percentage of seedlings in forest islands (Menacho, in prep). There is a need for developing practices which will ensure a sustainable production of this species. Some of the most important steps would include the reduction and control of burning, rotational grazing and improved harvesting methods for mature fruit. These measures would improve the productivity of this widespread species and ensure its future exploitation in Bolivia.

Geonoma deversa (jatata palm)

The jatata palm has a multistemmed habit with each individual bearing from 3-20 stems up to 4 m tall. This species grows in sandy, well-drained soils in the Andean foothills and in the adjacent piedmont region between 200-500 m elevation. This palm is used intensively for its leaves that are gathered for thatch construction. Five to fourteen leaves are generally gathered per stem. Roofs made from jatata leaves last about 15-20 years. This species is an important source of revenue for the Chimane, Tacanas and Mosetene ethnic groups that are settled on the piedmont and foothills of the Andes. Traditional harvesting practices entail the removal of the entire leaf crown of each stem. Stems that are less than 2 m tall are not harvested until flowering the next year (Sarmiento & Moraes, in prep.).

Current practices affect negatively the population of jatata, decreasing the abundance of mature individuals. The density of mature individuals in non-harvested plots is six times higher than harvested ones; both number of leaves per stem and number of stems are also higher. A comparison of harvested and non-harvested stands also showed a reduction in both seedlings and juveniles. Some stands are exploited more intensively than others and these become exhausted, forcing people to search at a greater distance from human settlements. There is overwhelming evidence that the jatata palm is very susceptible to over-exploitation. In addition, forest clearing by colonizers in neighbouring areas is also reducing the potential harvest of this economically important species. Sustainable exploitation would require a rotational use of assigned plots between 4-6 years, with a greater percentage of adult stems being saved for seed production.

Euterpe precatoria (asaí palm)

The asaí palm is an arboreal species with a single graceful slender stem and pinnate leaves. Mature individuals bear various inflorescences in different stages of maturation; each individual produces 1-4 inflorescences per year (Moraes 1994, Peña 1996). It grows in lowland rain forests at 150 m in the northeast Amazonian region and can be found up to 2,000 m in the Andes. It is adapted to rich soils in the montane forests and poorly drained or seasonally inundated soils along the alluvial Amazonian plain (Moraes 1994).

Mature individuals taller than 12 m are considered harvestable for palmito production; prior to this stage, meristems are elongate and considered unacceptable for commercial palmheart extraction. Density in lowland forest varies with reports ranging from 34 individuals per hectare (Peña 1996) up to 200 stems per hectare in some areas with poor drainage (T. Killeen, pers. comm.). The asaí palm is exhaustively exploited for its palmheart, which has a high commercial value in Brazil, Chile, and Europe. In order to harvest the heart, local gatherers must destroy the entire plant and clean the surrounding leaves until having a palmheart of ca. 50 cm long and 4-6 cm in diameter (Moraes 1994). This practice causes a high mortality of fruits and avoids regeneration of harvested stands. Due to its slow growth, a harvesting cycle of 10-14 years should be establish to allow a sustained regeneration of the asaí palm in northeastern Bolivia (Johnson 1995, Peña 1996). This procedure should also combine the development and management of other competitive palmheart producing species, like the pejibaye or tembé palm (*Bactris gasipaes*), both in the wild and under cultivation in agroforestry schemes (Moraes 1996b).

Parajubaea sunkha (the sunkha palm)

The sunkha palm is a tree with pinnate leaves that reaches up to m tall, and is endemic to the inter-Andean valleys of central Bolivia. It is a component of a deciduous forest formation or occurs on grassy and shrubby slopes between 1700 and 2200 m elevation (Moraes 1996c). Stands flower and produce fruits during the entire year; five infructescences per plant reach maturity in a year (Vargas & Moraes, in prep.). Each tree bears approximately 840-980 fruits per year, while 15-20 leaves are developed every two years.

Local residents use its fruits to produce a sweet drink, and harvest different parts of the leaf for its fibre. Each fibre measures 40 x 100 cm and is extracted from the base of each leaf. Exploitation of trees is usually restricted to individuals greater than 5-7 m tall.

About 70% of the stand was composed of mature individuals and a total absence of seedlings and juveniles was found in harvested areas; non-harvested plots show a more or less even percentage of juveniles vs. adults (Vargas & Moraes, in prep.). A comparison between harvested and non-harvested stands showed significant differences in the number of fruits produced (Vargas & Moraes, in prep.). Harvesting practices for fibre exploitation are based on a sequential extraction of leaves. Since each leaf subtends an inflorescence and a premature fruit drop prior to maturation, leading to fewer ripe fruits and a lower seed germination, this practice causes a reduction in the production of infructescences.

Most exploited stands are used in an agroforestry setting that includes maize crops and cattle. This practice leads to an evident destruction of seedlings, a high predation of seeds, and an inefficient dispersion of seeds (Vargas & Moraes, in prep.). Only few individuals are left unexploited by local peasants as seed trees; this causes a reduced crop for the next germinating season.

In order to keep populations productive over the long term, it is necessary to organize different management criteria: more individuals need to be left for seed production, rotation of harvested plots, the exclusion of domestic animals, and the establishment of plots where natural regeneration may occur in the absence of maize cultivation.

4 Essential criteria for palm management

Due to their numerous useful products, their widespread presence in different ecosystems and their perennial growth, wild palms are an important resource for development of agrosilvocultural systems (Pedersen & Balslev 1992). Palm resources can be exploited in a sustainable manner through the knowledge of the natural history of their individual species. In Bolivia, palm populations are currently being exploited without management of individuals to promote reproduction or germination.

There are many elements which help in the designing of long-term management schemes of palms in Bolivia. These elements can ensure high production in tandem with economic sustainability in order to improve local living standards, as well as decrease the genetic degradation of current populations. Management criteria can be summarized using the following themes that are integrated in pilot projects (Moraes 1996b):

Harvesting practices and utilization. Most exploitation activities are focused on the extraction of certain products and do not take into account the reproductive component and natural regeneration to assure the production of following generations. These predator harvesting practices may exhaust the resource under commercial conditions. In addition, several palms are highly susceptible to clearing and burning.

Profit distribution among gatherers, commerce and industries. The profit variation from gatherers of palm raw materials on one end of the scale through commerce with the final product on the other shows rising prices and low input back into management of the product (Moraes 1996b). In most cases, harvest activities are characterized by a traditional exploitation within subsistence economies. In contrast, intermediate-level commercializers inflate prices and overvalue their own role in selling the product to a local or regional market. The most lucrative role is represented by industrial interests because this sector obtains relatively high prices in external markets, depending on product quality and international demand. As a consequence, gatherers do not assume responsibility for keeping the production of palm resources high over a long-term period. This is due not only to a lack of technical capacity but also to a lack of financial incentives.

Conservation status of palm stands and production. In several cases, palm populations are endangered due to habitat destruction. This threat is increased when they are influenced by a natural sensitivity to certain environmental variations or have restricted geographic distribution.

The Bolivian National System of Protected Areas provides an opportunity to promote sustainable exploitation of several palm species under natural conditions. Most protected areas have objectives for conservation as well as for improving the quality of life of human residents.

Economic value and marketing. If there is no commercial demand for palm products, they have no economic value. Many palm products have only local importance and value is often obtained through their exchange for other products. Such examples include the fruits of *Attalea phalerata*, the fibres of *Parajubaea torallyi* and thatching produced by *Geonoma deversa*. These products have little commercial value in monetary terms, because their production is neither permanent nor stable. But there are some cases, such as the palmito of *Euterpe* *precatoria*, where a moderate to high demand and an established international market provide sufficient monetary incentives to finance sophisticated management plans.

Biology and natural renovability. Knowledge of the population structure of natural stands must be the basis for any management program. This information also can provide reliable data on the real production capacity of each stage of growth and development of a given species. There are virtually no exploitation and utilization programs being carried out in Bolivia. Most harvested stands have no regeneration programs, and no efforts are made to guarantee profitability over long periods.

There is a critical need for detailed studies on selected palm species, particularly on demography, distribution patterns, phenology, pollination and dispersal mechanisms, habit, seed viability, reproductive and productive capacity, related palms for similar products, and environmental conditions.

Human settlements and traditional knowledge. Several groups, such as ethnic groups, peasants and recent colonizers, depend on palm populations for part of their income or subsistence. Each group's degree of knowledge varies from recent and casual to a more profound understanding that has been developed over generations, depending on the group's cultural history and interaction.

Environmental policies and strategies for forest management. A framework consisting of decision-making involving local communities and a set of national priorities also includes regulations that define goals for forestry management of natural resources. In Bolivia, these national goals still need strategies and applicable programs to solve problems related to natural resources. Problems such as margination of use and management policies, chaotic exploitation without development plans, exhaustive utilization, and an inadequate territory ordination all combine to inhibit the sustainable development of palm resources.

Bolivia is an important centre of diversity of palm species due to its wide range of both altitudinal and ecological conditions. A high percentage of native palm species offers a wide range of uses, but there are few examples of sustainable management of palm stands in Bolivia. The traditional use and cultural relationships between ethnic groups and palm stands is not sufficient to reinforce historic ways of exploitation or ensure the sustainable use of these resources. Future harvesting practices should balance human needs with the production capability of palm populations. Background information on taxonomy, habitat preference, and population dynamics of individual species is particularly important for attaining this goal.

Technical information on the biology of palms must be combined with an understanding of the economic forces driving exploitation and the social conditions under which it is harvested. Only an integrated approach will yield the oft-cited goal of sustainable development.

Acknowledgements. Funding from the Instituto de Ecología in La Paz and grants supported by the International Foundation for Science (Stockholm, Sweden) and USAID to the National Herbarium of Bolivia enabled my fieldwork and studies in various research projects since 1986. I also appreciate the financial support of the organizers of this Symposium which funded my trip and stay in Bonn. My special thanks to Timothy Killeen from the Missouri Botanical Garden for his comments and suggestions on earlier versions of this manuscript.

References

Balée W (1989) The culture of Amazonian forests. Adv Econ Bot 7: 1-21

Johnson DV (1995) Sustainable management of assai boliviano (*Euterpe precatoria*) for palm heart production in the Tarumá forest concession Paraiso, Velasco province, Santa Cruz, Bolivia. BOLFOR, Santa Cruz

Menacho M (in prep.) Efecto del pastoreo sobre los renovales de las palmeras *Attalea phalerata* (motacú) y *Acrocomia aculeata* (totaí) en las islas de bosque de la Estación Biológica del Beni

Moore HE Jr (1973) The major groups of palms and their distribution. Gentes Herb 11: 27-141

Moraes RM (1992) Usos de palmas bolivianas. Proceedings Etnobotanica 92, Córdoba

Moraes RM (1994) Lineamientos para un plan de manejo del asaí (palmito): Un desafío para el Oriente boliviano. Suplemento Forestal, La Razón (28 octubre)

Moraes RM (1996a) Diversity and distribution of palms in Bolivia. Principes 40: 75-85

Moraes RM (1996b) Bases para el manejo sostenible de palmeras nativas de Bolivia. Dirección Nacional Conservación de la Biodiversidad (Ministerio de Desarrollo Sostenible y Medio Ambiente)-Tratado de Cooperación Amazónica (PNUD), La Paz

Moraes RM (1996c) Novelties in the genera *Parajubaea* and *Syagrus* (Palmae) from interandean valleys in Bolivia. Novon 6(1): 85-95

Moraes RM, Borchsenius F, Blicher-Mathiesen U (1996) Notes of the biology and uses of the motacú palm (*Attalea phalerata*, Arecaceae) from Bolivia. Economic Botany 50(4): 423-428

Pedersen HB, Balslev H (1992) The economic botany of Ecuadorean palms. In: Plotkin M, Famolare L (eds) Sustainable harvest and marketing of rain forest products. Conservation International, Island Press, Washington, DC, pp 173-191

Peña CM (1996) Palm heart production: sustainable management of *Euterpe precatoria* ("asai") in the Tarumá concession area, Santa Cruz, Bolivia. Tropical Research and Development, Inc., Gainesville

Sarmiento J, Moraes R M (in prep.) Uso tradicional de la jatata (*Geonoma deversa*) en el NE de Bolivia

Vargas I, Moraes R M (in prep.) Sustainable use of *Parajubaea sunkha*: biological cycle and harvesting practices

Diversity of mammals in Bolivia

Rainer Hutterer
Zoological Research Institute and Museum Alexander Koenig, Bonn, Germany

Abstract. At present, 4,700 species of extant mammals are recognized in the world, 1,100 of which occur in South America. Bolivia is home to one third of all South American mammal fauna. To date, 319 species of native mammals have been recorded; this number is expected to continue to increase in the future. New species and genera are still being described. Consequently, a complete picture of Bolivia's mammal fauna is not yet available. A summary of current knowledge is presented.

Diversidad de mamíferos en Bolivia

Resumen. Actualmente están reconocidas en el mundo 4,700 especies de mamíferos vivientes, de las cuales 1,100 se registran en Suramérica. Bolivia sirve de hogar a una tercera parte de toda la fauna de mamíferos suramericana. Hasta la fecha se han registrado 319 especies de mamíferos autóctonos y se espera que dicha cifra siga aumentando en el futuro. Nuevas especies y géneros están siendo descritos, por lo que aún no es posible ofrecer una imagen completa de la fauna de mamíferos de Bolivia. Se presenta un resumen de los conocimientos acumulados hasta ahora.

1 Introduction

Mammals are attractive animals, and they are an important source of food and clothing (Fig. 1). It therefore comes as no surprise that even very early human societies collected information on the character, behaviour and economic value of game mammals. People in South America have probably been doing this since the end of the Pleistocene (Roosevelt *et al.* 1996). In our modern world however only printed records count, and species are treated as "unknown" or "known" according to their absence or presence in formal taxonomic catalogues. With regard to mammals, we have a fairly complete record of all taxa described since Linnaeus in 1758 to 1993 (Wilson & Reeder 1993). Some 4,700 valid species of extant mammals have been described since 1758 (Cole *et al.* 1994), and most recent figures show that the number of mammalian species is steadily increasing. New speculation on the actual number of mammal species would have us believe there are 8,000 species (Morell 1996). Although this figure seems somewhat exaggerated, 5,000 to 6,000 may be entirely realistic. Of all mammal species,

Fig. 1. A herd of vicunas at the Reserva Nacional de Fauna Ulla Ulla, Bolivia, photographed by U. Hirsch in 1977
Ilustración 1. Una manada de vicuñas en la Reserva Nacional de Fauna Ulla Ulla en Bolivia, fotografiada por U. Hirsch en 1977

24% occur in South America. Of these, 80% are endemic to the Neotropics. A total of about 50 families, 309 genera and 1,100 species of extant mammals exist in South America (Cole *et al.* 1994).

Looking at the geographical areas where new species of mammals have been described in the last 10 years (Hutterer 1995), tropical regions clearly dominate. Bolivia is one of the 'hot spots' for recently discovered species (Fig. 2). This increase in knowledge is mainly the result of research conducted since 1982. The purpose of this article is to briefly review our current knowledge of Bolivia's native mammal fauna.

2 Recording Bolivian mammals

Pine's statement (1982) that "the mammal fauna of Bolivia may very well be the most poorly known in South America" is no longer valid. However, the continent's mammal fauna has not been studied in such detail that definitive figures can be given at this time. Many additional genera have since been found in Bolivia, such as *Andalgalomys* (Olds *et al.* 1987), *Catagonus* (Eisentraut

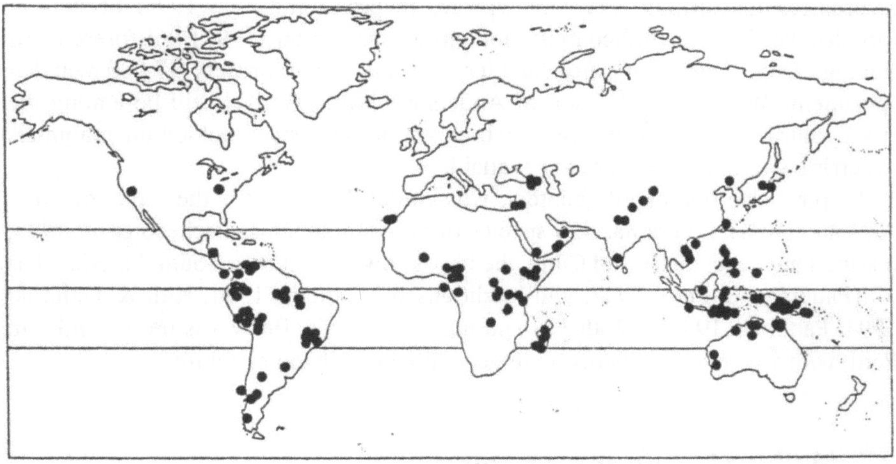

Fig. 2. World locations from which new species of mammals have been described between 1985 and 1995 (Hutterer 1995)
Ilustración 2. Lugares del mundo desde los que han sido descritas nuevas especies de mamíferos entre 1985 y 1995 (Hutterer 1995)

1986), *Chibchanomys* (Ergueta & Morales 1996), *Chironectes* (Cabot 1989), *Diphylla* (Anderson 1991), *Mesomys* (Emmons 1994), and *Pteronotus* (Ibañez & Ochoa 1989), to name a few. New species of *Abrocoma* (Glanz & Anderson 1990), *Akodon* (Myers & Patton 1989) and *Oxymycterus* (Hinojosa *et al.* 1987) were described for Bolivia, and a number of new taxa still await description (Anderson pers. comm.). New species and locality records were provided by Anderson (1991, 1993), Anderson & Webster (1983), Anderson *et al.* (1982, 1987, 1993), Christen & Geissmann (1994), Cook & Yates (1994), Cook *et al.* (1990), Hershkovitz (1990, 1992, 1994), Ibañez (1985), Ibañez *et al.* (1994), Myers *et al.* (1990), Olds & Anderson (1987), Patterson (1992a), Salazar *et al.* (1994), Torres *et al.* (1988), Wilson & Salazar (1990), and Yensen *et al.* (1994).

Mammal research in Bolivia dates back to about 1830. The study of bats in the country also began around that time (Anderson 1991, 1997). To date, 29 mammal taxa bear the name of Bolivia, in epithets like *boliviae, boliviensis,* and *bolivianus*. Of these, 8 are actually regarded as clearly seperate species, others as subspecies, some as synonyms. The most recently named species is *Abrocoma boliviensis* (Glanz & Anderson 1990).

Several species are known only from the territory of Bolivia. They include two primates (*Callicebus modestus, Callicebus ollalae*), two marsupials (*Marmosops dorothea, Monodelphis kunsi*), and nine rodents (*Abrocoma boliviensis, Akodon dayi, Akodon siberiae, Ctenomys goodfellowi, Ctenomys lewisi, Ctenomys steinbachi, Oxymycterus hucucha, Phyllotis wolffsohni,* and *Thomasomys ladewi*) (Anderson 1993, Ergueta & Morales 1996).

Anderson (1985) was the first to present a preliminary list of Bolivian mammals. Subsequent taxonomic changes and a critical evaluation of old and new records resulted in 319 acceptable species records (Anderson 1993, Ergueta & Morales 1996). In a forthcoming monograph, Anderson (1997) will present his findings from a study of pertinent literature and approximately 37,000 voucher specimens. With the publication of Anderson's work, Bolivia will be among the first countries in South America for which a comprehensive book on the mammals occurring within its boundaries is available.

To put these figures into context, we should also consider the fauna of other Andean countries. Peru has 426 species of native mammals (Pacheco *et al.* 1995, marine fauna excluded), and Chile, the best-known country in South America, has 92 (Patterson & Feigl 1987, with additions by Hutterer 1994, Kelt & Gallardo 1994, Patterson 1992b). With 319 and more mammals, Bolivia is more similar to Peru, with which it also shares some general geographical structures.

3 Mammals and their habitats

Ergueta & Morales provide a detailed biogeographical division of Bolivia in their book (1996). For the purpose of this paper, only three principal regions will be considered here.

Fig. 3. A chacoan peccary (*Catagonus wagneri*) kept at the Berlin Zoo in 1977 (F. Kleinschmidt, courtesy Zoological Garden Berlin)
Ilustración 3. Un pecarí de Chaco *(Catagonus wagneri)* en el Zoológico de Berlín en 1977 (F. Kleinschmidt, por cortesía del Zoological Garden de Berlín)

The Andean highlands (region altoandino and puna) have a very special mammal fauna, including "vicuña," "viscacha," and numerous rodent genera (*Andinomys*, *Auliscomys*, *Chinchilulla*, *Chinchilla*, *Eligmodontia*, *Galenomys*, *Microcavia*, *Octodontomys*, *Phyllotis*) and species.

The eastern Andean slopes with their intersecting valleys, and the forested yungas are extremely important for mammals. Work conducted by Patton *et al.* (1990) in Peru and Bolivia shows that the yungas are home to a mammal fauna all their own, and that the altitudinal separation plus the vertical separation by valleys of many of these forests has enhanced the evolution of local species of small mammals. Such an example is the mouse *Akodon siberiae* (Myers & Patton 1989) which is known only from the Siberia cloud forest at an elevation of 2,800 m from west of Comarapa, Cochabamba Department. The long-nosed mouse genus *Oxymycterus* is another element of cloud forest region on the eastern slopes of the Andes, as are large mammals like *Mazama chunyi* and *Tremarctos ornatus* (see maps in Tarifa 1996). It may be said that many more species of small mammals will be found in these yungas, including species new to science, in coming years.

It is possible to distinguish between two main lowland faunal provinces: the lowland temperate zone (chaco and eastern lowlands), and the lowland tropics (Amazon and Beni). The latter provides a home to typical Amazonian mammals such as pygmy anteaters, sloth, primates, many carnivores, tapirs and opossums. The temperate zone is home to a number of larger and widely distributed species, including marsh deer. The chaco is a well-defined section of the temperate zone, and Bolivia shares a small strip of it in the southern parts of Tarija and Santa Cruz Provinces. The dry thornbush of the chaco supports at least four mammals endemic to that region, all of which have been found in the Bolivian part as well: the armadillos *Cabassous chacoensis* and *Chlamyphorus retusus*, the tuco-tuco *Ctenomys conoveri* and the chacoan peccari, *Catagonus wagneri* (Fig. 3). Eisentraut (1983, 1986), who recorded this large peccari in Tarija, also mentioned rumours about the existence of another reddish species of peccari which he had already heard about in 1933. However, his later efforts to obtain a specimen were not successful.

4 Examples of current research

Two case studies out of many may be used to illustrate different approaches toward a better understanding of mammal diversity in Bolivia. The first deals with tuco-tucos (Fig. 4) which are small-to-large rodents of the genus *Ctenomys* that live in burrow systems and eat roots and other plant matter. Of the roughly 40 species known from South America, 8 are known from Bolivia, from the altiplano down to the chaco. Anderson *et al.* (1987) and Cook *et al.* (1990) identified the species using museum specimens and karyotypes, and defined their ranges in Bolivia. Cook and Yates (1994) went one step further and analyzed the allozyme

Fig. 4. A tuco tuco (*Ctenomys opimus*) in front of its burrow, photographed 1977 by U. Hirsch at Huancaroma, Bolivia
Ilustración 4. Un tuco tuco (*Ctenomys opimus*) frente a su madriguera, fotografiado en 1977 por U. Hirsch en Huancaroma, Bolivia

variation in 13 Bolivian populations. Their results indicate that several so-called species appear to be polyphyletic, and that 4 to 5 further biological species must be recognized in Bolivia.A revision of the entire group is therefore necessary and will eventually result in the recognition of 12 or more species.

Another aspect of current research in Bolivia involves the field of archaeozoology. Man has left traces of his presence in South America for 11,000 years now. These traces sometimes include the bones of vertebrates and certain mammals (Stahl 1995). The analysis of these remains can tell us much about the former presence of species and thus add a historical perspective to the study of mammals. In Bolivia, however, little to nothing has been done in this field thus far. Heiko Prümers of the German Archaeological Survey in Bonn and Wilma Winkler of the Instituto Nacional de Arqueología in La Paz have both conducted archaeological excavations in Pailón, a village east of Sta. Cruz de la Sierra on the right bank of the Rio Grande (Prümers & Winkler 1997). They discovered traces of an unknown culture dating from about 1,000 years ago, and they invited me to study the bones and other zoological remains found there. The species that were subsequently identified included 14 species of mammals, most of which had been used for food and ornamentation (Hutterer 1997). All of these species still occur in the lowlands of southeastern Bolivia. However, the tuco-tuco *Ctenomys conoveri* is confined today to the chaco region further south (Anderson *et al.* 1987). Marsh deer, white-lipped peccaries, tapirs and marsh rats were also present. Rich material obtained during a second expedition and now under study includes records of many more species of mammals. The bones of three species of *Ctenomys* from the same site and period were particularly noteworthy. Their presence contradicts the allopatric distribution pattern of the genus in Bolivia developed by Cook & Yates (1994). This data is the first to document Holocene fauna and faunal shifts in Bolivia. These findings may also shed some light on the role of native hunters (Redford & Robinson 1986) in Bolivia some 1,000 years ago. We know that a complete set of vertebrates, including giant sloths and giant vampire bats, vanished from South America during the Holocene, and current hypotheses hold that man played a crucial part in that process (Trajano & de Vivo 1991). More archaeozoological studies could increase our knowledge on the historical diversity in Bolivia, just as modern zoological studies could increase our knowledge on the extant mammal fauna.

5 Conservation

As in many other tropical countries, subsistance hunting of small and larger vertebrates is still common in Bolivia. Together with continuous habitat destruction, this may cause a critical decline of some species. A major step towards the conservation of mammals is the "Libro rojo de los vertebrados de Bolivia" (Ergueta & Morales 1996) and the legislation and education that go along with it. In this edition, the authors classified the larger mammals according to their conservation status for the first time. One species is regarded to be locally extinct (*Chinchilla brevicauda*), 10 species as endangered (*Callimicio goeldii*, *Chaetophractus nationi*, *Chrysocyon brachyurus*, *Tremarctos ornatus*, *Bassaricyon alleni*, *Pteronura brasiliensis*, *Catagonus wagneri*, *Lama guanicoe*, *Hippocamelus antisinensis*, *Mazama chunyi*), 15 species as vulnerable (*Alouatta caraya*, *Pithecia irrorata*, *Ateles paniscus*, *Saguinus imperator*, *Myrmecophaga tridactyla*, *Priodontes maximus*, *Lutra longicaudis*, *Felis pardalis*, *Panthera onca*, *Tapirus terrestris*, *Tayassu pecari*, *Tayassu tajacu*, *Vicugna vicugna*, *Ozotoceros bezoarticus*, *Blastoceros dichotomus*) and 37 further species as rare or insufficiently known. The latter category probably also applies to many smaller mammals which have not yet been evaluated due to a lack of data.

6 Conclusions

Our knowledge of Bolivian mammals has increased considerably since 1982, but we still do not know enough about mammal diversity in Bolivia. Although a solid framework already exists, we are still far from solving all our taxonomic problems or from having a set of distributional data from all over the country. The necessary methodology for such surveys is now readily available (Wilson *et al.* 1996, Voss & Emmons 1996). What is needed are more thorough inventories such as the one by Anderson *et al.* (1993) on the mammals of the Parque Nacional Amboró which was probably the first inventory of that kind in the country. Surveys of transects in the Andes like the one conducted by Patterson *et al.* (1996) on bats in Peru could also be promising. There is also much to be explored regarding the biology of most species (Pook & Pook 1981, for example) and the ecological interactions of mammals and plants (see Janos & Sahley 1995 for a nice example of the role rodents play in the dispersal of mycorrhizal fungi in Amazonian Peru). But with the assistance of existing programmes, many of our presently unsolved questions may be answered in the near future.

Acknowledgements. I am grateful to Sydney Anderson for providing unpublished information, and to Udo Hirsch and the Berlin Zoo for their photographs. Heiko Prümers kindly invited me to participate in the archaeological survey of the Pailón site.

References

Anderson S (1985) Lista preliminar de mamíferos bolivianos. Cuadernos VI, Zoología 3: 5-16

Anderson S (1991) A brief history of Bolivian chiropterology and new records of bats. Bull Amer Mus Nat Hist 206: 138-144

Anderson S (1993) Los mamíferos bolivianos: notas de distribución y claves de identificación. Instituto de Ecología, Colección Boliviano de Fauna. La Paz

Anderson S (1997) Mammals of Bolivia, taxonomy and distribution. Bull Amer Mus Nat Hist 231: 1-652

Anderson S, Koopman KF, Creighton GF (1982) Bats of Bolivia: an annotated checklist. Amer Mus Novit 2750: 1-24

Anderson S, Riddle BR, Yates TL, Cook JA (1993) Los mamíferos del Parque Nacional Amboró y la Región de Santa Cruz de la Sierra, Bolivia. Special Publication, The Museum of Southwestern Biology 2: 1-58

Anderson S, Webster DW (1983) Notes on Bolivian mammals 1: additional records of bats. Amer Mus Novit 2766: 1-3

Anderson S, Yates TL, Cook JA (1987) Notes on Bolivian mammals 4. The genus *Ctenomys* (Rodentia: Ctenomyidae) in the eastern lowlands. Amer Mus Novit 2891: 1-20

Cabot J (1989) Second record of *Chironectes minimus* (Marsupialia) in Bolivia. Mammalia 53: 135-136

Christen A, Geissmann T (1994) A primate survey in northern Bolivia, with special reference to Goeldi's monkey, *Callimico goeldii*. Int J Primatol 15: 239-274

Cole FR, Reeder DM, Wilson DE (1994) A synopsis of distribution patterns and the conservation of mammal species. J Mamm 75: 266-276

Cook JA, Anderson S, Yates TL (1990) Notes on Bolivian mammals 6. The genus *Ctenomys* (Rodentia, Ctenomyidae) in the highlands. Amer Mus Novit 2980: 1-27

Cook JA, Yates TL (1994) Systematic relationships of the Bolivian tuco-tucos, genus *Ctenomys* (Rodentia: Ctenomyidae). J Mamm 75: 583-599

Emmons LH (1994) New locality records of *Mesomys* (Rodentia: Echimyidae). Mammalia 58: 148-149

Eisentraut M (1983) Im Land der Chaco-Indianer. Biotropic-Verlag, Baden-Baden

Eisentraut M (1986) Über das Vorkommen des Chaco-Pekari, *Catagonus wagneri*, in Bolivien. Bonn zool Beitr 37: 43-47

Ergueta P, de Morales C (eds) (1996) Libro rojo de los vertebrados de Bolivia. Centro de Datos para la Conservaciòn, La Paz

Glanz WE, Anderson S (1990) Notes on Bolivian mammals. 7. A new species of *Abrocoma* (Rodentia) and relationships of the Abrocomidae. Amer Mus Novit 2991: 1-32

Hershkovitz P (1990) Titis, New World monkeys of the genus *Callicebus* (Cebidae, Platyrrhini), a preliminary taxonomic review. Fieldiana Zool NS 55: 1-109

Hershkovitz P (1992) The South American gracile mouse opossums, genus *Gracilinanus* Gardner and Creighton, 1989 (Marmosidae, Marsupialia): a taxonomic review with notes on general morphology and relationships. Fieldiana Zool NS 79: 1-56

Hershkovitz P (1994) The description of a new species of South American hocicudo, or long-nose mouse, genus *Oxymycterus* (Sigmodontidae, Muroidea), with a critical review of the generic content. Fieldiana Zool NS 79: 1-43

Hinojosa P F, Anderson S, Patton JL (1987) Two new species of *Oxymycterus* (Rodentia) from Peru and Bolivia. Amer Mus Novit 2898: 1-17

Hutterer R (1994) Island rodent: a new species of *Octodon* from Isla Mocha, Chile (Mammalia: Octodontidae). Z Säugetierkunde 59: 27-41

Hutterer R (1995) Unbekannte Säugetierwelt: Übersicht der seit 1985 beschriebenen Arten. Tier und Museum (Bonn) 4: 77-88

Hutterer R (1997) Archaeozoological remains (Vertebrata, Gastropoda) from prehispanic sites at Pailón, Bolivia. Beitr Allgemein Vergl Archäol 17: 325-341

Ibañez C (1985) Notas sobre distribución de quiropteros en Bolivia (Mammalia, Chiroptera). Hist Nat Argentina 5: 329-333

Ibañez C, Cabot J, Anderson S (1994) New records of Bolivian mammals in the collection of the Estación Biológica de Doñana. Doñana, Acta Vertebrata 21: 79-83

Ibañez C, Ochoa G J (1989) New records of bats from Bolivia. J Mamm 70: 216-219

Janos DP, Sahley CT (1995) Rodent dispersal of vesicular-arbuscular mycorrhizal fungi in Amazonian Peru. Ecology 76: 1852-1858

Kelt DA, Gallardo MH (1994) A new species of tuco-tuco, genus *Ctenomys* (Rodentia: Ctenomyidae) from Patagonian Chile. J Mammal 75: 338-348

Morell V (1996) New mammals discovered by biology's new explorers. Science 273:1491

Myers P, Patton JL (1989) A new species of *Akodon* from the cloud forest of eastern Cochabamba Department, Bolivia (Rodentia: Sigmodontinae). Occas Pap Mus Zool Univ Michigan 720: 1-28

Myers P, Patton JL, Smith MF (1990) A review of the *Boliviensis* group of *Akodon* (Muridae: Sigmodontinae), with emphasis on Peru and Bolivia. Miscell Publ Mus Zool Univ Michigan 177: 1-104

Olds N, Anderson S (1987) Notes on Bolivian mammals 2: Taxonomy and distribution of rice rats of the subgenus *Oligoryzomys*. Fieldiana Zool NS 39: 261-281

Olds N, Anderson S, Yates TL (1987) Notes on Bolivian mammals 3: A revised diagnosis of *Andalgalomys* (Rodentia, Muridae) and the description of a new subspecies. Amer Mus Novit 2890: 1-17

Pacheco V, de Macedo H, Vivar E, Ascorro C, Arana-Cardó R, Solari S (1995) Lista anotada de los mamíferos peruanos. Occas Pap Conserv Biol 2: 1-35

Patterson BD (1992a) Mammals in the Royal Natural History Museum, Stockholm, collected in Brazil and Bolivia by A. M. Ollala during 1934-1938. Fieldiana Zool NS 66: 1-42

Patterson BD (1992b) A new genus and species of long-clawed mouse (Rodentia: Muridae) from temperate rainforests of Chile. Zool J Linn Soc 106: 127-145

Patterson BD, Feigl CE (1987) Faunal representation in museum collections of mammals: Osgood's mammals of Chile. Fieldiana Zool NS 39: 485-496

Patterson BD, Pacheco V, Solari S (1996) Distributions of bats along an elevational gradient in the Andes of south-eastern Peru. J Zool Lond 240: 637-658

Patton JL, Myers P, Smith MF (1990) Vicariant versus gradient models of diversification: The small mammal fauna of eastern Andean slopes of Peru. In: Peters G, Hutterer R (eds) Vertebrates in the tropics. Zoologisches Forschungsinstitut und Museum Alexander Koenig, Bonn, pp 355-371

Pine RH (1982) Current status of South American mammalogy. In: Mares MA, Genoways HH (eds) Mammalian biology in South America, 27-37. Special Publ Ser Pymatuning Lab Ecol Univ Pittsburgh vol 6

Pook A, Pook G (1981) A field study of the socio-ecology of the Goeldi's monkeys (*Callimico goeldii*) in northern Bolivia. Folia Primatol 35: 288-312

Prümers H, Winkler W (1997) Archäologische Untersuchungen im bolivianischen Tiefland. Erster Bericht des Projektes Grigotá. Beitr Allgem Vergl Archäol 17:343-393

Redford KH, Robinson JG (1986) The game of choice: patterns of indian and colonist hunting in the Neotropics. Amer Anthopol 89: 650-667

Roosevelt AC, Lima da Costa M, Lopes Machado C, Michab M, Mercier N, Valladas H, Feathers J, Barnett W, Imazio da Silveira M, Henderson A, Sliva J, Chernoff B, Reese DS, Holman JA, Toth N, Schick K (1996) Paleoindian cave dwellers in the Amazon: the peopling of the Americas. Science 272: 373-384

Salazar, J, Campbell ML, Anderson S, Gardner SL, Dunnum JL (1994) New records of Bolivian mammals. Mammalia 58: 123-128

Stahl PW (ed) (1995) Archaeology in the lowland American tropics. Cambridge University Press, Cambridge

Tarifa T (1996) Mamíferos. In: Ergueta P, de Morales C (eds) Libro rojo de los vertebrados de Bolivia, Centro de Datos para la Conservaciòn, La Paz, pp 165-264

Trajano E, de Vivo M (1991) *Desmodus draculae* Morgan, Linares, and Ray, 1988, reported for Southeastern Brasil, with paleoecological comments (Phyllostomidae, Desmodontinae). Mammalia 55: 456-459

Torres MP, Rosas T, Tiranti SI (1988) *Thyroptera discifera* (Chiroptera, Thyropteridae) in Bolivia. J Mammal 69: 434-435

Voss RS, Emmons LH (1996) Mammalian diversity in Neotropical lowland rainforest: a preliminary assessment. Bull Amer Mus Nat Hist 230: 1-115

Wilson DE, Cole FR, Nichols JD, Rudran R, Foster MS (eds) (1996) Measuring and monitoring biological diversity: standard methods for mammals. Smithsonian Institution Press, Washington

Wilson DE, Reeder DA (eds) (1993) Mammal species of the world, 2nd edition. Smithsonian Institution Press, Washington

Wilson D, Salazar B J (1990) Los murcielagos de la Reserva de la Biósfera "Estación Biológica Beni", Bolivia. Ecología en Bolivia 13: 47-56

Yensen E, Tarifa T, Anderson S (1994) New distributional records of some Bolivian mammals. Mammalia 58: 405-413

Geoecology and biodiversity – Problems and perspectives for the management of the natural resources of Bolivia's forest and savanna ecosystems

Werner Hanagarth[1], Andrzej Szwagrzak[2]
[1] Institute of Ecology, La Paz, Bolivia
[2] Green Cross of Bolivia, La Paz, Bolivia

Abstract. A brief review of geoecology, biodiversity and biogeography is given, as well as a discussion of the problems and options for the natural resource management of Bolivia's forests and savannas.

The high level of biodiversity in Bolivia's forests and savannas is due to its geographical position that borders a tropical climate and is closely related to strong longitudinal and altitudinal gradients. Soil conditions and hydrology in combination with regional geology are strong determinants for biogeography and diversity. Bolivia is not located in the heart of any of the main neotropical biotas. The general high biodiversity of Amazonian ecosystems in juxtaposition with other very different floras and faunas is the reason for Bolivia's high diversity.

In many regions of Bolivia, low soil fertility, floods and geomorphology make it difficult to improve land utilization, but the chances are good for seeing management of forests and savannas implemented. In recent years, the Bolivian government has opened up to reform and is creating the foundation necessary for natural resource management. The low human population-forest and savanna ratio meets the prerequisites for the management of natural ecosystems. The past policy of "free" land has affected ecologically sensible planning up to the present. Timber logging, colonization and agro-business accelerated the exploitation of extant areas.

Nevertheless, there are options for the sustainable management of the country's forests, that also take various constraints and restrictions into consideration. In fact, Bolivia's savannas have the best prospects for implementing wildlife management programs based on a few economically promising species in the near future.

Geo-ecología y biodiversidad - Problemas y perspectivos para el manejo de los recursos naturales de los ecosistemas del bosque y la sabanas de Bolivia

Resumen. Se ofrece una breve visión de la geoecología, la biodiversidad y la biogeografía, así como una discusión sobre los problemas y alternativas para el manejo de los recursos naturales de los bosques y las sabanas de Bolivia.

El alto grado de biodiversidad en los bosques y sabanas bolivianos se debe a su posición geográfica fronteriza con un clima tropical y guarda una estrecha relación con fuertes gradientes longitudinales y altitudinales. Las condiciones del suelo y la hidrología, en combinación con la geología regional determinan decisivamente la biogeografía y la diversidad. Bolivia no se encuentra en el centro de ninguna de las principales biotas neotropicales. La alta diversidad de Bolivia es causada por la biodiversidad generalmente alta del ecosistema amazónico en yuxtaposición con otras floras y faunas muy diferentes.

En muchas regiones de Bolivia, el bajo grado de fertilidad del suelo, las inundaciones y la geomorfología dificultan una mejora del uso de la tierra, pero existen buenas perspectivas de cara a una implementación del manejo de bosques y sabanas. En los últimos años, el gobierno boliviano se ha mostrado abierto a reformas y está creando la base necesaria para el manejo de los recursos naturales. La baja proporción de población humana en bosques y sabanas responde a los prerrequisitos para el manejo de ecosistemas naturales. La política de tierra "libre" seguida en el pasado ha dificultado hasta ahora el ejercicio de una planificación preocupada por la ecología. La tala de madera, la colonización y la industria agraria aceleró la explotación de áreas aún existentes.

Sin embargo, existen posibilidades para el manejo sostenible de los bosques del país, posibilidades que contemplen también varias restricciones y limitaciones. De hecho, las sabanas de Bolivia cuentan con las mejores perpectivas para la implementación en un futuro próximo de programas de manejo de la vida salvaje basados en unas pocas especies económicamente prometedoras.

1 Introduction

About ten years ago, Bolivia was one of the biologically and ecologically least known countries in South America. Even in the early nineties, many parts of Bolivia's eastern Andean slope and its lowlands were largely unknown to science. Although this situation has since changed, it would be an exaggeration to say that today we have sufficient knowledge about the country's biodiversity. Numerous expeditions lead to the first-time registration of new plant and animal species in Bolivia and reports on species that are new to science.

There are few relatively complete long-term local inventories of biodiversity. Our knowledge regarding the distribution of endemism is meagre. The same is true of geoecological data. Too little is known of the soil properties of larger regions – information that is needed for evaluating land-use potential and for land-use planning.

This paper briefly reviews the relationship between the geoecology, biodiversity and biogeographical complexity of Bolivia's forests and savannas, and discusses some problems and options for natural resource conservation and utilization.

2 A brief review of the geography and geoecology of the Bolivian lowlands and the eastern Andean slope

The Bolivian territory encompasses three hydrological systems. Most parts of the lowlands and the eastern Andean slope belong to the Amazon drainage system. Only the southern-most areas drain into the La Plata basin. The smaller Andean highland (Altiplano) system is strictly endorrheic, without any drainage into the sea.

2.1 Area covered by forests and savannas

Forests cover still about 535,000 km² (CDC 1995) of Bolivia's territory. They are distributed throughout the eastern part of the country, with the exception of *Polylepis* open forests that occur in some parts of the Altiplano which can be found altitudes up to 5,000 m above sea level (Kessler 1993). Inner-Andean forests were destroyed in the years following the colonial period as the *ordonanzas* of the Spanish viceroy Francisco de Toledo (1563-1583) indicate (Heinrich & Equivar 1991). In recent decades, some 3% of the natural forest cover has been lost.

Natural savannas occupy 170,000 km² or more of the lowlands, primarily in the Beni savannas (Beck 1983, Hanagarth 1993), the Cerrado woodlands in the Chiquitania highlands (Killeen 1990) and the western-most parts of the Pantanal. The isolated mountain savannas in the Yungas mountain forests are probably the result of human influence. They displaced dry, semihumid and subhumid forests (Beck 1993).

2.2 Geology, soil and hydrology – Decisive factors for biota and diversity distribution

There is a marked relationship between the distribution of certain biota and regional geology, soil, hydrology and geomorphology. Bolivia's geographical location between the Andes in the west and the Brazilian shield in the east largely determines its geology.

The upper and mideastern Andean slope (1,000 - >6,000 metres above sea level) consists mainly of Paleozoic rocks. The terrain is extremely rugged and has deep valleys that are encompassed by high mountain chains. Although it is shallow and generally acidic, the soil is quite fertile. Cretacic, Tertiary and Pleistocene sediments as well as deeper, more nutrient-rich soils are much more widely distributed in the wider valleys found at lower altitudes.

The Precambrian Chiquitania highlands constitute the western-most part of the Brazilian shield. This ancient peneplain is located at altitudes of 200 to 300 metres above sea level. Much of it is covered by weathered lateritic and leached soils that are probably Tertiary in origin. A few scattered inselbergs and mountain chains reach altitudes ranging from 800 to 1,200 m above sea level (Killeen *et al.* 1990, Ibisch *et al.* 1995).

Located between both massifs, the Beni Chaco plain is part of the Pericratonic basin which extends from Venezuela to Argentina along the Andes. Most parts of this plain do not exceed an altitude of 200 metres above sea level. Exceptions are the subandean piemont and the Santa Cruz region which can reach more than 400 m a.s.l. This flat plain extends southward from northern Bolivia to the Chaco and Pantanal. Geologically speaking, it is divided into an older northern and a younger southern region by the Bala Rogagua line which runs through the middle of the Beni savannas. Large areas of the northern waterlogged Beni savannas are covered by highly weathered, extremely leached soils. Lateritic duricrusts, pisolithic and plinthic soils extend north to the Pando province in Bolivia and the Brazilian Acre province in Brazil which are covered by Amazonian forests.

Sediments from the southern Beni flooded savannas to the Chaco plain are from late Pleistocene and Holocene age, and are covered by slowly weathered soils. Alkaline soils are found even in the southern Beni savannas, where periodic floods cover areas of 80,000 km^2 or more, and occur more frequently with increasing proximity to the dry Chaco (Hanagarth 1993).

2.3 Latitudinal and altitudinal climate gradients

In addition to Bolivia's geological history, the country's diverse climates are vital to an understanding of the high degree of variability and heterogeneity found in Bolivia's forest and savanna ecosystems. The country's geographical location near the border to the tropics, and the structure of the Andes influence the movement of the intertropical convergence zone (ITCZ) which drifts south to southern Bolivia during the southern hemisphere's summer season and determines climatic conditions in the Bolivian lowlands, along the eastern Andean slope and in the Andean highlands. The ITCZ and low pressure centres in the Paraguayan Chaco cause extremely high precipitation levels, mainly in the southern part of the Beni Chaco plain and along the Andean slope.

In climatic terms, the eastern Andean slope is much more diverse than the lowlands. The most humid and perhumid area is located in the Chapare region, with annual precipitation exceeding 5,000 mm. The northwestern part of the Bolivian lowlands and the Andean slope is influenced by the southwestern Peruvian precipitation centre near Quince Mil and receives 7,000 mm or more of precipitation. Along the Andean slope, high altitudinal differences, "Massenerhebungseffekt", latitudinal gradients, mountain-valley wind systems and a complex orography that is related to the geology of this landscape are responsible for the region's high geoecological diversity which is a juxtaposition of arid and humid valleys, arid valley bottoms and perhumid mountain crests.

A general northwest-southeast precipitation gradient prevails in the lowlands. The eastern parts are generally drier than the western parts near the Andes, where humid and perhumid conditions dominate; a hot subhumid climate predominates in the centre of the country and in the northern lowlands. In the Santa Cruz region, a narrow west to east oriented semihumid-semiarid transition zone with strong rainfall gradients separates the arid Chaco from the tropical humid areas in the north.

Cold waves of polar origin, known as "surazos," cause the temperature to drop sharply to a absolute minimum of 6° to 7°C in the Beni province, and 0°C in the southernmost regions, and are occasionally accompanied by strong rainfalls (Hanagarth 1993). These cold waves affect most parts of the Bolivian territory, and possibly serve a selective function in regard to the distribution of tropical plant and animal species. Most parts of the country are influenced by strong El Niño phenomenon (Ronchail 1995, Hanagarth 1993).

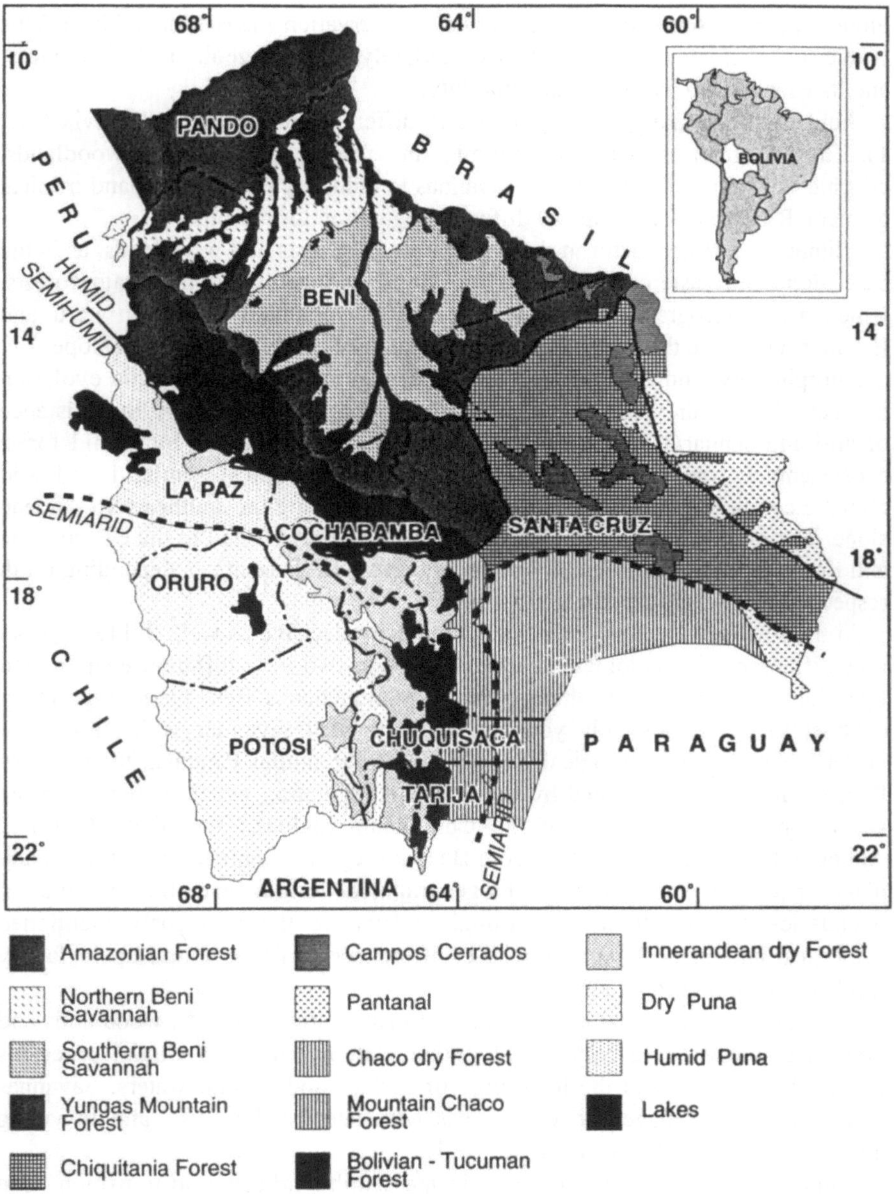

Fig. 1. Main Bolivian vegetation formations and biotas
Ilustración 1. Principales formaciones y biotas de vegetación en Bolivia

2.4 Biodiversity and biogeography

Bolivia's overall species richness is related to the country's biogeographical complexity. Bolivia is not located in the heart of any of the large biota. It can however claim to be the most important transitional country between the major biotic regions in southern South America (Conservation International 1991). This is due to its strong climatic gradients, markedly regional geological differences, and its edaphic and hydrological variability.

Aout twelve primary biogeographically different biotas exist in Bolivia (Fig. 1). They range from hot and perhumid rain forests, dry Chacoan woodlands, edaphic Cerrado and Pantanal-like savannas to cold, dry puna deserts and tropical glaciers. Ribera (1992) names 40 different vegetation formations.

Climate is a major factor in the distribution of forests and woodlands, as is the case along the eastern Andean slope. Humid and perhumid mountain forests, which are biogeographically related to similar mountain forests in Peru and Ecuador, dominate the northern and central parts of the eastern Andean slope. But geomorphology, and the effects of the Andes' geological and tectonic evolution interact with certain valley-mountain wind systems, thus permitting the existence of arid and semiarid inner-Andean valleys that are surrounded by humid forest ecosystems in the upper and middle Yungas. Subtropical semiarid and arid climatic conditions become increasingly important along the southeastern Andean slope. The mountain vegetation in this region is closely related to the Chaco biota and Bolivian Tucuman forest. It covers large parts of these areas, depending on th respective hygric and thermic conditions.

In the lowlands, factors other than climate are decisive as well. Table 1 shows how differences in geological evolution result in two very different edaphic and hydrological subregions in the apparently homogeneous Beni plain (Fig. 1). In contrast to the geologically younger southern Beni savannas which are more closely related to the geologically and ecologically similar Pantanal, the northern Beni savannas are dominated by Cerrado-like vegetation which is similar to the vegetation of the Cerrados of the Chiquitania highland and that of Brazil. Although the species diversity in both Beni subregions seems to be similar, many plant species are of different biogeographical origin and the distribution boundaries of both floras are strongly related to the previously mentioned geological and edaphic borders. These differences could be found up to family level (Hanagarth & Beck 1996).

Distribution of seasonal evergreen rain forests in the lowlands, woodlands and grasslands seems to be related with sedimentation caused by fluvial processes which alter the soil and the hydrology of surface and ground waters. Savannas predominate waterlogged areas, in areas with lateritic duricrusts and in young alluvial overflow plains with compact soils.

Climatic factors play a larger role in the semihumid-semiarid transition zone and particularly in the south. However, the distribution of Cerrado and Chaco dry forests still reflects the general influence of the region's geology and soils. Cerrado vegetation is generally found only in the old Precambrian shield. Chacoan forests are found only in areas having young Pleistocene and Holocene sediments that were deposited by rivers originating in the Andes. Although the sharply defined biogeographical boundaries of the northern Chaco and of the Chiquitanian semidecidous forest seem to be determined mainly by climate, but their occurrence within this climatic zone is generally related to geomorphology, certain drainage patterns and soils.

Table 1. Geoecological differences between the northern and southern Beni savannas

	Northern Beni	Southern Beni
Savanna type	Termite-earthworm-mound-savanna; proposed denomination: "Campos del Beni"	Levee-Isla-Park Savanna; proposed denomination: "Llanos de Mojos"
Main biogeographical relationship	Brazilian and Chiquitanian Cerrado	Brazilian Pantanal
Relief	Flat to moderately rolling high plains; glen-like, and river plain valleys	Generally flat to slightly rolling plains with deep riverbeds; comparable valleys do not exist
Geology, geomorphology	Old alluvial plain, partially dissected, large, poorly drained plains with zoogenic *sartenejales*, subrecent sedimentation present only near the Andes, erosional processes dominate in the centre	Young alluvial overflow plain, subrecent, and recent river levées, terraces, and platforms; erosional processes low, recent sedimentation near the Andes
Soils	Highly weathered, leached, very acidic to acidic (pH 3.5->5), presence of lateritic duricrusts, pisoliths, nodules, plinthites; acrisols, ferralsols, distric luvisols, and cambisols; contains no carbonates	Slightly weathered, moderately nutrient-rich, very acidic to alkaline (pH 3.4->9) laterites are absent, plinthite is scarce; hydromorphic soils, eutric luvisols, and cambisols, alkaline soils; rich in carbonates
Type of inundation	Large periodically waterlogged *sartenejales*, floods restricted to river flood plains, large swamps only west of the Beni river	Periodically deep (till up to over 2-5 m), large-scale floods (>80,000 km²), large swamps, duration of floods 3-4 months, in marshes 6-10 months
Surface waters of regional origin	Electric conductivity very low to low (4.2- 20(51)µS/cm); clear (white) waters dominate, "black" waters are rare	Electric conductivity low to high (very high) (13-130(600)µS/cm); in rivers white waters in dry season, "black" waters in rainy season, seasonal "black" and "brown-clear" water dynamic in swamps and creeks
Ground waters	Electric conductivity extremely low to low (5.5-40(50)µS/cm; fluctuation of water level high (up to 8 m)	Electric conductivity moderate to very high (135-1743µS/cm), fluctuation of water level low to moderate (1-5 m)

Sources: Hanagarth (1993), Hanagarth & Beck (1996), Hanagarth (unpubl.).

A biogeographical patchwork exists in the area surrounding the city of Santa Cruz where strong climatic gradients predominate. It is here that most Amazonian and Yungas elements reach their southernmost, Chaco and Bolivian-Tucuman forest species their northernmost, Cerrado species their westernmost, and Andean species their eastern-most distribution boundary (Hanagarth & Beck 1996).

2.5 Megadiversity is mainly the result of biogeographical juxtaposition

Species numbers are strikingly high, giving Bolivia a level of diversity that is comparable to that of the most diverse countries in South America as well as throughout the world. Moraes and Beck (1992) calculate that there are 18,000 to 20,000 plant species (Bryophyta: 1,200 species, Pteridophyta: 1,300 species, Gymnospermae: ca. 16 species, Dicotyledoneae and Monocotyledoneae: 17,000 species). Ibisch (1996) offers a checklist containing 1,054 Bolivian epiphyte species. About 14,000 plant species are actually known (Beck, pers. com.).

Birds are the best-studied groups of all animals. Arribas *et al.* (1995) registered 1,358 species (43% of all South American fauna). Compared with the data of Remsen and Traylor (1989), who mention 1,274 species, this represents an increase of 6.2% in six years. As more inventories are conducted in several lesser known areas near the Brazilian, Peruvian and Paraguayan borders, the number of bird species may rise to nearly 1,500. Although fish are little-known vertebrates, approximately 500 species are registered today and 200 more species may be expected when the Yungas rivers, the northern most Amazonian waters and the Bolivian Pantanal are studied more intensively. Almost 400 fish species were found in the Río Mamoré basin. New mammals were caught during several expeditions conducted in the past few years, raising the number of known native species to a total of 319. Reptiles and amphibians are the least known vertebrates. Ergueta & Sarmiento (1992) listed 220 reptile species, and De la Riva (1990) 112 amphibians. The number of species is increasing continually in both groups. Eleven new amphibian species have been described for Bolivia since 1990 (Erqueta & Morales 1996).

The level of our knowledge of endemism is still inadequate. Taxonomic revisions and more collections are needed. Previously published data regarding species numbers is not representative and requires correction. Ibisch (1996) mentions 298 (28%) confirmed or probable epiphyte endemics, distributed primarily in the eastern Andean slope. This is a contrast to the Beni savannas where actually 14 revised plant species seem to be true endemics (Hanagarth & Beck 1996). The level of our knowledge of fish, amphibians and reptiles does not permit any definitive conclusions. Today, 26 reptiles and 23 amphibians may be considered as endemic to Bolivia (Ergueta & Morales 1996). A recent review by Anderson & Tarifa (1996) yielded 15 endemic mammal species. Several potentially "new" species are still to be described.

Endemism seems to be very high in the Yungas mountain forests and in some transition zones with extremely pronounced rainfall and temperature gradients, such as those in the Cochabamba area, east of the Cordillera Real in the Department of La Paz, and along the extreme southern Andean slope, where climatic conditions were more stabile in the past. They seem not to have been as strongly affected by climate changes as most other areas were. The level of endemic diversity is considerably lower in the lowlands. However, this is not the last word on the subject. It seems that both species endemism and the number of species with a narrow distributional range areas are low. On the other hand, the number of subspecies endemics may be somewhat higher than is currently known.

Bolivian forests and savannas are considerably different in terms of their diversity. The greatest diversity has been found along the eastern Andean slope. Although alpha diversity is lower in the upper and middle Yungas than in the lower Yungas, beta diversity is high due to the orographical and local variations in the region's ecosystems. The northern part of the lower Yungas displays unusually high regional and local diversity due to the strong influence of the high general species richness of the upper Amazonian forests near the base of the Andes (Conservation International 1991) and as the result of the close juxtaposition of different floras and faunas. For example, it was possible to collect 988 plant species in a little more than one week and to register 403 bird species within two weeks in the Alto Madidi area. This, together with the region's suspected 140 mammal species, evidences a very high level of diversity. Remsen & Parker (1995) stress that the whole Madidi area may contain somewhat more than 1,000 bird species, and even several more when the adjacent fauna of the Tambopata-Candamo reserve is taken into account as well. They believe this area to have the greatest bird richness in the world.

Similar patterns of diversity may exist in many other areas along the Andean slope southward to Santa Cruz, where Clarke & Sagot (1996) were able to register more than 830 bird species in and around the Amboró National Park where biogeographical diversity of species and ecosystems may be higher than in the north due to high beta diversity. Although diversity decreases further south, endemism seems to remain high in the Bolivian-Tucuman and inner-Andean dry valleys (Ibisch 1996).

A strong north to east species-richness gradient as well as a slighter west to east gradient exists in the Bolivian lowlands.

It is estimated that diversity in the Amazonian rain forests near the base of the Andes is extremely high, with more than 2,000 plant species and over 800 tree species per 10,000 km^2 (Ibisch 1996). Conservation International (1994) and Duellman & Koechlin (1991) cite more than 550 bird species over the space of a few thousand hectares. 92 mammals more than 20 lizards, 47 snakes, five turtles, four crocodilians and almost 80 amphibians are expected in one locality in the neighbouring Peruvian Tambopata-Candamo reserve. Species numbers of mammals may increase to about 130 (including >50 bat species). There may be 500 ant species, some 1100 diurnal butterfly species (Lepidoptera) and 300 carabid species (Carabidae, Coleoptera), evidencing the high local diversity of many insect groups. The areas inhabited by several endemic bird species of the southwestern Amazonian Inambari endemism centre extend south to Santa Cruz. The eastern parts of the Bolivian Amazon region seem to have a biogeographical influence of the Rondônia centre and a lower species richness. A high level of diversity may exist along the base of the Andes and south to Santa Cruz, even though cold waves may keep some tropical plant and animal species from existing in the area's southern humid and perhumid forests. In contrast, southern elements may have increasingly greater influence.

The transition zones of forests and savannas draw sharp biogeographical borders for many species, and diversity is lower in savannas, compared to the rain forests.

Because they are more arid, Chiquitanian semideciduous forests may have only 200 to 400 tree species. Bolivian inselbergs which mostly occur as granitic outcrops are the most southeastern outcrops on the continent and are probably quite isolated from other neotropical inselbergs. Their species composition differs from the surrounding vegetation. Compared to Venezuela and Brazil, these inselbergs have lower levels of general and gamma diversity which can perhaps be attributed to their marginal and isolated location within a climatic transition. It appears that endemism is lower in Bolivian inselbergs than in other neotropical inselbergs (Ibisch *et al.* 1995).

There is somewhat more information regarding local and regional diversity in the Beni savannas and the Chiquitanian cerrado. Long-term botanical studies of two localities of the Beni savannas yielded 523 and 799 plant species respectively (Hanagarth & Beck 1996). Killeen (1990) collected 609 species in the Chiquitanian cerrados of Concepción. A total of 284 birds, 68 mammals and about 25 amphibians has been registered in the inundated savanna of Espiritu (Hanagarth & Specht in prep., Aguirre *et al.* 1996) as well as 257 bird species in Concepción (Davis 1993). This data tallies with comparable studies in other neotropical savannas. The same is true of many other faunal groups. That is to say, the number of ant species, including those species living in gallery and island forests, does not exceed 200 (Verhaagh, pers. com.), compared to more than 500 species in the rain forest.

The Chacoan dry forest has the lowest diversity. Based on their numbers, flora and fauna species are less diverse. Nevertheless Chacoan endemism is high. Abundances, among several mammal species for instance, may occur in resource-rich areas.

3 Outlooks for natural resource management

There is tremendous potential for the Bolivian economy in the high biodiversity of the country's forests and savannas and their biogeographical differentiation. At the same time however, this potential poses a tremendous challenge to efforts to implement the biodiversity convention and to realize sustainable development because it is not possible to apply general resource-use strategies. Considering how much must be done in the political, social and economic fields, Bolivia is confronted with problems that are directly or indirectly related to the destruction of natural resources. Poverty is no longer the only factor that accelerates the exploitation of valuable resources. It has been joined by the interests of powerful groups such as agro-business and the logging industry.

The Bolivian government expressed its interest in sustainable development in the "Cumbre de las Americas", the second western hemisphere meeting of politicians on energy, water, and biodiversity that was realized in December 1996 in Santa Cruz de la Sierra. Time will tell whether it is possible to create the foundation necessary to sustainable development at national and international level.

3.1 Status of Bolivian conservation

In counterpoint to the past (Hanagarth 1990), the Bolivian government has reformed several laws and passed a number of new ones such as the environmental law in the last eight years. The new law on the conservation of biodiversity has not yet gone into force. Several laws are essential to the implementation of sustainable development and natural resource management (see Sánchez de Lozada, this volume). Hopefully, the decentralization process supports these activities, but there is the possibility that local and regional bodies will favour the exploitation of resources for short-term economic gains. Efficient long-term monitoring systems are needed.

3.2 Protected areas

In recent years, Bolivia has made substantial progress in the conservation sector, with the help of international cooperation. The National System of Protected Areas (SNAP, Sistema Nacional de Areas Protegidas) consists of 26 conservation units (Sánchez de Lozada, same volume) Other areas have been proposed and their inclusion in the system is still pending. Several areas require a redefinition and/or need manageable boundaries (or limits). A few protected areas have management committees. The SNAP protects approximately 10% of Bolivia's entire territory.

The conservation and rigorous protection of several protected areas' natural key ecosystems may be very important for the long-term maintenance of many species. Nevertheless, conservation should not be restricted solely to these areas. Protected areas must be the heart of "source areas."

Representative examples of Amazon forests, Chiquitania semidecidous forests, Pantanal, the dry forests of the inner-Andean valleys, and the Beni savannas still need to be protected. The dry forests in the southeastern Bolivian Andes and in the western parts of the Chaco also need protection due to their high endemism and to the fact that several endemics exist in areas outside existing protected areas, where humans have lived for hundreds of years or where recent colonization could threaten some endemics. Our lack of knowledge about endemics makes it necessary for land-use planning activities to draw upon scientific studies for developing conservation and utilization action plans in such areas.

3.3 Land-use potential and constraints

The regional options for land use in Bolivia are as varied as the country's flora and fauna are diverse. The orographically, climatically and pedologically highly diverse landscape of the eastern Andean slope limits intensive land utilization in that region. Poor soils and periodic inundations hamper intensive land use in large parts of the lowlands. Several studies conducted in the Santa Cruz department, in the Pando and in the Beni prove that using large areas of land that are not appropriate for intensive production due to low soil fertility and other constraints is problematic. Although Furley (1990) holds that some earlier fears concerning the impact of possible agricultural developments on the soil resources of the Brazilian Amazon may have been premature, he establishes that maintaining tropical soil fertility requires sophisticated management methods. The number of strategies for improving or initiating cultivation, ranching and other forms of land use depend on how well they can be adapted to local conditions.

Bolivia's social, economic and infrastructure problems leave little room for establishing a basis for implementing intensive agricultural systems. Highland peasants have neither the know-how or experience necessary to manage the soil or natural lowland ecosystems. Agro-businessmen generally do not show any interest in appropriate land-use management. Similar patterns can be found among long-established inhabitants and, to some degree, among indigenous people who also contribute to the overexploitation of natural resources nowadays. Different strategies for the implementation of sustainable development and natural resource management have to be developed for all four groups. It is urgently necessary that we learn from the negative effects of the earlier, destructive policy of "free" land which created numerous problems for regional planning and complicated the implementation of reforms needed for successful resource management.

The regulation of population movements is one of the most important prerequisites for planning and implementing colonization programs in areas with high land-use potential. New plans are needed that go beyond establishing an infrastructure and assigning plots of land, that extend to assessing peasants in agroforestry and other land use systems, and integrate adjacent areas for natural resource management. If government development organizations and NGOs fail to take research findings from colonization areas elsewhere in the humid tropics into account, they will continue to implement plans that are at odds with field results. Each colonization area has its own special properties and has to be understood as an experimental area where ecologically and economically feasible farming strategies emerge only slowly (approximately 10 years) (Moran 1990). Future colonization projects have to be geared toward minimizing loss of nature and toward balanced growth that causes a minimum of ecological damage.

3.4 Forest management: Problems and options

Despite several land-use constraints, Bolivia still has good chances (at least in theory) for implementing resource management systems in extant natural forests and savannas. Any future success depends on appropriate policies, education, regional planning, control of land occupation and the monitoring of migration, assessment and other factors. Bolivia must develop clear-cut strategies that reconcile the need to utilize the land and the need to conserve nature on the basis of ecological, conservational and bioeconomical know-how.

Since the logging industry presently holds logging concessions for approximately 37% of Bolivia's forest land, it is a highly crucial factor and must be taken into account in any discussion of the management of natural forests. There are approximately 190 logging companies in Bolivia. Forty of them held logging concessions for 2.2 million hectares in the Beni department alone. In 1995, logging concessions had been granted for some 200,000 km^2 (about 18% of the national territory). Most logging areas are located in the Santa Cruz province (64%), in the Beni (20%) and in the Amazonian La Paz province (11%)(CIMAR, unpubl.). Approximately 443,000 m^3 and 477,600 m^3 were harvested in 1993 and 1994 respectively, primarily in Santa Cruz, Cochabamba and Beni (INE-SNIC 1996). About 160 timber species are exploited throughout the entire country. Fifty to 80% of all cut timber is mahogany (mara, *Swietenia macrophylla*). Illegal "motosierristas" (chain-saw loggers) exploit many other areas. It is worth noting that only a few larger areas are not yet affected by timber exploitation. Both logging companies and "motosierristas" use only 10 to 50% of a tree's usable timber. Today's primary forest exploitation is destructive (Fig. 2) and corresponds to a "cut, cash and run strategy. Such strategies are closely related to bioeconomical and market problems (see also the analysis by Rice *et al.* 1996).

Following extensive debate, Bolivia's new forest law was passed in 1996. Under this new law, 40-year forest concessions are transferred to companies that have to meet stringent requirements and obligations. The law allows indigenous peoples and other traditional groups, to use the forest in their native areas. Areas rich in rubber (gen. *Hevea*), Brazil nuts (*Bertholletia excelsa*) and palms (mainly *Bactris gasipaes, Attalea phalerata, Euterpe precatoria*) are transferred preferably to the peasant communities that have traditionally used them. The Ministry of Sustainable Development and Environment is politically responsible for this. The logging industry is not in a position to implement sustainable forest management. Experience is meagre and logging companies do not possess the necessary technical and economic know-how. Further, there are no incentives for implementing sustainable management. Technical assistance may be needed. The governments of Bolivia and the United States recently applied for the inclusion of mahogany in the CITES Appendix II because its status has become critical.

Existing logging concessions are not incorporated into general regional planning. Corresponding reforms are urgently necessary in light of the experience of other tropical countries which shows that conditions are most favourable for the sustainable management of natural forests when (1) there is little competition

Fig. 2. Legal logging of mahogany *(Swietenia macrophylla*) in a lowland forest
Ilustración 2. Tala legal de mara (caoba; *Swietenia macrophylla)* en un bosque de las tierras bajas

over use of the forest, (2) large tracts of forest still exist and (3) the quality of the forest products is so high that they can be sold for high prices (Prabhu *et al.* 1993).

3.5 Use of biodiversity by local people

Since it is highly unlikely that local inhabitants and people living in adjacent areas will accept the situation, when logging companies have exclusive rights for utilization of large forests concessions, forest protection policy must include, as one of its most important components, the promotion of social forest management programs. In addition to some form of certification or a "green" label for the sustainable production of timber, natural-forest management needs a "red" label as well, to signal that it is "socially sustainable." Tests for applying such methods, integrating the needs of local people, are being conducted in many parts of the tropics (Poffenberger 1990, Plotkin & Famolare 1992, Kiernan *et al.* 1992). Socially sustainable forest management needs to include other ways of utilizing forests than just logging, because of economic necessity. Many Bolivian forests are used in diverse ways. For instance, the Chácobo Indians who live in the northern Beni use 82% of the tree species. In the vicinity of the Ríos Blanco y Negro Wildlife Reserve, three indigenous tribes reported using a total of 305 medicinal plant species (Halloy 1996). The value of sustainable "unimproved" forests is several times greater than the value of the same land after it has been converted to cattle ranching or plantations (Plotkin & Famolare 1992). But given our abysmal lack of knowledge of natural products and their potential uses, it is still hard to answer the essential question of how far high biodiversity is reflected in a high diversity of useful plants. Many non-timber products have only a "local price" for subsistence but no market price because their potential use is unknown.

The cultivation of pharmaceutical useful plants for industrial use may be just one option. Their utilization for health-care purposes at local and regional level is another alternative. It is however doubtful that such products could be subject to long-term demand as several conservationists and politicians hope. The forests' potential is diversified by the harvest of fruits, racines and nuts. Good marketing strategies are important instruments for introducing new products to national and international markets. Forests rich in Brazil nuts and palms (see also Moraes, same volume) must be placed under management before their last natural resources suffer any more negative impacts. We run the risk of losing natural capital that is needed to initiate successful projects. Bioeconomical assessments of rubber and other promising products that permit the implementation of low-cost initial projects are needed.

Although there is potential for improved market-oriented production that uses various enrichment/refining-systems, natural forest management, conserving biodiversity of an acceptable level, support a relative low human carrying capacity. This includes that each managed area needs to be clearly zoned in a way that indicates areas with different levels of utilization and conservation, as well as the type and degree of intervention required, applying the source-sink model, considering the creation of a corridor network, permitting the exchange of populations, among the partial areas. An array of alternatives such as agroforestry, cattle ranching and ecotourism has to be evaluated before their introduction, in order to buffer the human impact they involve.

Traditional communities and indigenous peoples have fundamental know-how, but their current forest utilization systems suffer destructive changes when they are adapted to market demands and Western culture. They are not capable of dealing with the complex socio-economic, cultural and ecological changes of a rapidly-changing environment on their own. It is therefore necessary to develop strategies for implementing management systems that integrate the specialized knowledge of traditional resource users into scientific knowledge and vice versa.

3.6 Wildlife utilization in forests

The future of the forests and the people who inhabit them will also depend in part on how successfully we incorporate wildlife management into the broad array of land-use options. If large tracts of wilderness do not begin to produce significant returns in the form of forest and wildlife products for local peoples within the next few years, the pressure that leads to spontaneous colonization and larger development projects will quickly gain the upper hand (Freese & Saavedra 1991). Even though most game species seem to be reaching their natural population limits in some remote areas of Bolivia, there is a strong tendency to overhunt, even in regions with low human density. Even in the Pando province which is almost completely forested (0.6 inhabitants/km², INE-SNIC 1996), large mammals were decimated in the region, probably by Brazil-nut gatherers. Several regions are still suffering from the effects by the trade in animals and hides that occurred between the 1950s and 1980s. The fauna in areas covered by logging concessions has been depleted to feed the loggers, and in several other areas that villages and towns can access, monkeys and other game mammals are the target of intense commercial hunting. So-called "vaciamientos," which means "abnormal absence of populations of game species in one or more areas" caused by excessive hunting are also found in sparsely populated protected areas such as the Biological Station Beni and the indigenous territory Isiboro Sécure.

As in other countries, large mammals are more endangered than smaller ones. Since monkeys have low reproduction rates, they disappear more quickly than larger animals with higher reproduction rates and smaller mammals and birds (Robinson & Redford 1991, Ojasti 1991). For many species, productivity depends on their density and on the carrying capacity of the ecosystems in which they live. In Bolivia, only a few studies have been conducted on the qualitative and semiquantitative status of game species, but it has been proven that the most commonly hunted game species are usually the largest members of their taxonomic group as well as the largest species in the forest. Hunting has a strong effect on such species. The effects of moderate and 'heavy' hunting were reviewed by Redford (1993), who stresses that moderate hunting reduces non-primate mammalian game populations to 19.3% of the levels reported at similar, unhunted sites, and 'heavy' hunting can even reduce them to 6.3%. A similar comparison made for game birds shows that their original density falls to 26.5% under the effects of moderate hunting and to 5.4% in the case of 'heavy' hunting.

Comparisons of data indicate that non-commercial subsistence hunting may affect the populations of some game species when human density is about or more than 1.3 inhabitant/km², causing some local thinning and/or a general population decrease. Townsend (1996) confirms that several species are overexploited in the Sirionó territory of Ibiato (1.35 inhabitants/km²) in the eastern Beni province. Cuellar (1997) documents that the Guaraní Indians living in the territory of Akae (which measures 33 km² and has about six inhabitants/km²) in the northwestern Chaco have reported a severe depletion of game species. Stearman (1990) and Robinson & Redford (1991) doubt that the Yuquí Indians' hunting of certain game species is sustainable.

Although Bolivia's indigenous territories suffer the negative effects of overhunting today, they may have the best options for implementing long-term management of game species, as well as the larger protected areas. Native people utilize game in more diversified ways than peasants and colonists do (Redford 1993). But there are serious long-term constraints such as very rapid human population growth and the growing amount of settled territory, the erosion of cultural integrity, the restricted diversity and heterogeneity of ecosystems and their carrying capacity for game species. Townsend (1996) concludes that the Sirionó population may double through natural growth in about 10 years' time, and Cuellar (1997) reports annual growth of about 4 % in the Akae community which means a doubling in about 18 year's time. Game management has to include source areas where ecological conditions permit high reproduction rates and sink areas – where reproduction rates are lower as a result of unfavourable conditions – which may be the most important hunting areas. Establishing resource management in such territories involves widely diverse political and economic interests which makes conflict inevitable. When indigenous territories must have a role in biodiversity utilization and conservation, then they have to be designated for long-term sustainability, must be large enough. Moreover, planners together with native communities have to look for economical alternatives to reduce or to moderate the pressure on game species. Townsend (1996) ascertained that the territory of Ibiato, which covers an area of 340 km^2 (459 Sirionó inhabitants in 1993) with 161 km^2 of forest and 179 km^2 of savanna woodlands, grassland and swamps, comprises only one third to one half of the minimum area needed for sustainable game management. Management must integrate the surrounding areas. The Isiboro Sécure, in Pilón Lajas and in the Chacoan Kaa-Iya territories may have the best chances for the implementation of game management with high subsistence and economical benefits.

3.7 Restrictions on and options for integrating wildlife management in the Beni savannas

Natural resource management in Bolivian savannas is different from forest management in several ways. Since the Beni savannas are the largest and most promising savannas for specific wildlife management, the following comments will focus mainly on their potential for natural resource management. Land in the Beni savannas is generally used for private low-level cattle ranching. In contrast to agriculture which must deal with economical and ecological constraints, cattle ranching could be improved by about 30% to 50% in the southern Beni savannas with ecologically sound management methods.

Few other regions can match the benefits that the Beni savannas offer for the implementation of wildlife management systems. Human population density is low (1.3 inhabitants/km^2, INE-SNIC 1996). Most parts are periodically flooded wetlands that have many small and large swamps (some covering more than 300 to 400 km^2) and comprise one of the most important ecosystems for waterfowl, wading birds and migrant birds of South America. Although crocodiles and some other species have been decimated as the result of commercial hunting since the 1950s and 1960s, it would be possible to implement game management systems in a relatively short time, in addition the black caiman (*Melanosuchus niger*) and the londra *(Pteronura brasiliensis)* have not yet been able to recover their populations and swamp deer populations *Odocoileus dichotomus*, Fig. 3) are currently being reduced by hunting and possibly by the transmission of cattle diseases. In the years since the Bolivian government's "veda total" (total

Fig. 3. The Beni savannas are important habitats for the swamp deer (*Odocoileus dichotomus*)
Ilustración 3. Las sabanas de Beni constituyen hábitats importantes para el ciervo de los pantanos *(Odocoileus dichotomus)*

prohibition of hunting) in 1986, the caiman or "lagarto" populations (*Caiman yacare*) have increased well in several extant swamps and rivers. Capibaras or "tapiguaras" (*Hydrochaeris hydrochaeris*) are very numerous because they are only occasionally hunted for hides and their meat is not consumed. Ornamental and commercial fish are abundant, sometimes with more than 200,000 individuals/hectare in swamps (Sarmiento & Hanagarth unpubl.). Since most terrestrial savanna mammals depend in part on forests (Redford & Da Fonseca 1986, Aguirre *et al.* 1996) they are less abundant in open grassland savannas. As a result, peccaries, tapirs and other mammals do not offer interesting opportunities for economically-oriented management. Conservation plans are nevertheless needed. Mammals are more abundant in forest-savanna transition zones (Townsend 1996).

Inundation savannas are fragile ecosystems. For this reason, it is important that decision-makers determine how they are to be changed and utilized on a long-term basis. Although inundation savannas are highly productive ecosystems, there are some constraints that restrict their management. The following options are available for the development of wetland savannas: (1) Conversion of the wetland into a radically different ecosystem in order to create new conditions; (2) Strict conservation of the wetland as a whole to profit from it (3) Maintenance of the wetland ecosystem, its functions and the benefits it provides while allowing a

Fig. 4. *Caiman yacare* (a, above) and *Rhea americana* (b, below) are promising species for cattle and wildlife ranching in the southern Beni

Ilustración 4. *Caiman yacare* (a, arriba) y *Rhea americana* (b, abajo) son especies prometedoras para ganadería y manejo de vida silvestre en el sur de Beni

certain level of resource use; and (4) Maintenance of the greater part of the wetland and conversion of the rest (Roggeri 1995).

In light of the fact that the geomorphology and ecology of these savannas is very complex, options 3 and 4 are to be recommended. A basin-wide management plan that integrates the bordering regions of the Santa Cruz and Cochabamba departments into cross-sectoral and cross-departmental planning and coordination activities is needed. Poorly planned drainage projects in the Santa Cruz region and adjacent regions could cause flood peaks in the central-southern Beni savanna that greatly exceed those experienced during the enormous floods of 1982 and 1992, and thus put towns, villages, cattle breeding operations and natural resources at threat.

Drainage measures and natural erosion processes prove that although local alteration of water regimes changes the composition of pastures in a favourable way in one area, it may have a catastrophic effect in others. Ecologically important areas such as the Biological Station Beni Biosphere reserve could be adversely affected when large swamps, that play a key role in the region's ecology are drained, a measure that would drastically alter plant and animal communities. Road construction projects can cause severe changes. Drained areas may suffer edaphic aridity that causes water and feed shortages during the dry season affecting the regional ecology and economy. Because these savannas are extremely flat, drainage measures can result easily in uncontrollable, chaotic and unfavourable erosion processes along artificial channels.

These savannas offer excellent options for the integration of wildlife management into cattle ranching operations, which would enhance the economic value of large parts of this region particularly if this is accomplished in conjunction with the controlled harvesting of three pre-selected wildlife species which could represent the most efficient economic value. For this purpose, the abundant caiman (*Caiman yacare*, Fig. 4a), ñandu (*Rhea americana*, Fig. 4b) and capibara (*Hydrochaeris hydrochaeris*) currently offer the best basis for implementing management systems. The farming of neotropical species has not proven to be more economical than ranching to date because the higher overhead costs reduce any profit (Terborgh *et al.* 1986). In the case of caiman, ranching may be the only economical method for managing a species whose hides are not as valuable as those of crocodiles and alligators, which can also be found in other tropical regions and compete with the neotropical caiman (Magnusson 1984, Thorbjarnarson 1991). The population of the more valuable benian black caiman remains very small, and they may need rigorous protection for many years to come.

Since passage of the "veda total" in 1986, the caiman population in some areas has recovered considerably better than in the years before. The ñandu population is especially large in the wetlands of the central-southern Beni savanna, but excessive egg gathering takes a heavy toil on their numbers. Meat and valuable hides could be produced in the management of the ñandu. Capibaras remain abundant because their meat is still rarely consumed. Examples from Venezuela and Argentina prove that it is both possible and profitable to integrate the capibara into open-habitat management systems. All three species have high reproduction rates. For instance, the reproductive efficiency of capibara is six times that of cattle and it produces 2.6 times more meat in Venezuela (Ojasti 1991). Whether management is successful depends on the development of markets for meat and the integrated utilization of the hides and derivative products. These three species offer an opportunity for implementing of wildlife management systems within a short time, and could serve to change the minds of politicians who advocate cattle and wildlife ranching.

The swamps, ponds, river and marshes in the cattle ranches around the Biological Station Beni, in the Isiboro Sécure and Ibiato territories as well as in some large swamps may contain areas with potential for management projects. Concepts and plans are needed – that benefit not only ranchers, but small farmers and landless peasants who may resort to poaching in order to access these resources, should game management be placed exclusively in the hands of the ranchers (see also Ojasti 1991).

3.8 The "common property" problem

The 'tragedy of the commons' where communally-held resources tend to be overexploited is likely to prevail. This "common property" problem is of great importance for the management of natural resources (Endres & Querner 1993). At the root of this problem is non-existent or insufficient regulation of property, harvesting and hunting. Control mechanisms must prevent the overexploitation not only of game species, but of timber and non-timber products as well. Optimal long-term utilization of natural resources has to be oriented to the maximum annual yield which is used together with other components to determine a resource's bioeconomically optimal use. Without such regulations, overexploitation and even local or regional extinction may happen. Voluntary restraint on the part of individuals does not have much success. Plotting "common property" may be a remedy, but does not take into account that resource stock moves freely between plots. In this case the user's right in one plot does not give him the right to use a certain resource unit; he is given only the possibility of accessing defined quantities within defined spatial limits. It is absolutely necessary to determine the intensity of use and form of utilization which in turn have to be adjusted to the carrying capacity of the ecosystems and to the dynamics and regenerative capacity (and migration) of each resource. A wealth of experience proves that the "common property" problem exists and that efficient solutions are decisive for the successful management of natural resources. Cost-benefit analyses that also take the biological-economic, social and cultural potential of an area into consideration are indispensable for the development of a conservation-oriented economy.

4 Final comments

The high geoecological and biological diversity of Bolivia is a true challenge that does not permit us to recommend general "prescriptions" for the sustainable management of natural resources. Various examples have been given which could be implemented within a relatively short time, demonstrating to politicians and people that there are opportunities for sustainable resource management. It must however be recognized that the conservation and use of biological resources is much more complicated than the management (or manipulation) of abiotic resources or biological elements in an altered or man-made environment. Managing biological diversity successfully with long-term aims depends on whether politicians complete the task of creating the various necessary foundations soon. This means working to develop a socio-economic and ecological conscience at national and international level. Globalization in its present form does not seem to be compatible with numerous prerequisites for the management of diversity and sustainable development. If the necessary political measures are not implemented, this dream may burst like a soap bubble. The question is: Is there a serious desire for integrating the conservation and utilization of biodiversity into sustainable development? Scientists can offer usable and practical know-how, and managers have to integrate the knowledge and experience that exists at international and local level into projects. Politicians have to create the national and international basis for success.

References

Aguirre LF, Hanagarth W, de Urioste RJ (1996) Mamíferos del refugio de Vida Silvestre Espíritu, Dpto. Beni, Bolivia. Ecología en Bolivia 28: 29-44

Arribas MA, Jammes L, Sagot F (eds) (1995) A Birdlist of Bolivia. Asociación Armonía, Santa Cruz, Bolivia

Beck S (1983) Vegetationsökologische Grundlagen der Viehwirtschaft in den Überschwemmungssavannen des Río Yacuma (Departamento Beni, Bolivien). Diss Bot 80, J Cramer, Vaduz

Beck S (1993) Bergsavannen am feuchten Ostabhang der bolivianischen Anden - Anthropogene Ersatzgesellschaften? Scripta Geobotanica 20: 11-20

Boom, B (1989) Use of plant resources by the Chácobo. Adv Econ Bot 7: 78-96

Centro de Datos para la Conservación (1995) Recursos forestales y características de uso. CDC-Bolivia, La Paz

Conservation International (1991) A Biological Assessment of the Alto Madidi Region (and adjacent areas of Northwest Bolivia). Rapid Assessment Program, RAP Working Papers 1, Conservation International Publications, Washington

Conservation International (1994) The Tambopata-Candamo Reserved Zone of Southeastern Peru: A Biological Assessment. Rapid Assessment Program, RAP Working Papers 6, Conservation International Publications, Washington

Clarke R, Sagot F (1996) Guía para observadores de aves en el mejor lugar del mundo: Area protegida Amboró, Bolivia. Armonía/BirdLife International, Santa Cruz, Bolivia

Cuellar RL (1997) Aprovechamiento de la fauna silvestre en la comunidad Akae, sector Charagua Norte, Cordillera, Dpto. de Santa Cruz, Bolivia. Tesis de maestría, Fac Cienc Puras y Nat, UMSA, La Paz

Davis SE (1993) Seasonal Status, Relative Abundance, and Behavior of the Birds of Concepción, Departamento Santa Cruz, Bolivia. Fieldiana, Zoology, New Series, No. 71, Publ. 1444

Duellman WE, Koechlin JE (1991) The Reserva Cuzco Amazónico, Peru: Biological Investigations, Conservation and Ecoturism. Occ Papers, Mus Nat Hist, Univ Kansas, 142: 1-38.

Endres A, Querner I (1993) Die Ökonomie natürlicher Ressourcen. Wissenschaftliche Buchgesellschaft, Darmstadt

Ergueta P, Morales C de (eds) (1996) Libro rojo de los vertebrados de Bolivia. CDC-Bolivia, La Paz

Freese CH, Saavedra CJ (1991) Prospects for Wildlife Management in Latin America and the Caribbean. In: Robinson JG, Redford KH (eds) Neotropical Wildlife Use and Conservation. The University of Chicago Press, Chicago, London, pp 430-444

Halloy, S. (1996) Developing Plants as a Conservation Tool in Bolivia. Medicinal Plant Conservation, IUCN, vol. 2: 7

Hanagarth W (1990) The Nature conservation Problems of a Developing Country, taking Boliva as an Example. Natural Resources and Development 31: 65-84

Hanagarth W (1993) Acerca de la geoecología de las sabanas del Beni, noreste de Bolivia. Editorial Instituto de Ecología, La Paz

Hanagarth W, Beck SG (1996) Biogeographie der Beni-Savannen (Bolivien). Geographische Rundschau GR 48 H 11: 662-668

Heinrich F, Eguivar MR (1991) El medio ambiente en la legislación boliviana. La Paz

Ibisch PL (1996): Neotropische Epiphytendiversität - das Beispiel Bolivien. Martina Galunder Verlag, Wiehl

Ibisch PL, Rauer G, Rudolph D, Barthlott W (1995) Floristic, biogeographical, and vegetational aspects of Pre-Cambrian rock outcrops (inselbergs) in eastern Bolivia. Flora 190: 299-314

INE-SNIC (1996) Bolivia en cifras 1990-1995. Instituto Nacional de Estadística - Secretaría Nacional de Industria y Comercio, Cumbre de las Américas, Santa Cruz, Bolivia

Kessler M (1995) *Polylepis*-Wälder Boliviens: Taxa, Ökologie, Verbreitung und Geschichte. Diss Bot 246, J Cramer, Gebr. Borntraeger Verlagsbuchhandlung, Berlin

Kiernan M, Perl M, McCaffrey D, Buschbacher RJ, Batmanian G (1992) Pilot natural forest management initiatives in Latin America: lessons and opportunities. Unasylve 169 (43): 16-23

Killeen T, Louman T, Grimwood T (1990) La ecología paisajística de la región de Concepción y Lomerío en la provincia Ñuflo de Chávez, Santa Cruz, Bolivia. Ecología en Bolivia 16: 1-45

Magnusson WE (1984) Economics, developing countries and the captive propagation of crocodilians. Wildl Soc Bull 12: 194-197

Moraes M, Beck SG (1992) Diversidad florística de Bolivia. In: Marconi M (ed) Conservación de la diversidad biológica en Bolivia. Centro de Datos para la Conservación, La Paz, pp 73-111

Moran EF (1990) Private and Public Colonisation Schemes in Amazonia. In: Goodman D, Hall A (eds) The Future of Amazonia - Destruction or Sustainable Development? Macmillan Press, London, pp 70-89

Ojasti J (1991) Human Exploitation of Capybara. In: Robinson JG, Redford KH (eds) Neotropical Wildlife Use and Conservation. University of Chicago Press, Chicago, London, pp 236-252.

Plotkin M, Famolare L (eds) (1992) Sustainable harvest and marketing of rain forest products. Conservation International, Island Press, Washington D.C.

Poffenberger M (ed) (1990) Keepers of the forest - Land management alternatives in Southeast Asia. Kumarian Press, West Hartford, Connecticut

Prabhu BR, Weidelt HJ, Leiner S (1993) Erfahrungen und Möglichkeiten einer nachhaltigen Bewirtschaftung von artenreichen tropischen Regenwäldern - eine Untersuchung anhand von vier Fallbeispielen. BMZ Forschungsbericht 109, Weltforum Verlag, Bonn

Redford KH (1993) Hunting in Neotropical Forests: A Subsidy from Nature. In: Hladik CM, Hladik A, Linares OF, Pagezy H, Semple A, Hadley M (eds) Tropical Forests, People and Food: Biocultural Interactions and Applications to Development. Man and Biosphere Series, Vol. 13, pp 227-246

Redford KH, Da Fonseca GAB (1986) The Role of Gallery Forest in the Zoogeography of the Cerrado's Non-volant Mammalian Fauna. Biotropica 18(2): 126-135

Ribera MO (1992) Regiones ecológicas. In: Marconi M (ed) Conservación de la diversidad biológica en Bolivia. Centro de Datos para la Conservación (CDC-Bolivia), La Paz, pp 9-71

Remsen JV, Traylor MA (1989) An Annotated List of the Birds of Bolivia. Buteo Books, Vermillion, South Dakota

Remsen JV, Parker III TA (1995) Bolivia has the opportunity to create the planet's richest park for terrestrial biota. Bird Cons Intern 5: 181-199

Rice RE, Gullison RE, Reid JW (1996) Can Sustainable Management save Tropical Forests? unpubl. Draft

Robinson JG, Redford KH (1991) Sustainable Harvest of Neotropical Forest Animals. In: Robinson JG, Redford KH (eds) Neotropical Wildlife Use and Conservation, University of Chicago Press, Chicago, London

Rodan BD, Newton AC, Verissimo A (1992) Mahogany conservation: Status and Policy Iniciatives. Environmental Conservation, vol. 19, No. 4: 331-342

Roggeri H (1995) Tropical Freshwater Wetlands - A Guide to Current Knowledge and Sustainable Management. Kluwer Academic Publ., Developments in Hydrology 112: 84-119

Ronchail J (1995) Variabilidad interanual de las precipitaciones en Bolivia. In: Ribstein P, Francou B, Coudrain-Ribstein A, Mourguiart P (eds) Eaux, glaciers & changements climatiques dans les Andes tropicales. Bull Institut Français d'Études Andines vol. 24 N° 3, Lima: 369-378

Stearman AM (1990) The Effects of Settler Incursions on Fish and Game Resources of the Yuquí, a Native Amazonian Society of eastern Bolivia. Human Organization 49: 373-385

Terborgh J, Emmons LH, Freese C (1986) La fauna silvestre de la Amazonia: El despílfarro de un recurso renovable. Bol de Lima 46: 77-85

Thorbjarnarson, JB (1991) An Analysis of the Spectacled Caiman (*Caiman crocodilus*) Harvest Program in Venezuela. In: Robinson JG, Redford KH (eds) Neotropical Wildlife Use and Conservation. University of Chicago Press, Chicago London, pp 217-235

Townsend, WR (1996) Nyao Itõ: Caza y pesca de los Sirionó. Instituto de Ecología-FUNDECO, La Paz

Priorities for conservation in Bolivia, Illustrated by a continent-wide analysis of bird distributions

Jon Fjeldså , Carsten Rahbek
Centre of Tropical Diversity, Copenhagen, Denmark

Abstract. This paper reviews Bolivian avifauna from a continent-wide perspective. In a computer-based analysis with a 1° geographical resolution, humid Andean slopes exceed the Amazon rain forests and lowland savannas in species richness. The biota of hydrologically unstable lowlands are dominated by phylogenetically old species which take advantage of high levels of local habitat turnover and are therefore widespread. According to an analysis of complementarity of species ranges, a very small number of well chosen conservation areas is needed to cover all Amazonian species. A much higher number of conservation areas is needed in the Andes due to the complex patterns of endemism in that region. This difference is even more evident in Gap Analyses which pre-select (1) grids which are already well protected and (2) grids which are virtually uninhabited. It is found that the existing reserve network protects few species which are not already relatively safe in uninhabited areas, and largely fails to protect the most unique aggregates of endemic species in the Andes. These aggregates comprise species of rapidly radiating groups together with some relic taxa, the latter indicating local ecoclimatic stability. The human population pressures are particularly strong near places with peak concentrations of endemics (compared to areas with high species richness but low endemism), possibly because of ecoclimatic predictability, together with a stable water supply from cloud forest ridges. Such ecosystem functions are now threatened by habitat conversion. A balanced regional conservation strategy should comprise concentrated actions in centres of endemism (in Bolivia especially in the montane basins of La Paz and Cochabamba, and in Chuquisaca) and improvements at macropolitical level for sustainable management of widespread biota.

Prioridades de conservación en Bolivia, ilustradas por un análisis realizado en todo el continente sobre la distribución de las aves

Resumen. El presente documento estudia la avifauna boliviana desde una perspectiva que abarca todo el continente. En un análisis basado en datos informáticos con un 1° propósito geográfico, las húmedas laderas andinas presentan una mayor riqueza de especies que la selva tropical amazónica y las sabanas de las tierras bajas. La biota de las tierras bajas hidrológicamente inestables está dominada por especies filogenéticamente viejas que se aprovechan de los altos niveles de transformación del hábitat local y están, por lo tanto, muy diseminadas.

Según un análisis de complementariedad de la extensión de las especies, se necesita un número muy reducido de áreas de conservación bien seleccionadas para cubrir todas las especies amazónicas. En los Andes se requiere un número considerablemente mayor de áreas de conservación debido a los complejos patrones de endemismo en esa región. Esta diferencia resulta incluso mucho más evidente en *Gap Analyses* que preseleccionan (1) las áreas que ya están bien protegidas y (2) las áreas que están virtualmente deshabitadas. Parece que el sistema de áreas protegidas existente protege a pocas especies que aún no se encuentran demasiado seguras en áreas deshabitadas y deja desprotegido al conjunto más importante y singular de especies endémicas de los Andes. Este conjunto auna especies de grupos de expansión rápida y algunos taxa superviviente, indicando esto último estabilidad ecoclimática local. La presión de la población humana es especialmente fuerte cerca de los lugares que muestran una alta concentración de especies endémicas (comparado con áreas con una gran riqueza de especies pero bajo endemismo), quizá a causa de la posibilidad de predicción ecoclimática, junto con un abastecimiento de agua estable procedente de las montañas con bosques de neblina. Tales funciones del ecosistema se encuentran ahora amenazadas por la transformación del hábitat. Una estrategia regional equilibrada de conservación debería incluir acciones concentradas en centros de endemismo (en Bolivia especialmente en los valles de La Paz y Cochabamba y en Chuquisaca) y mejoras a nivel macropolítico para fomentar un manejo sostenible de la biota general.

1 Preface

The conservation of key areas for biodiversity often conflicts with poverty-driven pressures on nature or with national development strategies. It is therefore essential to designate land for conservation and development as efficiently as possible. Good data on biodiversity distribution is needed to identify representative networks of conservation areas. However, urgency and the prohibitive cost of complete biodiversity inventories force us to seek acceptable compromises on accuracy and proxy data. Birds are frequently used because they are more completely charted than any other groups of organisms and are thought to be fairly representative, at least in terms of patterns on large geographical scales (see ICBP 1992, Burgess *et al.* in press).

During recent years, most conservationists have shifted their attention from selected species (mainly spectacular mammals and birds) to ecosystems, especially lowland rain forests. The Amazon lowlands are recognized as the richest of all biomes, and receive special attention. This biome has just more than 1,000 bird species (our data), and up to 233 species per km² (Terborgh *et al.* 1990). However, the magnitude of this species pool is partly due to the biome's enormous area (Rahbek 1997). This paper, which reviews data from a more comprehensive study (Fjeldså & Rahbek in press), evaluates the myths about the conservation value of the lowland rain forests by examining the variation in species richness and endemism in a 1° grid system for the entire South American continent and by identifying fully representative minimum sets of areas which include all species. While the majority of lowland birds are relatively widespread (and therefore inhabit various localities where they may survive with minimal conservation effort), the costs of conservation rise as we try to protect the entire assortment of species. Balmford & Long (1994) demonstrated a world-wide correlation between avian endemism and deforestation rate. We will show that peak concentrations of endemism are often located immediately adjacent to some of the principal centres for human development. In order to formulate balanced conservation strategies, we will relate this observation to a hypothesis concerning the underlying processes (Fjeldså 1994, Fjeldså & Lovett 1997, Fjeldså *et al.* 1997, Roy *et al.* 1997). In the process, we will show that the probably most critical areas for conservation are in specific regions in the Andes where dense human populations depend on ecosystems which are unique in terms of endemic species. Although the causal nexus underlying the correlation between endemism and human populations needs to be explored in detail, we suggest that the principal cause is local ecoclimatic stability. Thus, predictable conditions supported the evolution of unique biological communities and also facilitated the establishment of stable human cultures.

2 Materials and methods

Baseline maps for South American birds were prepared over a ten-year period through close collaboration between the Academy of Natural Sciences of Philadelphia and the Danish research group (see Fjeldså & Rahbek 1997 and in press). These maps include verified records and assumed range boundaries which are conservative but assume that species are continuously present between collecting points where habitat maps and satellite images suggest fairly uniform habitats. The data are entered in a 1° grid using the WORLDMAP computer programme (version 3.18/3.19), a PC-based graphical tool designed for fast, interactive assessment of priority areas for conserving biodiversity (Williams 1994). The program accommodates distributional data for large numbers of species and calculates the endemism, as a rarity score, by adding the inverse range sizes of all species present per cell as a percentage of the total rarity scores for all taxa (see Williams 1994 for details). We regard the 1° grid scale as the finest resolution permissible considering the large collecting gaps which exist in certain parts of the continent. However, a more intensive sampling and fairly detailed GIS modelling permit a finer resolution (15'x15') for the tropical Andean montane zone.

Since not all of these datasets have been fully updated or checked for errors, we present only overview maps for the breeding ranges of 1,201 species with a 1° scale which we consider fairly representative in view of their range of ecological adaptations (Fig. 1 a).

Using WORLDMAP we identified conservation priorities using the principles of complementarity of species ranges (Austin & Margules 1986). We identified irreplaceable grids, new flexible grids, and flexible grids from ties (see Williams 1994) of which the first two make up the minimum set of areas needed to keep all species. In order to suggest realistic conservation areas we assume that a species needs to be maintained in at least three grids in order to prevent global extinction. Irreplaceable grids are then determined by those species whose total range falls within 1-3 grids. New flexible grids are those which could be exchanged for other areas, although this will often require larger sets of areas

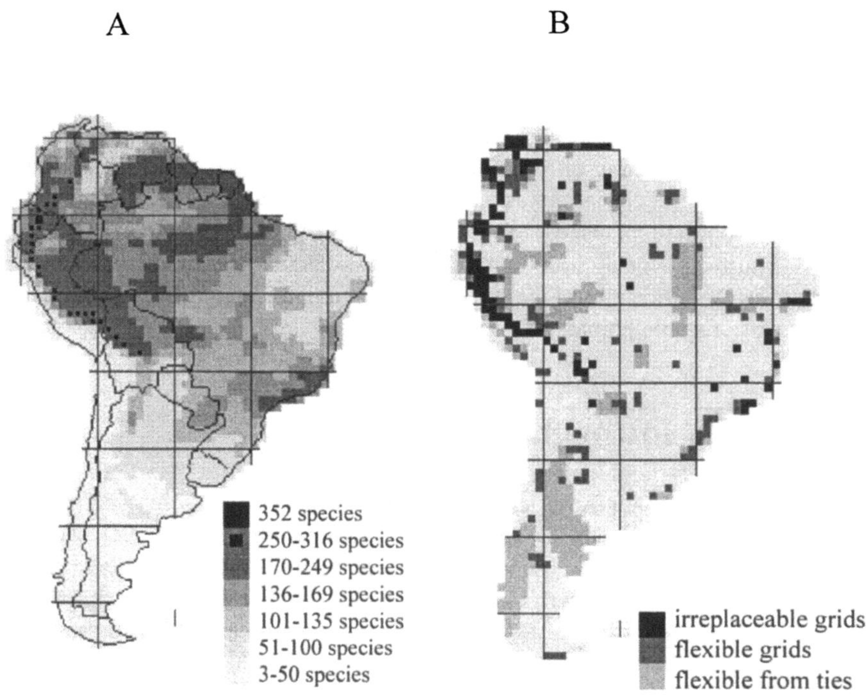

Fig. 1. Species richness of South American birds (A; a sample comprising a total of 1201 species: 1, tinamous [Tinamidae, 46 species] ground-living granivores and frugivores birds inhabiting closed as well as open habitats; 2, parrots [Psittacidae, 118 species], mainly forest-adapted frugivores; 3, hummingbirds [Trochilidae, 241 species], nectarivorous birds of forest and scrub habitats; 4, ovenbirds and woodcreepers [Furnariidae, incl. Dendrocolaptinae; 279 species], insectivores of scrub and open land, with some subgroups of scansorial forest birds; 5, New World flycatchers [Tyrannidae sensu lato, incl. Cotinginae and Piprinae, 458 species], which comprise fly-catching and foliage-gleaning insectivores as well as frugivores, inhabiting a wide habitat range; and 6, icterids [Icteridae, 59 species], which inhabit forests as well as grasslands and marshes. B is the minimum set of areas needed for three representations of all species

Ilustración 1. Riqueza en especies de pájaros suramericanos (A); una muestra que comprende un total de 1.201 especies: 1) tinamús [Tinamidae, 46 especies], pájaros terrestres granívoros y frugívoros que se encuentran en hábitats tanto cerrados como abiertos; 2) loros [Psittacidae, 118 especies], en su mayoría frugívoros adaptados a los bosques; 3) colibríes [Trochilidae, 241 especies], pájaros que se alimentan de néctar y habitan en los bosques y hábitats de monte bajo; 4) horneros y trepadores [Furnariidae, incluyendo Dendrocolaptinae, 279 especies], insectívoros de monte bajo y zonas abiertas con algunos subgrupos de pájaros trepadores de bosques; 5) cazadores de moscas del Nuevo Mundo [Tyrannidae sensu lato, incluyendo Cotinginae y Piprinae, 458 especies], entre los que se incluyen insectívoros cazadores de moscas y recolectores de follaje además de frugívoros, que se encuentran en hábitats de amplio alcance y 6) ictéridos [Icteridae, 59 especies], que habitan en los bosques, así como en praderas y pantanos. B es la cantidad mínima de áreas necesarias para tres representaciones de todas las especies.

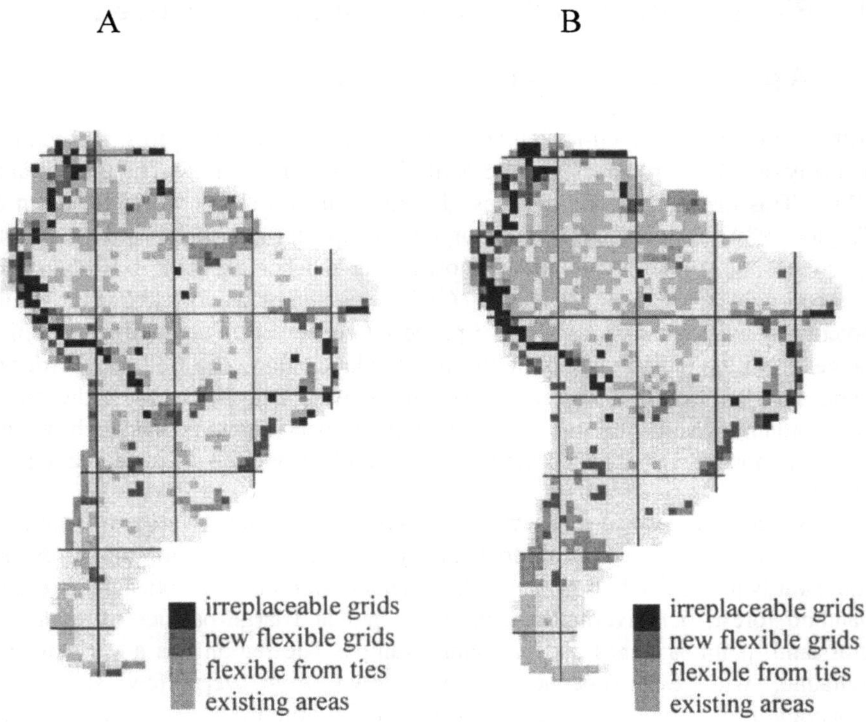

Fig. 2. Minimum sets of New Conservation Areas needed to supplement the existing network of well-protected areas (A) and to protect species which do not live in uninhabited areas (where they may survive at a minimum conservation cost) (B).

Ilustración 2. Cantidad mínima de Nuevas Áreas de Conservación necesarias para complementar la red existente de áreas bien protegidas (A) y para proteger especies que no viven en áreas deshabitadas (donde pueden sobrevivir con un coste mínimo de conservación) (B)

than the minimum set. Flexible areas from ties are the most likely alternatives to identified flexible grids if biodiversity management is impossible in the latter due to conflicts with other interests.

We supplemented the identification of a near-minimum set of potential conservation areas (Fig. 1 b) with two Gap Analyses in which we pre-select grids where the species are relatively safe (Fig. 2 a and b). The first of these analyses considers existing protected areas (from IUCN 1992, and later updates; see Fjeldså and Rahbek in press for precise criteria). We identified 252 such grid units, hereafter referred to as the existing well-protected areas. The second analysis considers human population pressures. Because these are hard to quantify using existing demographic data for political units, we instead adopted some simple criteria which could be applied to the existing 1:500,000 and 1:1,000,000 maps (Fjeldså and Rahbek in press for criteria). We defined a set of 343 grids which we refer to as "uninhabited" areas for convenience.

3 Biogeographic patterns and conservation needs

3.1 A general outline of Bolivian avifauna

Although completely land-locked and therefore lacking marine birds, Bolivia has as many as 1385 bird species, or 43% of all South American avifauna (Armonia 1995). This places it among the most diverse countries in the world (Remsen & Traylor 1989, Armonia 1995). Although a few species are known only from old specimens, no species are definitely known to have become extinct in Bolivia.

According to Remsen & Parker (1995), as many as 1088 species could potentially exist within a 10,000 km² proposed conservation area (Madidi) which ranges from the puna down to the tropical lowlands along the Bolivian border to Peru. This number (if correct) corresponds to what can be found in the entire Amazonian lowland! The area has the potential to become the planet's richest park for birds and presumably for other terrestrial biota as well (Remsen & Parker 1995).

This richness of species reflects the diversity of habitats. Northern Bolivia is covered by tropical lowland rain forest, as typical Amazonian communities extend southwards to Prov. Ichilo in Santa Cruz. This region also has extensive areas of bamboo forest, rich riverine habitats and small and large patches of "pampas" grassland, palm savannas and pantanal swamps. The region has a very diverse avifauna, with two endemic species known from very few sites (*Ara glaucogularis*, *Turdus haplochrous*). A very large proportion of the non-swampy savanna habitats has now been converted into cattle ranches. In the south, this region gives way to the semideciduous Chiquitania forests which grade into the Chaco in southern Bolivia, western Paraguay and northwestern Argentina. The lowlands of Santa Cruz also have scattered hills with low scrubby woodland as found in the uplands (Campos Cerrados) of southern Brazil.

The sub-Andean zone is extensively forested, but this habitat is rapidly dwindling because of timber extraction and the slash-and-burn culture that follows it. Endemic birds of this zone comprise *Simoxenops striatus*, *Myrmotherula grisea*, *Hylopezus auricularis* and *Hemitriccus spodiops* (and probably *Popelairia letitiae*, known only from two nineteenth-century specimens).

The higher Andean slopes have montane cloud forest, which is wet and rich in epiphytes in the north (Yungas of La Paz and Cochabamba). In the Valles region of southern Cochabamba, Chuquisaca and Tarija, the forest is semideciduous, with evergreen forest (including considerable tracts of *Podocarpus*) on certain ridges. The rich avifauna of the humid tropical Andes is fairly intact all the way to Chapare in Cochabamba (with *Aglaeactis pamela*, *Grallaria erythrotis* and *Schizoeaca harterti* endemic to the Bolivian part). Twenty species drop out at the La Paz canyon but nearly 200 cloud-forest birds have their southern boundary within the short distance between Chapare and adjacent Santa Cruz, and only few cloud-forest birds exist further south. However, the Valles region (and adjacent northwestern Argentina) have many unique species (e.g., *Penelope dabbenei*, *Amazona tucumana*, *Otus hoyi*, *Eriocnemis glaucopoides*, *Veniliornis frontalis*, *Scytalopus zimmeri* and *Cinclus schulzi* in semi-evergreen montane forests; *Ara rubrogenys*, *Myiopsitta (monacha) luchsi*, *Oreotrochilus adela*, *Ocethorhynchus harterti*, *Mimus dorsalis*, *Oreopsar bolivianus*, *Poospiza boliviana* in deciduous forests and dry land in montane basins and *Asthenes heterura*, *Saltator rufiventris* and *Poospiza garleppi* in thickets and remnant patches of *Polylepis* forests that exist in the high parts). Nowhere in South America do so many lowland species enter highland habitats as in this part of the Andes, and it seems that dispersal along the dry valleys, and subsequent isolation in the Cochabamba basin or on surrounding ridges, has been an important element for the evolution of the avifauna of the high Andes (Fjeldså 1992, Fjeldså & Maijer 1996). Two endemic species are known from rainshadow valleys in La Paz (*Asthenes berlepschi* and the newly discovered *Cranioleuca henricae*).

The avifauna of high-altitude grasslands and semideserts is rather poor. This zone has been severely changed by overgrazing and burning, with only tiny remnants of natural vegetation left (Fjeldså & Kessler 1996). The ornithologically richest habitats in the highlands are the wetlands, which range from reed-fringed lakes with a large assortment of grebes, ducks and waterhens (Lakes Titicacas and Oruro, and Lake Alalay in Cochabamba town) to highland bogs with geese and ducks, and various types of salty and alkaline lakes in the southwest with flamingo colonies.

The highest species richness in South America, on the 1° scale, is not in the tropical lowlands (as often believed) but along the eastern slope of the tropical Andes region (Fig. 1 a; a similar pattern being found also for other taxonomic groups, see e.g. Barthlott *et al.* 1996). Avian richness peak in the equatorial Andes, but is almost uniform all the way to Cochabamba in Bolivia. However, the endemic species cluster together in quite specific areas (Fjeldså 1995, Wege & Long 1995).

The minimum set of potential conservation areas (with three representations of each species) is 192 grids (86 irreplaceable, 106 flexible; see Fig. 1 b). Because of the complex patterns of endemism, 40% of these grids are in the tropical Andes region, southwards to Cochabamba, and 16% in southeastern Brazil. The entire avifauna of the Amazon Basin can be covered (with three representations) in 25 grids (12%), and this minimum set comprises ten grids placed across major rivers (with different allospecies occupying the opposite banks) and nine grids on the ecotones to adjacent biomes. Two irreplaceable grids relate to species known from single sites.

3.2 Processes underlaying the biogeographic pattern

Haffer (1974) explained the tremendous number of species in the tropical lowland forests with their isolation in ecological 'refuges' which were permanently forested even during the coldest periods of the Pleistocene. These cold periods were the result of interactions between climatic oscillations caused by orbital changes and a general global cooling following the Pliocene period (see Bennet 1990 and Hooghiemstra *et al.* 1993). However, the ages of species were inferred from assumptions about a Pleistocene speciation mechanism, and little attention had been given to whether the patterns could be the result of post-speciation redistribution.

A new "model hypothesis" (Fjeldså 1994, Fjeldså & Lovett 1997) was based on comparisons of distributions of species of different ages (using the comprehensive DNA-DNA hybridization data by Sibley & Ahlquist 1990). This comparison shows that hydrologically unstable plains (including the Bolivian lowlands) are dominated by widespread species which mainly represent deep phylogenetic branches. As many as 80% of the younger species of the Amazon Basin also exist in adjacent biomes (e.g., in the Andes, Brazilian Atlantic Forests, or in gallery forests in the Cerrados, see Silva 1995), and so need not be Amazonian by origin. The tropical lowland areas can therefore be seen as a 'melting-pot' where taxa of potentially diverse origins accumulate rather than as centres of Pleistocene diversification. This interpretation agrees with detailed knowledge that has been obtained in recent years about marine ingressions, tectonically induced flooding cycles and landscape turnover caused by river shifting and meandering (Kalliola *et al.* 1993). Thus, the majority of Amazonian species exist over tremendous areas, as they find patches of suitable habitat that arise locally as a result of the functional heterogeneity of this kind of biome. A particularly interesting situation is described for the Beni plains (Hanagarth 1993, Silva 1994). During the Tertiary, there was continuous hilly terrain from the base of the Bolivian Andes across southern Brazil, probably with rather uniform biota. This biogeographic track was interrupted in the early Pleistocene period when a geological subsidence led to the formation of a sedimentation basin which now appears as hydrologically unstable plains (Hanagarth 1993). These plains are nutritious, characterised by frequent river-shifting and high habitat turnover, and includes Pantanal-like swamp habitats. A completely new ecological regime arose which favoured communities of widespread and to some extent nomadic birds, and which isolated the avifaunas of the non-hydromorphic landscapes in the Andes and southern Brazil.

The DNA divergence data associates the intensive speciation burst in the Pliocene and Pleistocene with the tropical Andes and in adjacent sub-Andean forelands. Undoubtedly, much of the speciation here can be associated with the formation of physical barriers. However, many local assemblies of unique species may also reflect intrinsic properties of certain areas. It has been documented that places with peak concentrations of neoendemics correlate with the presence of aggregates of biogeographic relicts (see, e.g., Fjeldså 1995). The latter would seem to indicate low local rates of extinction which in turn suggests low levels of ecoclimatic change. Preliminary analyses of ecoclimatic data (Best 1992, Fjeldså 1995, Fjeldså *et al.* 1997) indicate that the most important cause could be local climatic moderation leading to persistent mist formation and cloud forests on certain slopes, and thus a predictable water supply in adjacent basins.

3.3 Avian diversity in pre-selected well-protected and "uninhabited" areas

The networks of areas which we considered well-protected or "uninhabited" (see Figs. 1 b and c) comprise large numbers of grids in the tropical lowlands and in southern Chile. The overlap between the two pre-selected sets of 252 and 339 grids is 101 grids. The main differences between the two sets are the enormous extension of uninhabited land in certain parts of the Amazon-Guianas, the few "uninhabited" areas in the tropical Andes region and a total lack of such areas in southeast Brazil.

As demonstrated by Fjeldså & Rahbek (in press), many of the larger reserves in the Andes are situated where endemism is decidedly low compared with nearby areas. This is also the case for the proposed "richest park for terrestrial biota" near the Bolivian/Peruvian border (Remsen & Parker 1995; see Fig. 2) and for the large and species-rich Amboro conservation area. Although rich in species, the "uninhabited" grid cells have even lower levels of endemism than the protected areas (Fjeldså & Rahbek in press).

3.4 Suggested new conservation areas

The differences between the Amazon lowlands and the Andes are very clearly illustrated in Fig. 2. Since most Amazonian species range well into the "uninhabited" hinterlands (away from rivers or ecotones), only very few new conservation initiatives are needed to supplement existing reserves or "uninhabited areas" (3 grids in Fig. 2 b). Comparing Figs. 1 b and 2 a, it is clear that the conservation needs in the Andes (and in southeastern Brazil) were only marginally reduced by the pre-selected conservation areas. Altogether, the 252 pre-selected protected areas must be supplemented with at least 135 new areas (63 irreplaceable, 72 flexible) if the goal of three representations of each species is to be reached. The 339 "uninhabited" areas need to be supplemented with 157 grids (78 irreplaceable, 79 flexible). What is most remarkable is the 115-grid overlap between these suggested new networks (124 overlapping grids if we examine the data matrix for flexible ties which would serve the same aim). Thus, the priorities for new conservation investments remain largely the same whether we pre-select "uninhabited" or formally protected areas (and even if we do not pre-select any areas; compare with Fig. 1 b). Proposed new conservation areas have very high endemism rates compared with the pre-selected protected and "uninhabited" areas ($P<0.000000001$ for both sets using students' t-tests; Fjeldså & Rahbek in press), whereas species richness is not significantly different.

The key problem involved in conserving all South American birds is therefore related to the fact that aggregates of narrowly endemic species are found in regions which lack large uninhabited areas, and many of them are located immediately adjacent to centres of past high cultures or to areas with currently dense human populations in the Andes (Fig. 3 b). Although the causal nexus underlying this spatial relationship between endemism and human settlement is

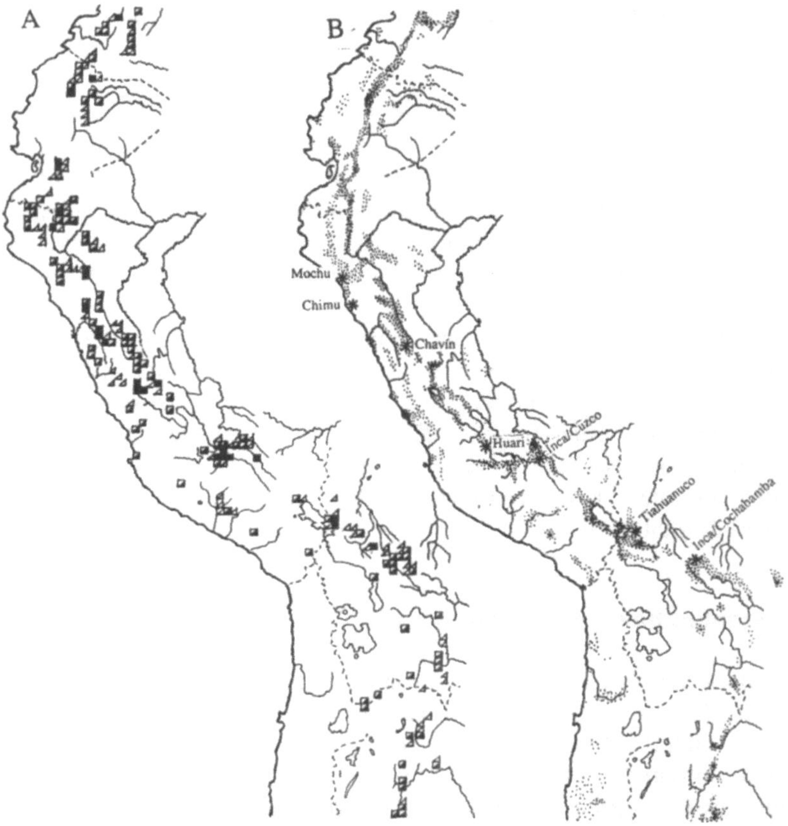

Fig. 3. A minimum set of areas for conservation of Andean species (A), compared with human population pressures (B). The minimum set (irreplaceable [ ] and flexible areas [ ], with the most likely flexible ties indicated [ ]) was identified from a database containing the distributions of 731 Andean montane species (WORLDMAP, 15' resolution), and with the requirement that each species should be represented in three grid units; the clusters being determined mainly by aggregates of endemic species. In B, the stippling shows dense human settlement and infrastructure (based on 1:500,000 maps from Defense Mapping Agency Aerospace Center, St.Louis, Miss.), dots show larger towns, * marks centres of past high cultures, the Huari centre actually extending from the * in Ayacucho to * in Cuzco (I. Schjellerup pers. comm.)

Ilustración 3. Una cantidad mínima de áreas de conservación de especies andinas (A), comparada con presiones debidas a la población humana (B). La cantidad mínima (áreas irreemplazables [ ] y flexibles [ ], con los vínculos más flexibles indicados [ ]) fue identificada por una base de datos que contenía las distribuciones de 731 especies andinas de zonas montañosas (WORLDMAP, resolución 15') y con el requisito de que cada especie habría de ser representada en tres unidades de la reja; los grupos se determinan principalmente por agregados de especies endémicas. En **B**, el granulado muestra densos asentamientos humanos e infraestructura (basado en mapas de 1:500,000 del Defense Mapping Agency Aerospace Centre, San Luis, Miss.), el punteado refleja ciudades grandes, el * marca centros de importantes culturas del pasado, el centro de Huari se extiende actualmente desde el * en Ayacucho hasta el * en Cuzco (I. Schjellerup comentario personal)

still unexplored, we believe that the assumed ecoclimatic stability of these places is a crucial factor.

The network of Bolivian conservation areas is quite adequate for lowland birds (except for *Ara glaucogularis* and *Turdus haplochrous*). However, much more attention must be paid to the dry habitats with endemic species in the montane basins (especially in the Cochabamba Basin and in the hills around Río Caine, but also in the Sorata and Cotacajes basins in La Paz) and for semihumid montane forest in Chuquisaca (Fig. 3 a).

4 Discussion

It is apparent that the existing network of protected areas (Fig. 2 a) was not planned in a way that explicitly considers biogeographic patterns, diversity-maintaining processes or conservation needs. Furthermore, many widespread species were redundantly conserved in reserves situated in regions with virtually no people. Such areas have few endemics and therefore few species to lose! A realistic conservation strategy should consider factors and processes that determine endemism and human settlement.

We do not dispute the uniqueness of the Amazon area, or its overall value for maintaining global biodiversity and moderating the global climate (see Gash *et al.* 1996). However, because its species pool is partly a consequence of its enormous area and heterogeneity in space and time, we fail to recognize precise conservation goals here. Furthermore, global persistence of species is indeed possible despite considerable amounts of local elimination. The need for intensive management is also reduced because of the high general resilience of unstable areas such as plains characterized by river shifting, periodic flooding or other instability. This may be illustrated by comparing the conclusions of different studies on selective logging and forest fragmentation (Danielsen 1997). While these activities had markedly negative effects in centres of endemism, assumed to have been continually forested (Willis 1979, Karr 1982, Kattan *et al.* 1994, Leck 1979, Thiollay 1992, and Canaday 1997 for South America), the impact was moderate or unclear in areas characterized by great ecological shifts (Johns 1991 and Bierregaard *et al.* 1992 for Amazonian sites which were directly affected by past river shifting). We should therefore consider alternative ways for conserving the high biodiversity of lowland plains which acknowledge its interrelationships with dynamic processes operating over large areas.

For tropical lowlands we suggest that macropolitical decisions may be overall more effective than formal protection of specific sites. High priority should be given to better international agreements on forests, climate, tariffs and trade, removal of perverse subsidies and promoting economic planning that takes all the environmental values into account. Support could also be given to colonists to establish stable communities, and the land tenure systems could be adapted to local market situations, as suggested by Beaumont & Walker (1996). In the Bolivian lowlands, traditional site-specific protection might still be needed to secure more natural grassland and palm savannas.

Centres of endemism seem to be threatened to a much larger extent by dense rural populations that often have few alternatives. In much of the tropical Andes, the highest population pressures are found just below the mist zone at 3,000-3,500 m. Many local human populations have already virtually destroyed the natural vegetation in this zone, which now causes soil degradation and an irregular flow of water. The centres of origin of earlier high cultures in the Andes region (Mochu, Chimu, Chavín, Huari, Inca, Tiahuanuco, and also Tairona near Nevado Santa Marta, Colombia) were immediately adjacent to mountain slopes with peak concentrations of endemic species (Fig. 3). There is strong evidence that even very early human cultures exerted strong pressures on the Andean habitats (Fjeldså & Kessler 1996). The centres of the Peruvian Huari and Inca cultures have severely modified and degraded habitats; nevertheless the tiny patches of montane forest which remain have good populations of endemic species (Fjeldså & Kessler 1997). Earls (1991) argues that famines and social tensions resulting from ecosystem degradation were driving forces for the development of the Andean cultures. The Incas' well-regulated land management was terminated by the Conquista when Spanish landlords enforced new and ecologically very inappropriate land use methods. Many areas which may initially have been covered by species-rich montane forests have now been turned into eroded and dry wastelands. In some districts so little cloud forest is left that important water catchment functions may have been lost. The situation is particularly severe in many parts of the Colombian Andes, in southwest Ecuador, in the montane basins of northern and central Peru, and in the dry montane basins and prepuna habitats of Bolivia, where only vestiges of the natural vegetation are left. A critical situation may arise in the Cochabamba area because of irrigation projects and a lack of investment to protect the natural vegetation of the primary water catchment zone.

Although the cost of developing a fully representative network of conservation areas will be high, this investment is indeed relevant, since actions to protect unique local bird communities will relate directly to ecosystem functions which are also important for people. The approach does not necessarily have to entail formal protection, as it may often be more efficient to provide local communities massive support in the form of better land tenure systems and methods which would diminish the need to continue devastating methods of land use. Investment is particularly needed to allow mist vegetation to regenerate on the higher ridges, and could perhaps be induced through agreements to regulate traditional grazing and burning in return for assistance for obtaining higher yields on other parts of the land. Although methods of land use must be changed, we will also emphasize that any such change is unlikely to be successful unless the traditional territorial organization (aunokas) is accepted as the norm for regulating fallow cycles, crop rotations, integration of agriculture and livestock and access to different types of land.

Acknowledgements. This analysis was made possible by the long-term financial support for data gathering and analysis provided by the Danish Natural Science Research Council (currently grant No. 11-0390) and the Danish Environmental

Research Programme (DIVA project). P. Williams kindly provided the WORLDMAP software and did us an enormous additional service by programming it according to our specific wishes. For the provision of distributional data we thank mainly R. Ridgely and the Academy of Natural Sciences of Philadelphia, but also BirdLife International, N. Krabbe, S. Maijer and J. M. Cardoso da Silva; we also thank Ivan Olsen and Casper Paludan for their major computer efforts.

References

Armonia (1995) Lista de las Aves de Bolivia. A Birdlist of Bolivia. Asociación ARMONIA, Santa Cruz, Bolivia

Austin MP, Margules CR (1986) Assessing representativenes. In: Usher MB (ed.) Wildlife conservation evaluation. Chapman & Hall, London, pp 45-67

Balmford A, Long A (1994) Avian endemism and forest loss. Nature 372: 623-624

Barthlott W, Lauer W, Placke A (1996) Global distribution of species diversity in vascular plants: towards a world map of phytodiversity. Erdkunde 50: 317-327

Beaumont PM, Walker RT (1996) Land degradation and property regimes. Ecol. Econ. 18: 55-66

Bennet KD (1990) Milankovitch cycles and their effects on species in ecological and evolutionary time. Paleobiology 16: 11-21

Best BJ, (ed) (1992) The threatened forests of south-west Ecuador. Biosphere Publications, Leeds, U.K.

Bierregaard RO Jr, Lovejoy TE, Kapos V, Santos AAd, Hutchings RW (1992) The biological dynamics of tropical rainforest fragments. BioScience 42: 859-866

Burgess N, de Klerk H, Fjeldså J, Rahbek C (in press) A preliminary assessment of congruence between biodiversity patterns in Afrotropical forest mammals and forest birds. Ostrich

Canaday C (1997) Loss of insectivorous birds along a gradient of human impact in Amazonia. Biol Conserv 77: 63-77

Danielsen F (1997) Stable environments and fragile communities: does history determine the resilience of avian rain-forest communities to habitat degradation. Biodiver Conserv 13: 423-434

Earls J (1991) Ecologia y Agronomia en los Andes. Hisbol, La Paz

Fjeldså J (1992) Biogeographic patterns and evolution of the avifauna of the relict high-altitude woodlands of the Andes. Steenstrupia 18: 9-62

Fjeldså J (1994) Geographical patterns for relict and young species of birds in Africa and South America and implications for conservation priorities. Biodiv Conserv 3: 207-226

Fjeldså J (1995) Geographical patterns of neoendemic and older relict species of Andean forest birds: the significance of ecologically stable areas. In: Churchill SP, Balslev H, Forero E, Luteyn JL, (eds) Biodiversity and conservation of neotropical montane forests. The New York Botanical Garden, New York, pp 89-102

Fjeldså J, Kessler M (1996) Conserving the biological diversity of Polylepis woodland of Peru and Bolivia. A contribution to sustainable natural resource management in the Andes. NORDECO, Copenhagen, pp 1-250

Fjeldså J, Maijer S (1996) Recent ornithological surveys in the Valles region, southern Bolivia - and the possible role of Valles for the evolution of the Andean avifauna.

Centre for Research on Cultural and biological Diversity of Andean Rainforests (DIVA) Tech Rep 1. Kalø, Denmark, pp 1-62

Fjeldså J, Ehrlich D, Lambin E, Prins E (1997) Are biodiversity "hotspots" correlated with ecoclimatic stability? A pilot study using NOAA-AVHRR remote sensing data. Biodiver Conserv 13: 401-422

Fjeldså J, Lovett JC (1997) Geographical patterns of old and young species in African forest biota: the signifance of specific montane areas as evolutionary centres. Biodiver Conserv 13: 325-347

Fjeldså J, Rahbek C (1997) Species richness and endemism in South American birds: implications for the design of networks of nature reserves. In Laurance WF, Bierregaard R, Moritz C (eds) Tropical forest remnants: ecology, management and conservation of fragmented communities. University of Chicago Press, Chicago, pp 466-482

Fjeldså J, Rahbek C (in press) Continent-wide conservation priorities and diversification processes. In Mace GM, Balmford A, Ginsberg JR (eds) Conservation in a changing world. Integrating processes into priorities for action. Cambridge Univ. Press, Cambridge

Gash JHC, Nobre CA, Roberts JM, Victoria RL (1996) Amazonian Deforestation and Climate. John Wiley & Sons, Chichester

Haffer J (1974) Avian speciation in tropical south America: with a systematic survey of the toucans (Ramphastidae) and jacamars (Galbulidae). Publ Nuttall Ornith Club 14: 1-390

Hanagarth W (1993) Acerca de la geoecologia de las sabanas del Beni en el norest de Bolivia. Instituto de Ecología, La Paz

Hooghiemstra H, Melica JL, Berger A, Shackleton NJ (1993) Frequency spectra and paleoclimatic variability of the high-resolution 30-1450 ka Funza I pollen record (Eastern Cordillera, Colombia). Quaternary Sci Rev 12: 141-156

ICBP (1992) Putting biodiversity on the map: global priorities for conservation. ICBP, Cambridge, U.K.

IUCN (1992) Protected areas of the World: A review of national systems. Volume 4: Nearctic and Neotropical. IUCN, Gland, Switzerland and Cambridge

Johns AD (1991) Responses of Amazonian rain forest birds to habitat modification. J Trop Ecol 7: 417-437

Kalliola R, Puhakka M, Danjoy W (1993) Amazonia Peruana. Vegetacíon húmeda tropical en el llano subandino. ONERN, Lima

Kattan GH, Alvarez LH, Giraldo M (1994) Forest fragmentation and bird extinctions: San Antonio eighty years later. Conserv Biol 8: 138-146

Leck CF (1979) Avian extinctions in an isolated tropical wet-forest preserve, Ecuador. Auk 96: 343-352

Pacheco LF, Simonetti JA, Moraes M (1994) Conservation of Bolivian flora: representation of phytogeographic zones in the national system of protected areas. Biodiver Cons 3: 751-756

Rahbek C (1997) The relationship among area, elevation, and regional species richness in Neotropical birds. Am Nat 149: 875-902

Remsen JV Jr, Traylor MA Jr (1989) An Annotated List of the Birds of Bolivia. Buteo Books, Vermillion, South Dakota

Remsen JV Jr, Parker TA III (1995) Bolivia has the opportunity to create the planet's richest park for terrestrial biota. Bird Cons Intern 5: 181-199

Roy SM, Silva JMC, Arctander P, García-Morena J, Fjeldså J (1997) The role of montane regions in the speciation of South American and African birds. In: Mindell DP (ed.) Avian molecular evolution and systematics. Academic Press, New York, pp 321-339

Sibley CG, Ahlquist JE (1990) Phylogeny and classification of birds: A study in molecular evolution. Yale University Press, New Haven, Connecticut

Silva JMC (1994) Can avian distribution patterns in northern Argentina be related to gallery forest expansion-retraction caused by Quaternary climatic changes? Auk 111: 495 499

Silva JMC (1995) Birds of the Cerrado Region, South America. Steenstrupia 21: 69-92

Williams PH (1994) WORLDMAP. Priority areas for biodiversity. Using version 3. Privately distributed computer software and manual, London, U.K.

Willis EO (1979) The composition of avian communities in remenescent woodlots in southern Brazil. Papéis Avulsos Zool, S. Paulo 33: 1-25

Amphibian species diversity in Bolivia

Jörn Köhler, Stefan Lötters, Steffen Reichle
[1] Zoological Research Institute and Museum Alexander Koenig, Bonn, Germany
[2] Institute of Geography, University of Bonn, Germany

Abstract. To obtain a first tentative impression of amphibian diversity patterns in Bolivia, distributional data of 166 known species was compiled and linked to eco-geographical zones. These findings confirm that the general worldwide tendency for species richness to decrease as latitude and/or altitude increase also occurs in large parts of Bolivia in regard to amphibians. The perhumid Yungas play a special role, harbouring a remarkable variety of species for its size, including almost all endemic taxa.

Diversidad de especies de anfibios en Bolivia

Resumen. Para recibir una primera impresión aproximada de los patrones de diversidad de anfibios en Bolivia, se recopilaron datos distributivos de 166 especies conocidas y se asociaron con zonas ecogeográficas. En la mayor parte de Bolivia se confirmó la tendencia general hacia una disminución de la riqueza de especies en anfibios al aumentar la latitud y la altitud. Las siempre húmedas Yungas desempeñan un papel especial, ofreciendo refugio a una considerable variedad de especies relativa al tamaño del área, incluyendo casi todos los taxa endémicos.

1 Introduction

From a faunistic point of view, Bolivia has some of the least explored areas in the Neotropics. This is especially true for vertebrates, with the exception of birds. Bolivian herpetofauna in particular was only fragmentarily known until the end of the 1980s because previous investigative efforts had been very limited. Regarding Bolivian amphibians, a useful checklist that includes 112 species was first provided by De la Riva (1990). Since then numerous frog species have been recorded for the first time or newly described for Bolivia. Other taxa recognized in De la Riva's (1990) list have been or will be synonymized. To date, the number of valid amphibian species known to occur in Bolivia has increased to a total of 166 (not including some unpublished records in scientific collections that have already been recognized), which provides a more precise idea of the country's amphibian fauna. Nevertheless, there is undoubtedly a large number of species that have yet to be discovered.

Table 1. Amphibian taxa currently known from Bolivia (unpublished results included)

Family	Genus	Number of species
Bufonidae	*Atelopus*	1
	Bufo	13
	Melanophryniscus	1
Centrolenidae	*Cochranella*	2
	Hyalinobatrachium	1
Dendrobatidae	*Colostethus*	2
	Epipedobates	3
Hylidae	*Gastrotheca*	2
	Hemiphractus	1
	Hyla	30
	Osteocephalus	4
	Phrynohyas	2
	Phyllomedusa	8
	Scinax	11
	Sphaenorhynchus	1
Leptodactylidae	*Adenomera*	3
	Chacophrys	1
	Ceratophrys	2
	Eleutherodactylus	14
	Ischnocnema	1
	Lepidobatrachus	1
	Leptodactylus	19
	Lithodytes	1
	Odontophrynus	2
	Phrynopus	5
	Phyllonastes	2
	Physalaemus	6
	Pleurodema	3
	Pseudopaludicola	2
	Telmatobius	8
Microhylidae	*Chiasmocleis*	2
	Dermatonotus	1
	Elachistocleis	2
	Hamptophryne	1
Pipidae	*Pipa*	1
Pseudidae	*Lysapsus*	1
	Pseudis	1
Ranidae	*Rana*	1
Plethodontidae	*Bolitoglossa*	1
Caeciliidae	*Caecilia*	1
Siphonopidae	*Siphonops*	2
Total: 12	41	166

Bolivia encompasses a remarkable range of eco-geographical zones within its borders; its territory reveals extreme relief changes in conjunction with different biomes. As to be expected, each of these zones harbours a distinct amphibian fauna. To obtain an impression of diversity patterns we cross-linked the distribution of amphibian species to eco-geographical zones. However, available data on the distribution of Bolivian amphibians is far from complete and is not sufficient to conduct a detailed analysis. Therefore, this work should be regarded as tentative.

2 Bolivian amphibians – a short survey

The anurans (162 species) comprise the most specious group of Bolivian amphibians by far. Only three caecilians and one salamander species (i.e., *Bolitoglossa altamazonica*) are known. The actual number of amphibians known to occur in Bolivia is more than 180 when the several new species that have not yet been described and recognized records in museum collections are also taken into account. Using the classification system of Frost (1985) and Duellman (1993), Bolivian amphibians comprise 12 families with 41 genera (Table 1). Presently, 27 species (ca. 16 %) are endemic to Bolivia.

In comparison, 315 amphibian species had been recognized in neighbouring Peru prior to 1993 (Rodríguez *et al.* 1993). Given that both countries are almost the same size, the obvious differences in species richness can be attributed to differences in the amount of investigative effort and to the fact that Peru has a larger portion of perhumid life zones which are generally rich in species.

Diversity in Bolivian amphibians is also evidenced in their reproduction. In addition to the most common reproductive mode – egg-laying in bodies of water followed by the aquatic development of tadpoles – there is also egg deposition on vegetation above water with aquatic development of tadpoles, different types of foam nest production as well as terrestrial direct development in leptodactylid frogs, terrestrial egg-laying followed by parental tadpole transport to water in dendrobatid frogs, development of the young on the back of the female, *inter alia*. According to Crump's (1974) classification of anuran reproductive modes, nine out of ten defined modes can be found among Bolivian frogs.

3 Spatial patterns of species richness

Bolivia's eco-geographical zones were comprehensively defined, described, and mapped by Beck *et al.* (1993) and by Ibisch (1996). Due to a lack of detailed data on the distribution of Bolivian amphibians, a simplified classification with only eight zones was used for analyzing spatial patterns of amphibian diversity and is presented in this paper. We evaluated relative quantities of species in each zone, with a base of 166 species equalling 100%. Approximately one third of all amphibian species occur in more than one zone. Relative rates of endemism were also determined for each zone using the absolute number of species occurring in the respective zone as a baseline of 100%. The results are shown in Fig. 1.

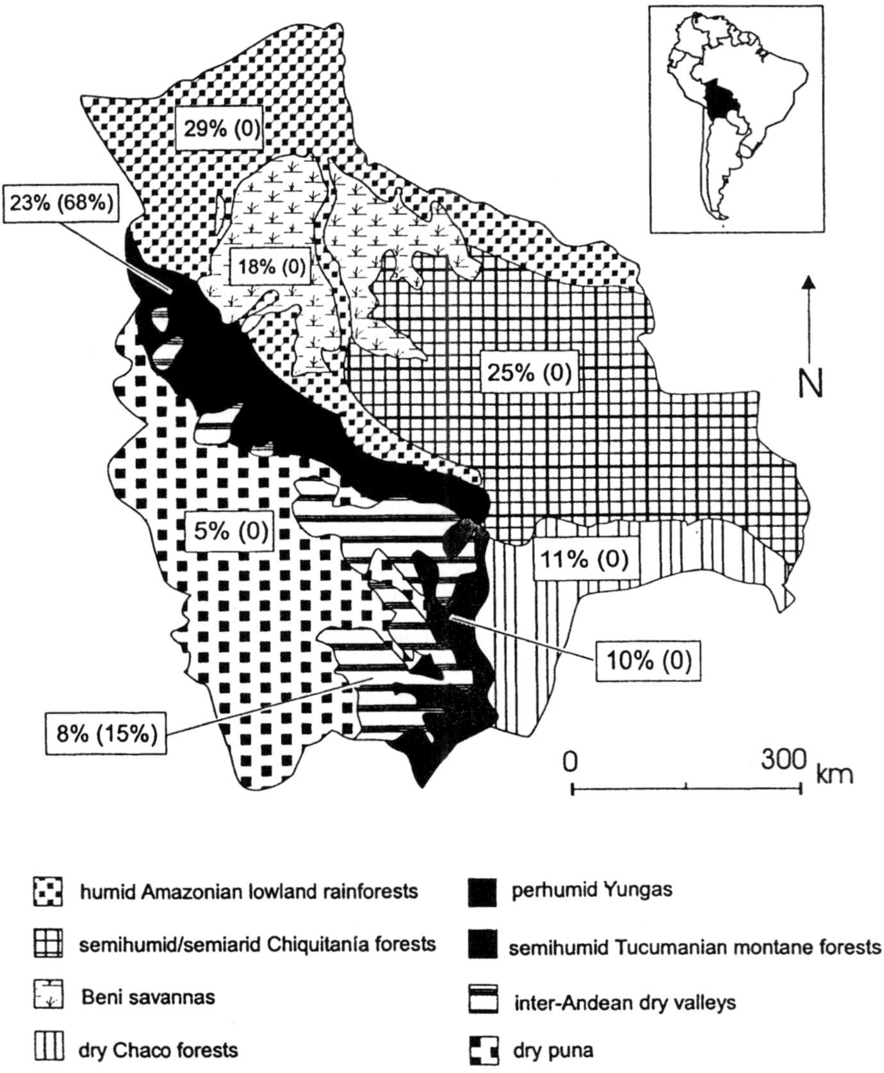

Fig. 1. Map of Bolivia showing eight eco-geographical zones and relative amphibian species richness. Percentages of species richness refer to 166 species (i.e., 100 %). The percentage of endemic taxa relative to the absolute species number of each zone is provided in parentheses

Ilustración 1. Mapa de Bolivia en el que se reflejan ocho zonas ecogeográficas y la riqueza relativa de especies de anfibios. Los porcentajes de la riqueza de especies se refieren a 166 especies (esto es, 100%). El porcentaje de taxa endémicos relativo al número absoluto de especies en cada zona aparece entre paréntesis

The level of species richness in the lowlands progressively decreases the further one travels from the humid Amazonian rain forests to the semihumid and semiarid Chiquitanía forests and Campos Cerrados south to the dry Chaco forests. However, this north-south gradient is interrupted by the Beni savannas. High levels of amphibian diversity are found in the perhumid Yungas; in addition almost all endemic taxa occur there. By contrast, the other eco-geographical zones in the Andes (i.e., the semihumid Tucumanian montane forests, the inter-Andean dry valleys and the dry puna) show comparatively low species diversity.

4 Discussion

As a rule, there is a negative correlation between species richness in amphibians and decreasing precipitation and temperatures. Thus, there is also a similar relationship between species richness and an increase in latitude or altitude (Duellman & Trueb 1986). The amphibians of Bolivia, at least in the lowlands, also reflect this pattern (Fig. 1). Even though the lowland zones diminish in size as one progresses from the north (i.e., humid Amazonian rain forests) to the south (i.e., dry Chaco), it is suggested that the decrease in species richness is representative, since these zones exhibit relatively homogenous physio-geographical structures. However, this general pattern is interrupted by the Beni savannas (Fig. 1). This could be explained by their biogeographical affinities to the Brazilian Campo Cerrado (northern part) and the Pantanal (southern part) respectively, rather than their lesser relationships to other Bolivian zones (Hanagarth & Beck 1996).

The general correlation between species richness and altitude and latitude mentioned above can only be partially confirmed in regard to the eco-geographical zones of the Bolivian Andes. As shown in Fig. 1, there is a recognizable pattern that indicates decreasing amphibian diversity as one progresses from the semihumid Tucumanian montane forests into the inter-Andean dry valleys and up to the high-Andean dry puna. However, this general pattern does not apply to certain areas of the Andes because species communities of inter-Andean valleys can be very distinct from one another. There are areas containing a diverse amphibian fauna (perhaps due to habitat richness and/or historical climatic conditions), whereas others display a more uniform amphibian fauna. For example, recent surveys have revealed diverse communities which also show an unusual species composition (Köhler *et al.* 1995a, Harvey 1996): In the montanous vicinity of the village Samaipata in the Departamento Santa Cruz, at 1600-1900 m a.s.l., both lowland (e.g. *Leptodactylus gracilis*) as well as high-Andean taxa (e.g. *Pleurodema cinereum*) were found sympatrically with montane forest species (e.g. *Hyla marianitae*, *Cochranella nola*). Such a phenomenon may also occur in other localities where different eco-geographical zones meet.

Although the perhumid Yungas may be comparatively minor in terms of species richness when compared to humid lowland forests (Fig. 1), the authors regard them as the most diverse eco-geographical zone.

The Yungas exhibit a remarkable fragmentation of habitats (due to relief barriers and drastic changes in climatic conditions within short distances). For this reason, area size must be considered when comparing them to more homogenous lowland zones (although species compositions within the lowlands vary from locality to locality, especially where different zones meet). As a result, communities are not only species rich but their composition also varies greatly within small areas. One example is "La Siberia," an area at the border of the Departamentos Cochabamba and Santa Cruz which appears to harbour more taxa than neighbouring sites in the Yungas. Despite this, it is possible that collecting gaps play a role as well. The high rate of endemic species in the Yungas (Fig. 1) is most probably the result of orogenesis, recent relief, and historical climatic conditions (Köhler *et al.* 1995b). There is apparently a distributional border between the Yungas de La Paz and the Yungas de Cochabamba because certain taxa are known to occur in only one or the other of the two regions. The Yungas' endemism as well as its species richness make it a very special area that is gravely in need of protection.

Glaw and Köhler (in press) have found a general increase in species description rates in amphibians, especially in the tropics. This is partly due to recently intensified surveys in previously uninvestigated areas, such as major parts of Bolivia. Another reason is the growing use of modern techniques which particularly include the analysis of mating calls in anurans (which allow for identifying sibling species), and genetic and biochemical studies.

Acknowledgements. We are indebted to the following persons and institutions who kindly supported our work in Bolivia or who helped us in other ways (in alphabetical order): James Aparicio (La Paz), Wolfgang Böhme (Bonn), Colección Boliviana de Fauna (La Paz), Ignacio De la Riva (Madrid), Lutz Dirksen (Bonn), Patricia Ergueta (La Paz), W. Ronald Heyer (Washington), Pierre L. Ibisch (Bonn), Karl-Heinz Jungfer (Fichtenberg), Albert Meyers (Bonn), Carmen Miranda (La Paz), Juan Carlos M. Quiroga (La Paz), Hinrich Rahmann (Stuttgart), Andreas Schlüter (Stuttgart), Sennheiser electronic KG (Wedemark), Stuttgarter Museum für Naturkunde (Stuttgart), Gustav Peters (Bonn), Zoologisches Forschungsinstitut und Museum Alexander Koenig (Bonn).

References

Beck SG, Killeen TJ, García E (1993) Vegetación de Bolivia. In: Killeen TJ, García E, Beck SG (eds) Guía de árboles de Bolivia. Herb Nac Bolivia, Miss Bot Gard, La Paz, pp 6-24

Crump ML (1974) Reproductive strategies in a tropical anuran community. Univ Kansas Mus Nat Hist Misc Publ 61: 1-68

De la Riva I (1990) Lista preliminar comentada de los anfibios de Bolivia con datos sobre su distribución. Boll Mus reg Sci nat Torino 8(1): 261-319

Duellman WE (1993) Amphibian species of the world: additions and corrections. Univ Kansas Spec Publ 21: 1-372

Duellman WE, Trueb L (1986) Biology of amphibians. McGraw-Hill, New York

Frost DR (1985) Amphibian species of the world. A taxonomic and geographical reference. Allen Press, Lawrence

Glaw F, Köhler J (in press) Amphibian species diversity exceeds that of mammals. Herpetological Review

Hanagarth W, Beck SG (1996) Biogeographie der Beni-Savannen (Bolivien). Geogr Runds, Braunschweig, 48(11): 662-668

Harding KA (1983) Catalogue of New World amphibians. Pergamon Press, Oxford

Harvey MB (1996) A new species of glass frog (Anura: Centrolenidae: *Cochranella*) from Bolivia, and the taxonomic status of *Cochranella flavidigitata*. Herpetologica 52(3): 427-435

Ibisch PL (1996) Neotropische Epiphytendiversität - das Beispiel Bolivien. Galunder Verlag, Arch naturwiss Diss 1, Wiehl

Köhler J, Ibisch PL, Dirksen L, Böhme W (1995 a) Zur Herpetofauna der semihumiden Samaipata-Region, Bolivien. I. Amphibien. Herpetofauna 17(98): 13-24

Köhler J, Dirksen L, Ibisch PL, Rauer G, Rudolph D, Böhme W (1995 b) Zur Herpetofauna des Sehuencas-Bergregenwaldes im Carrasco-Nationalpark, Bolivien. Herpetofauna 17(96): 12-25

Rodríguez LO, Córdova JH, Icochea J (1993) Lista preliminar de los anfibios del Perú. Publ Mus Hist Nat UNMSM (A), Lima 45: 1-22

Part 5.2

Sustainable use of biodiversity - case studies from the Andes

Land use, economy and the conservation of biodiversity of high-andean forests in Bolivia

Michael Kessler
Systematic-Geobotanic Institute, University of Göttingen, Germany

Abstract. Centuries of overexploitation of high-Andean ecosystems in Bolivia have led to the destruction of approximately 90% of the high-Andean forests (98-99% in the eastern highlands), making them one of the most threatened ecosystems in the Neotropics. The main reasons for the destruction are unsustainable land-use methods, in particular burning and overgrazing. Other factors leading to the destruction of Andean ecosystems are human population growth and an increasing inclusion of rural communities in the country's market economy, which creates the need for cash income and, therefore, production surplus. Due to the small size of relict forest patches, and since all patches are used by local people, the conservation of high-Andean forests is not possible in reserves, but has to be accomplished in cooperation with the rural population. In order to preserve high-Andean forests (and any other Andean ecosystem), it will be necessary to create a sustainable basis for human development. This will necessitate (a) establishing land-use systems with a clear spatial separation of land-use units (fields, pastures, forests), (b) slowing human population growth, (c) putting strong emphasis on the education of children and adults, and of women in particular, (d) strengthening local subsistence and non-monetary economies, (e) making sure that national and international governmental and non-governmental aid organizations provide economic and technical support for the above, and (6) ensuring that national governments pass supporting legislation.

Uso del suelo, economía y conservación de la biodiversidad en los bosques altoandinos de Bolivia

Resumen. Siglos de sobreexplotación de los ecosistemas de los altos Andes de Bolivia han llevado a la destrucción de aproximadamente un 90% de los bosques altoandinos (98-99% en las tierras altas del este), convirtiéndolos así en uno de los ecosistemas más amenazados de la zona del Neotrópico. La causa principal de esta destrucción son los métodos insostenibles de uso del suelo, en particular la quema y el sobrepastoreo. Otros factores que conducen a la destrucción de los ecosistemas andinos son el crecimiento demográfico y la mayor inclusión de las comunidades rurales en la economía de mercado del país, creando la necesidad de obtener ingresos en efectivo y, por consiguiente, un excedente de producción. Debido a la reducida extensión de los retazos de bosque autóctono y a que todos ellos son usados por la población local, no resulta posible la conservación en reservas de los bosques altoandinos, sino que ha de conseguirse en cooperación con la población rural. Con vista a preservar los bosques de los altos Andes (y cualquier otro ecosistema andino) será necesario crear una base sostenible para el desarrollo humano. Esto requerirá (1) establecer sistemas de uso del suelo con una clara división espacial de las unidades de uso del suelo (campos, pastos, bosques); (2) ralentizar el crecimiento demográfico; (3) poner mayor énfasis en la educación de niños y adultos, y de las mujeres en particular; (4) fortalecer la subsistencia local y las economías no-monetarias; (5) garantizar que las organizaciones gubernamentales y no-gubernamentales nacionales e internacionales de ayuda proporcionen apoyo económico y técnico para los aspectos anteriormente mencionados y (6) asegurar que los gobiernos nacionales adopten la legislación de apoyo correspondiente.

1 Introduction

Today's landscape in the high Andes of Bolivia, including the Altiplano and the surrounding mountain ranges, is dominated by open grasslands, bushy steppes – *tholares* – and huge salt pans. The inhabitants of this region generally belong to the Aymara and Quechua cultures. They live in scattered houses and villages, growing mainly potatoes and raising livestock (sheep, cattle, some llamas and alpacas). In the lower-lying valleys – the *valles* – where most of the Bolivian population is concentrated, flat areas are subject to intensive agriculture, while the shrub-covered mountain slopes are used as pastures. With the exception of planted *Eucalyptus* and *Pinus*, trees are mostly absent throughout these regions.

Small patches of native high-elevation forests are however found occasionally (Fig. 1). These forest patches, dominated by species of the genus *Polylepis* (Rosaceae), reach elevations of up to 4,200 m in the eastern cordillera and 5,200 m on volcanoes near the Chilean border, making them the highest woody plant formations on earth (Jordan 1983, Kessler 1995a). Until a few decades ago, researchers believed that the occurrence of such small, isolated forest patches high above the closed tree line (ca. 3,500 m in Bolivia) was due to special microclimatic conditions which restrict tree growth to sheltered ravines, boulder slopes, and rock faces (e.g., Koepcke 1961, Walter & Medina 1969, Rauh 1988). Recent ecological, biogeographic, and phytosociological evidence has shown, however, that these patches are remnants of once much more widespread forests (Fig. 1) which have been almost completely destroyed by thousands of years of human impact (Fjeldså 1992, Lægaard 1992, Kessler & Driesch 1993, Ibisch 1994, Hensen 1995, Kessler 1995a, b, Ellenberg 1966). Although some sizeable forests remain in the western cordillera, only 1-2% of the original area remains forested in the eastern highlands today (Kessler & Driesch 1993, Kessler 1995b).

The plant and animal communities of the native high-elevation forests of Bolivia and Peru, while not especially notable in terms of species numbers, represent a unique set of taxa that have adapted to extreme ecological conditions. Some 35 bird species live predominantly in *Polylepis* forests and six of these inhabit only this habitat (Fjeldså 1992). Although no complete enumeration of plant taxa exists (see Hensen 1993, 1995 on the Cochabamba region), many wild varieties of cultivated species such as potatoes (*Solanum* spp.), oca (*Oxalis tuberosa*), papalisa (*Ullucus tuberosus*) or mashua (*Tropaeolum tuberosum*) occur in native high-elevation forests (Fjeldså & Kessler 1996).

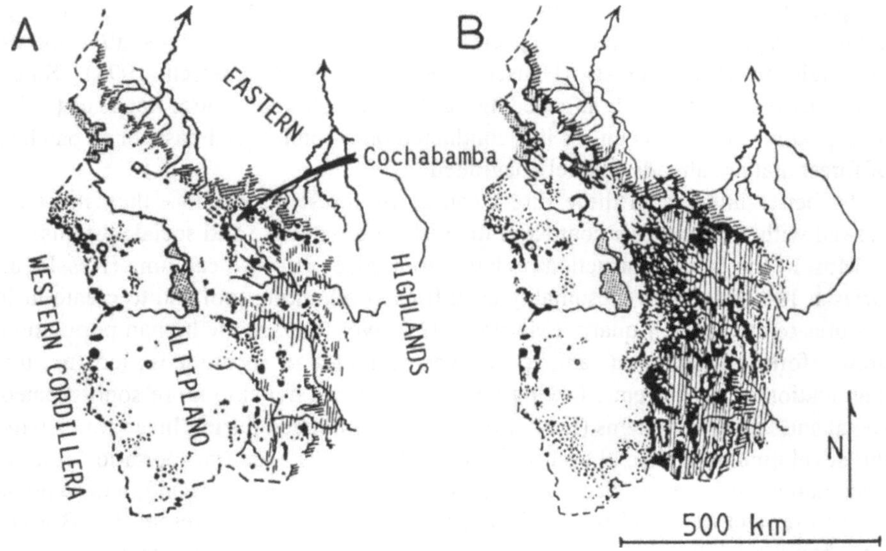

Fig. 1. Current (A) and natural (B) distribution of Andean forests (> 2,000 m) in Bolivia and geographic regions mentioned in the text. Black areas denote *Polylepis* forest, horizontal hatching humid montane forest, and vertical hatching arid montane forest. Note that on map A the actual extent of forested areas is exaggerated due to scale and because many remaining forests are severely degraded. The extent to which the northern and central Altiplano would naturally be forested is unclear

Ilustración 1. Distribución común (A) y natural (B) de los bosques andinos (>2.000m) en Bolivia y las regiones geográficas mencionadas en el texto. Las zonas negras indican bosques *Polylepis*, el sombreado horizontal indica bosques montañosos húmedos y el sombreado vertical bosques montañosos áridos. Nótese que en el mapa A la actual extensión de áreas boscosas aparece exagerada debido a la escala y a que muchos de los bosques que quedan se encuentran altamente degradados. No está claro en qué medida podría ser forestado naturalmente el norte y el centro del Altiplano

I will discuss the reasons for the catastrophic decline of the natural forests in the high Andes of Bolivia, assess the effect of this destruction on natural and human communities and put forward some possible solutions.

2 Present-day land use

The two main factors directly influencing the distribution of *Polylepis* forests are overgrazing and the use of fire. Much of the Bolivian highlands is burnt on a yearly basis to enhance the growth of pasture grasses or, in many cases, for no apparent reason (Seibert 1983). While fires generally do not kill adult trees, they very effectively destroy seedlings and young trees, thus preventing tree regeneration, creating senescent forests and eventually leading to the disappearance of forests (Lægaard 1992, Kessler & Driesch 1993). Grazing animals, especially non-native species, have a similar though usually not as thorough effect on forests (Hensen 1993, Kessler & Driesch 1993). Since *Polylepis* resprouts readily after being cut, timber extraction does not lead to the disappearance of forests unless it is conducted on a commercial basis or in patches of forest that are already severely degraded.

To better understand these direct causes of forest destruction, they must be viewed within the broader context of the history of land use and social systems.

Most of the forest destruction probably took place in pre-Incan time (Kessler & Driesch 1993). Hunters presumably used fires as a hunting tool and to create lush pastures to attract their quarry (Martin 1992). Over time, as the human population grew, forest destruction and primitive agricultural techniques led to the degradation of ecosystems, forcing the inhabitants to develop more sophisticated social and land-use systems (Earls 1991). This culminated during Incan times with the development of "social security systems," long-distance transportation routes, sustainable land-use practices with extensive irrigation and terracing, and explicit laws to protect trees and forests (Ansión 1984). The area of present-day Bolivia was conquered by the Incas only a few decades before the arrival of the Spaniards. As a result, these systems had not yet been fully established there by the time of the *conquista*. The Spaniards, in turn, disrupted the entire social system and introduced a variety of new land-use techniques, such as deep-plowing, new crops and animals, and allowed traditional systems to fall into neglect (Posnansky 1983). In the last few decades, increasing inclusion of rural communities into the national market and growing mechanization have added new, complex facets to the country's agricultural economy.

As a result of this turbulent history, present-day land-use practices represent a mixture of native and foreign techniques, with little or no long-term experience as to their sustainability. In fact, numerous ecological studies have shown that the current land use is not sustainable (e.g., Ellenberg 1979, 1981, LeBaron *et al.* 1979, Posnansky 1983, Kessler & Driesch 1993, Morales 1994, Schad 1995, Fjeldså & Kessler 1996). Some examples of widespread, unsustainable land-use techniques include:

1. The extensive use of fire, which, while enhancing the growth of grasses on a short-term basis, leads to the depletion of soil nutrients (Zech & Feuerer 1982)
2. The use of non-native grazing animals, particularly sheep, goats, and cattle, which destroy the protective vegetation cover and promote soil erosion (Ellenberg 1981, Schad 1995)
3. Agriculture on steep slopes, often with furrows directed downhill, which promotes erosion
4. The use of non-native trees, and *Eucalyptus* in particular, for reforestation projects which, while often giving excellent short-term yields, leads to the depletion of soil water and nutrient levels in the long run (Poore & Fries 1987, Lisanework & Michelsen 1993)
5. Increasing human population density which leads to ever shorter fallow periods (Schad 1995) and to chronic overstocking of pastures (Augstburger 1990)

What are the long-term effects of these and other unsustainable land-use practices and how do they affect ecosystems and local inhabitants?

3 The effects of unsustainable land use

Today and probably over the last several thousand years as well, every suitable piece of land in the Bolivian highlands is and has been inhabited by humans (excluding the wet Yungas montane forests which were largely uninhabited until recently). Thus, it is impossible to find any ungrazed pastures or undisturbed forests in the Bolivian highlands today. Based on the current relictual distribution of forests it seems probable that about 95,000 km² of the Bolivian highlands (about 35% of the total area) would be under natural conditions forested (Kessler 1995a, b). Of these, only about 10% is forested today, mostly in the western cordillera. In the eastern highlands 1-2% of the forest are preserved; this figure is less than 0.1% on the Altiplano (Kessler 1995a, Ellenberg 1996). In the east in particular, much of the remaining forest is subject to intensive human pressure and is heavily degraded. Every year, the area covered by forest is reduced.

Although as far as we know no plant or animal species inhabiting high-elevation forests in Bolivia has ever become extinct, several are considered to be endangered, including the Andean Deer (*Hippocamelus antisiensis*), six bird species (Collar *et al.* 1992), and possibly wild varieties of crop plants (Fjeldså & Kessler 1996). On a local scale – the area used by a rural community for example – the total destruction of native forests has marked consequences for local biodiversity. About 35% of the plant species found by Hensen (1991a) in the community of Chorojo (Cochabamba region) were restricted to native *Polylepis* forest. Many of these species were of local importance as medicinal plants, natural vegetables, fodder, or for ritual purposes. While native high-elevation forests in the Cochabamba region are inhabited by 25-45 bird species, including all locally threatened and endemic species, plantations of *Pinus* and *Eucalyptus* contain only 2-12 widespread species (T. Hjarsen, pers. comm.).

Possibly more important than the decline of biodiversity is the loss of ecosystem functions provided by natural forests (Fjeldså & Kessler 1996): Forests comb moisture out of clouds and thus increase precipitation; they have a high water catchment capacity and regulate water run-off; they prevent soil erosion and hold sediments and nutrients; and they provide wood for fuel, construction material and non-timber products. Fields adjacent to native forest or surrounded by rows of native trees are more productive than fields on open slopes (Fjeldså & Kessler 1996). Thus, the living conditions of local people are considerably worse once native forests are destroyed.

Reduced water catchment capacity and increased soil erosion are probably the most important consequences of forest destruction, especially on a larger scale (Schulte 1994). The soil in about two thirds of the Bolivian highlands is currently considered to be severely eroded (Mansilla 1984). Soil erosion will also lead to pollution of streams and rivers and to the silting of water reservoirs.

The cumulative effect of these unsustainable land-use practices, of which forest destruction is only a single – albeit crucial – component, is an overall decrease in agricultural productivity in degraded areas. This condition forces inhabitants to use their land ever more intensively which then leads to even greater depletion of resources (Ibisch 1993). This vicious circle, coupled with a growing human population, generally leaves at least part of the community no other alternative than to emigrate to new agricultural land (usually in the lowlands) or to urban areas.

4 An economic perspective

Besides land-use and population development factors, the degree of sustainability of human development in the Bolivian highlands (and probably elsewhere as well) also depends on the region's economic development. Originally, Andean communities were mostly self-sustaining and had less economic exchange with other communities than is the case today where most *comunidades* depend to a growing degree on the sale of agricultural products at local markets and on cash income from relatives working in the city or in the coca fields in the Yungas. To pay for many of today's commodities (e.g., transportation, electricity, clothing), local inhabitants are increasingly dependent on monetary income. The Bolivian government is attempting to further strengthen economic links in the country.

While the improvement in the local inhabitants' living conditions must be regarded as a positive development, the increasing importance of the market economy is not entirely without problems. By creating incentives for the *campesinos* to increase their productivity, a market economy also increases the pressure on already over-used lands and consequently leads to even more rapid degradation of ecosystems.

Furthermore, our current market economy is inherently unsustainable: By depending on continuous growth (i.e., interest) in a physically and economically limited world, a market economy inevitably induces periodic economic collapses which are frequently accompanied by social instability (Gesell 1984, Creutz 1986, King 1987). Any attempt to achieve sustainable development within the framework of this economic system is therefore doomed to fail. This applies to a wide variety of cases which target the conservation of biodiversity by creating a market value for biodiversity (e.g., Nader, this volume), setting up trust funds to finance conservation schemes (e.g., Sánchez de Lozada, this volume) or simply by creating economic incentives for local people in order to achieve a specific conservation goal (e.g., Fjeldså & Kessler 1996). Each of the above approaches will be ineffective or unworkable if economic crisis strikes. Financial aid on part of the "developed" countries is currently of critical importance to conservation and development programmes in Bolivia, but will likely decrease in the future as constraints on national budgets grow. Sustainability is based on the wise use of reliably renewable resources. Money is not this type of resource.

When summarizing the factors affecting the development of natural and human communities in the Bolivian Andes, we are confronted with a complex

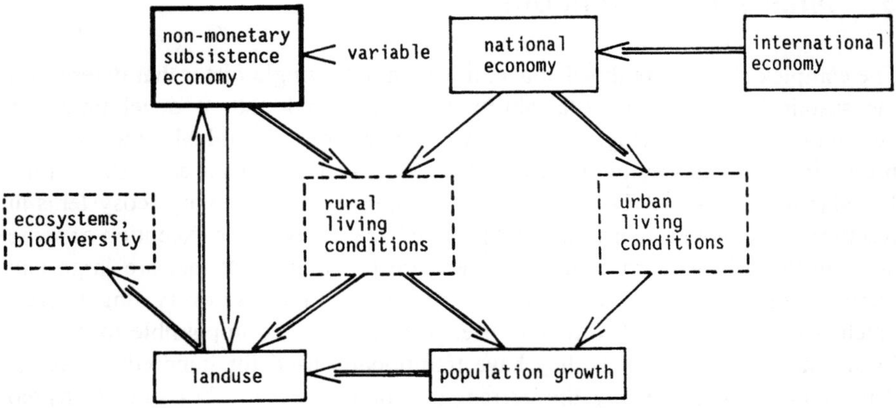

Fig. 2. Schematic representation of the main relations between factors determining the sustainability of human development (closed boxes) and their effects on natural ecosystems and human living conditions (open boxes) in the Bolivian Andes. Closed boxes with narrow borders denote crucial factors which could potentially be sustainable but currently are not; the box with the bold-faced border denotes the non-monetary subsistence economy which is intrinsically sustainable since it is not based on growth. Double and single arrows show strong and slight influences, respectively. In more mechanized, large-scale agricultural systems (e.g., in the lowlands around Santa Cruz), the relation between land use and the national economy (and *vice versa*) is stronger. See text for further explanations
Ilustración 2. Representación esquemática de las principales relaciones entre factores determinantes de la sostenibilidad del desarrollo humano (cajas cerradas) y sus efectos sobre los ecosistemas naturales y las condiciones de vida humanas (cajas abiertas) en los Andes bolivianos. Las cajas abiertas con bordes estrechos indican factores decisivos que potencialmente podrían ser sostenibles pero que por lo general no lo son; la caja con los bordes de superficie pronunciada muestra la economía de subsistencia no-monetaria intrínsecamente sostenible, ya que no se basa en el crecimiento. Las flechas dobles y simples muestran respectivamente influencias fuertes y débiles. En sistemas agrícolas más mecanizados y de gran escala (p. ej. en las tierras bajas alrededor de Santa Cruz), la relación entre el uso de la tierra y la economía nacional (y *vice versa*) es más fuerte. Para más explicaciones véase el texto.

network of interrelationships (Fig. 2). The environmental quality of the ecosystems in this area depends largely on (currently unsustainable) land-use practices which in turn are mostly determined by the living conditions of the local inhabitants, the development of rural economies, and local population development. These factors are in turn increasingly dependent on the development of Bolivia's national economy, on economic and technical input from governmental and non-governmental organizations, and on the government's political decisions.

5 What are the solutions?

The complex situation outlined above, in which not a single factor that determines the sustainability or non-sustainability of human and natural development is currently sustainable (Fig. 2), clearly reveals how formidable the task of preserving the biodiversity of high-Andean forests (or any other ecosystem in the Bolivian highlands) will be. The traditional approach of preserving ecosystems in reserves would not be appropriate in this case because (with the exception of some areas in the arid western cordillera) there are no forests left that are large and remote enough to warrant establishing a reserve and because every single forest patch is currently subject to human use. It would also be impossible to remove local inhabitants from areas they have traditionally used for hundreds of years. This point is illustrated by the current situation in Parque Nacional Carrasco (Cochabamba). This park was established in 1991 without consulting or even informing the local communities, generating strong opposition to the park among local residents. The *campesinos* have even threatened to burn native forests in order to regain their land (T. Tarifa, pers. comm.). Conservation of high-Andean forests can be achieved only with the cooperation of local communities. This will require change on many levels.

Current land-use practices that are unsustainable need to be changed so that they are sustainable. Technically, this would primarily involve the spatial separation of land with different uses. At present, the same piece of land is generally subject to two or three different uses (planting, grazing, timber extraction) with, at most, a temporal separation. These uses involve conflicting management techniques which lead, *inter alia*, to grazing and the extraction of bushes for firewood during fallow periods – activities which promote soil erosion and prevent the land from recuperating. It is therefore essential that land units be spatially separated. This would allow each of them to be managed in a profitable and sustainable manner. Traditional and modern techniques for sustainable agriculture such as irrigation, terracing, agroforestry, intercropping, etc. are presented, *inter alia*, by Ellenberg (1981), Posnansky (1983), and Fjeldså & Kessler (1996), and are being implemented to an increasing degree in the Bolivian Andes. To reduce soil erosion, Schad (1995) recommended keeping cattle and sheep on well-watered pastures with closed vegetation cover or in *corrales*.

Traditional examples of such separated land-use units exist at scattered localities in Bolivia (Liberman 1990, Hensen 1991b, Kessler & Driesch 1993). Some communities selectively protect small forest patches as sources of firewood (Kessler 1995b). In such cases, these communities also control burning and grazing in the forest. Thus, optimal management is automatically implemented once spatial separation has been achieved. Although it is naturally difficult to change long-standing land-use techniques, gradual changes can be implemented. The rural population urgently needs to improve its land-use techniques; it also displays a cautious willingness to adopt necessary measures (Ibisch 1996). In fact, many communities are extremely interested in adopting novel land-use techniques but are cautious since they have already experienced repeated failures with non-practicable measures that aid organizations without enough practical knowledge recommended in the past (Ibisch 1996). Many of the necessary changes contain at least some traditional components which can be stressed in order to make it easier for the people involved to accept the changes.

As outlined above, conservation of Andean ecosystems must be pursued in cooperation with local rural communities, particularly by increasing the long-term value of healthy, natural ecosystems and agricultural systems. The most straightforward approach would be to increase the monetary income generated from such ecosystems, such as in the case of forests by establishing a local charcoal industry (the medium-term economic and long-term ecological feasibility of which has been shown by Fjeldså & Kessler 1996). However, since the current economic system is inherently unsustainable, the long-term sustainability of such approaches appears doubtful. What will happen, for example, to the 150,000 hectares of *Eucalyptus* plantations in the Bolivian Andes that were established largely to produce timber on a commercial basis, if the timber market collapses even temporarily? Rather than increase the market value of the land, a sounder and more sustainable approach would be to strengthen the land's non-commercial value to local people by increasing agricultural productivity, providing ample firewood, protecting watersheds and securing a safe, clean water supply – in short, by improving the local population's living conditions. Given their current economic and social situation, most Andean communities need financial and technical support to improve land use and raise living conditions to a level that will enable *campesinos* to improve their living conditions on their own. This, rather than creating additional income, should be the primary aim of aid organizations. Many *campesinos* are at least partially aware of the ecological consequences of unsustainable land-use practices but are trapped in a system of extreme poverty in which outside aid is often aimed at providing certain services rather than improving living conditions (Birgegard 1992).

At first glance, there seems to be a contradiction between the conclusion that financial aid is necessary for the development of Bolivian highland communities and that this aid should not strengthen the *comunidades'* economic dependence on the national or international market. However, this contradiction is only superficial. It is the result of pragmatically dealing with two facts: (1) Due to historical reasons, Bolivia's highland communities are unable to achieve sustainable development on their own today and (2) the current economic system is unsustainable. The subsequent conclusion is straightforward: The opportunities offered by the current economic system must be used to achieve sustainable development via the creation of structures that are as independent as possible of the market economy. An example to illustrate this point: Numerous aid organizations are currently helping communities on the Altiplano to build greenhouses. In many cases, this requires some economic input from the *comunidades* which in turn means that these often have to obtain loans. This makes it necessary to grow cash crops (ranging from medicinal herbs to strawberries!) in the greenhouses and creates an economic dependency on the national market which is likely to persist even after the loans have been paid back. Although this is a rational strategy in terms of our current economic system, it does not make sense when viewed from the standpoint of sustainable economy.

Wouldn't it be much more rational (and ecologically sensible) to have the *campesinos* grow food for their own consumption (and, of course, some surplus if possible)? This might not increase the monetary income of the communities but it would improve their living conditions. Similar examples could be given for a wide variety of well-meant aid programmes ranging from fishing trout in Lake Titicaca for the markets of La Paz to producing alpaca sweaters for the international market. It is certainly not the responsibility of the Bolivian *campesinos* to change the world's unsustainable economy. But do we really do them a favour if we increase their dependence on this economic system?

Several steps must be taken if high-Andean forests and other Andean ecosystems are to be preserved and human living conditions to be improved in the process. Governmental and non-governmental aid organizations – in other words, those already working with a large number of highland communities – need to revise their policies to emphasize:

1. Educating children and adults, and women in particular, especially about environmental topics
2. Implementing sustainable land-use systems, which would especially include the spatial separation of fields, pastures and forests
3. Strengthening local economies which are potentially sustainable
4. Supporting and implementing population-control programmes. In light of the fact that a large family is the primary form of social security in many highland communities, the only way this long-term goal can be achieved is by improving living conditions and developing alternative social security systems
5. Strengthening local community organizations in order to ensure the applicability and medium-to-long-term success of aid programmes

At government level, it is necessary to support the development of sustainable land use and local economies by:

1. Creating the necessary legislative foundation and incentives. It should be emphasized that some laws, which recently went into force such as the *Ley de Participación Popular* (law on popular participation) and the *Ley INRA* (law on land ownership) already contain some of the requisite elements. However, a more specific legal framework is needed
2. Acknowledging the value of subsistence economies which represent an essential (and potentially sustainable!) component of the nation's economy even though they do not directly contribute to the GNP
3. As an ideal case, establishing a sustainable foundation for the nation's economy by adopting an interest-free and inflation-free monetary system (see, e.g., Onken 1983, Suhr 1983, Gesell 1984, Laistner 1986, King 1987). Such steps can be taken independently of other countries' decisions

The crucial changes needed to improve living conditions for Bolivia's highland population and to preserve Andean ecosystems are not all that technical in nature. We know what has to be and what could be done. Now, aid organizations, Bolivia's government and international organizations must change their strategies to favour long-term sustainable development over short-term profit and to give preference to implementing real solutions rather than simply treating symptoms. It must also be mentioned here that some organizations are already pursuing this type of change.

Acknowledgements. I would like to thank H. Ritter for introducing me to Gesell's theories of sustainable economy and J. Fjeldså for many stimulating discussions. P. Driesch, S.K. Herzog and P. Ibisch revised earlier drafts of this manuscript.

References

Ansión I (1984) El Arbol y el Bosque en la Sociedad Andina. Instituto Nacional Forestal y Fauna - FAO, Lima

Augstburger F (1990) La ganadería y los equilibrios ecológicos. In: COTESU, Desarrollo y medio ambiente. COTESU, Cochabamba, pp 25-27

Birgegard L-E (1992) Las actividades forestales no son respuesta a la deforestación. Bosques, Arboles y Comunidades Rurales 14 :35-37

Collar NJ, Gonzaga LP, Krabbe N, Matroño N. A, Naranjo LG, Parker III TA, Wege DC (1992) Threatened Birds of the Americas. The ICBP/IUCN Red Data Book. 3rd ed., part 2. ICBP, Cambridge, U.K.

Creutz H (1986) Wachstum bis zur Krise. Basis Verlag, Berlin

Earls J (1991) Ecología y Agronomía en los Andes. Hisbol, La Paz

Ellenberg H (1979) Man's influence on tropical mountain ecosystems in South America. J Ecol 67: 401-416

Ellenberg H (1981) Desarrollar sin destruir. Instituto de Ecología, La Paz

Ellenberg H (1996) Páramos und Punas der Hochanden Südamerikas, heute großenteils als potentielle Wälder anerkannt. Verh Ges Ökol 25: 17-23

Fjeldså J (1992) Biogeographic patterns and evolution of the avifauna of relict high-altitude woodlands of the Andes. Steenstrupia 18: 9-62

Fjeldså J, Kessler M (1996) Conserving the Biological Diversity of *Polylepis* Woodlands of the Highlands of Peru and Bolivia. A Contribution to Sustainable Natural Ressource Management in the Andes. NORDECO, Copenhagen

Gesell S (1984) Die natürliche Wirtschaftsordnung. 10. Aufl. Rudolf Zitzmann Verlag, Nürnberg

Hensen I (1991a) La Flora de la Comunidad de Chorojo, su Uso, Taxonomía Científica y Vernacular. AGRUCO, Cochabamba

Hensen I (1991b) El Bosque de Kewiña de Chorojo ¿Ejemplo de un Sistema Agroforestal Andino Sostenible? AGRUCO, Cochabamba

Hensen I (1993) Vegetationsökologische Untersuchungen in *Polylepis*-Wäldern der Ostkordillere Boliviens. PhD thesis, Syst -Geobot Inst, Univ Göttingen

Hensen I (1995) Die Vegetation von *Polylepis*-Wäldern der Ostkordillere Boliviens. Phytocoenologia 25: 235-277

Ibisch P (1993) Estudio de la Vegetación como una Contribución a la Caracterización del la Provincia de Arque (Bolivia). Cuaderno Científico 1, PROSANA, Cochabamba

Ibisch P (1994) Flora y Vegetación de la Provincia de Arque, Departamento de Cochabamba, Bolivia. Parte III: Vegetación. Ecología en Bolivia 22: 53-92

Ibisch P (1996) "Reparieren" von degradierten Agrar-Ökosystemen in den Anden zwischen Theorie und Praxis. Beispiel: Provinz Arque, Bolivien. Kritische Ökologie 14: 9-15

Jordan E (1983) Die Verbreitung von *Polylepis*-Beständen in der Westkordillere Boliviens. Tuexenia 3: 101-112

Kessler M (1995a) *Polylepis*-Wälder Boliviens: Taxa, Ökologie, Verbreitung und Geschichte. Dissertationes Botanicae 246, Cramer, Berlin Stuttgart

Kessler M (1995b) Present and potential distribution of *Polylepis* (Rosaceae) forests in Bolivia. In: Churchill SP, Balslev H, Forero E, Luteyn JL (eds) Biodiversity and Conservation of Neotropical Montane Forests. New York Botanical Garden, Bronx. pp 281-294

Kessler M, Driesch P (1993) Causas e historia de la destrucción de bosques altoandinos en Bolivia. Ecología en Bolivia 21: 1-18

King JL (1987) On the Brink of Great Depression II. Future Economic Trends, Goleta, CA

Koepcke H-W (1961) Synökologische Studien an der Westseite der peruanischen Anden. Bonner Geogr Abh 29

Lægaard S (1992) Influence of fire in the grass páramo vegetation of Ecuador. In: Balslev H, Luteyn J (eds) Páramo. An Andean Ecosystem under Human Influence. Academic Press, London, pp 151-170

Laistner H (1986) Ökologische Marktwirtschaft. Verlag Max Huber, Ismaning

LeBaron A, Bond LK, Aitken SP, Michael L (1979) An explanation for the Bolivian highlands grazing-erosion syndrome. J of Range Management 32: 201-208

Liberman M (1990) Estudio de un sistema agrosilvopastoril en la cordillera oriental andina de Bolivia. In: Agroecología y Saber Campesino. AGRUCO-Cochabamba, Bolivia, y PRATEC-Lima, Perú, pp 95-124

Lisanework N, Michelsen A (1993) Allelopathy in forestry systems: the effect of leaf extracts of *Cupressus lusitanica* and three *Eucalyptus* spp. on four Ethiopian crops. Agroforestry Systems 21: 63-74

Mansilla HCF (1984) Nationale Identität, gesellschaftliche Wahrnehmung natürlicher Ressourcen und ökologische Probleme in Bolivien. Beitr zur Soziologie und Sozialkunde Südamerikas 34, Fink Verlag, München

Martin PS (1982) Prehistoric Overkill: The Global Model. In: Martin PS, Klein RG (eds) Quaternary Extinctions. A Prehistoric Revolution. 2nd ed. Univ. of Arizona Press, Tucson

Morales CB (1994) (ed) Huaraco, Comunidad de la Puna. Instituto de Ecología, La Paz

Onken W (1983) Ein vergessenes Kapitel der Wirtschaftsgeschichte. Schwanenkirchen Wörgl und andere Freigeldexperimente. Zeitschr f Sozialökonomie 58/59: 3-20

Poore MED, Fries C (1987) Efectos ecológicos de los eucaliptos. Estudios FAO, Montes 59, Rome

Posnansky M (1983) Los efectos sobre la ecología del Altiplano de la introducción de animales y cultivos por los Españoles. In: Ecología y Recursos Naturales en Bolivia. Centro Portales, Cochabamba, pp 13-22

Rauh W (1988) Tropische Hochgebirgspflanzen. Wuchs und Lebensformen. Springer, Heidelberg

Schad P (1995) Einfluß traditioneller Bewirtschaftungsmethoden auf Kennwerte der Bodenfruchtbarkeit in Gebiet von Charazani (Bolivianische Hochanden). Alfons Kasper, Bad Schussenried

Schulte A (1994) Dürre oder montane Desertifikation? Ursachen und Folgen der Boden- und Waldzerstörung in Bolivien. Entwicklung und ländlicher Raum 5/94: 26-28

Seibert P (1983) Human impact on landscape and vegetation in the Central High Andes. In: Holzner W, Wegner MJA, Ikurima I (eds) Man's impact on vegetation. W. Junk, The Hague, pp 55-65

Suhr D (1983) Geld ohne Mehrwert. Knapp, Frankfurt/M

Walter H, Medina E (1969) La temperatura del suelo como factor determinante para la caracterización de los pisos subalpino y alpino en los Andes de Venezuela. Bol Soc Ven Cien Nat 28(115/116): 201-210

Zech W, Feuerer T (1982) Geoökologische Studien im Callawayagebiet, Bolivien. Gießener Beiträge zur Entwicklungsforschung, Reihe I, Band 8: 131-144

Using Lake Titicaca's biological resources – problems and alternatives[1]

Wolfgang Villwock
Zoological Institute and Zoological Museum, University of Hamburg, Germany

Abstract. A brief introduction of the geological history of the Altiplano and its water systems is followed by a description of the existing faunistic situation (autochthonous and endemic fishes of the genus *Orestias* versus exotic species, represented primarily by salmonids). Eco-social as well as ecological consequences and evolutionary aspects are discussed. In conclusion, four recommendations are made regarding a potentially sustainable use of natural resources in the Lake Titicaca region.

Empleo de los recursos biológicos del Lago Titicaca: Problemas y alternativas

Resumen. A una breve introducción sobre la historia geológica del Altiplano y sus sistemas hidráulicos le sigue una descripción de la situación de la fauna existente (peces autóctonos y endémicos del género *Orestias* frente a especies exóticas, representadas principalmente por samónidos). Se discuten consecuencias tanto ecosociales como ecológicas y aspectos evolutivos. En conclusión, se presentan cuatro recomendaciones respecto a un uso potencialmente sostenible de los recursos naturales en la región del Lago Titicaca.

1 Introduction

A discussion of the problems involved in and possible alternatives for the use of Lake Titicaca's biological resources today requires at least some knowledge of the region's geological history – in other words, of the Altiplano of Bolivia, Chile and Peru (Fig. 1). The manifold aspects of the present situation can scarcely be understood without this knowledge. For example, the lake's very unstable ecological balance is the product of its extreme geographic position and climate, which in turn is the product of the region's geological history.

Meanwhile, the lake's more recent history is quite well known (Moon 1939, Monheim 1956, Kött *et al.* 1995). It was created during the last 20 million years when the area was forced upward at an accelerated rate from about 800 m above sea level to its present average altitude of >4,000 m (Seyfried *et al.* 1995) (Fig. 2).

[1] funded by DAAD, DFG and GTZ

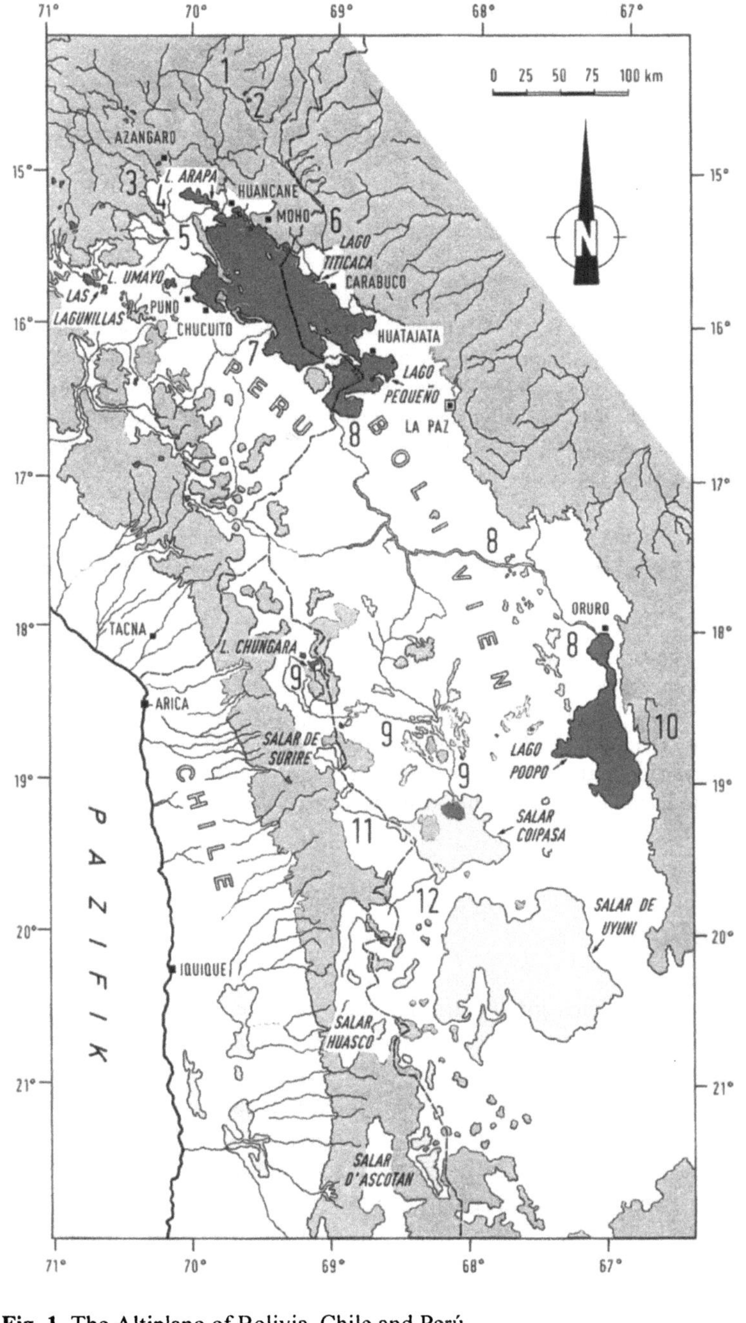

Fig. 1. The Altiplano of Bolivia, Chile and Perú
Ilustración 1. El Altiplano de Bolivia, Chile y Perú

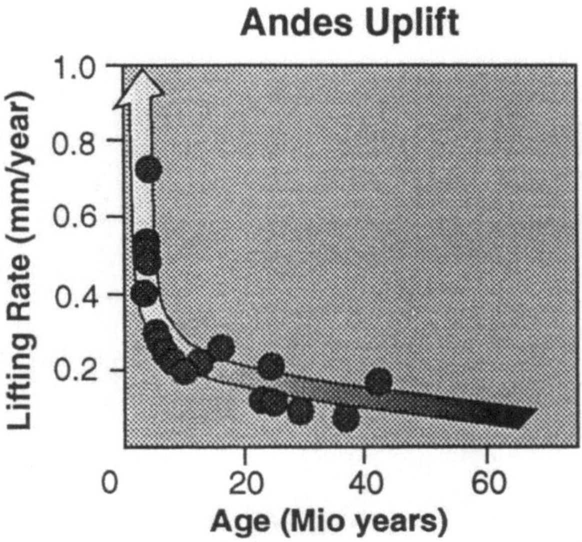

Fig. 2. Geological development of the Andes (from Seyfried *et al.* 1994; changed by Stiewe, Hamburg)
Ilustración 2. Desarrollo geológico de los Andes (de Seyfried *et al.* 1994, modificado por Stiewe, Hamburgo)

Lacustrine deposits show that this development originated with an enormous precursor of the present lake system which had a rich and evidently tropical flora and fauna (Ochsenius 1974). Both died out with the exception of a very small amount of remaining (plants and) animals as the area was forced upward. The primary reason for this was probably the change in climate from tropical to at least semi-arid and cool.

At the end of this process – some 200,000 years ago – the Lake Titicaca region was covered by another, an even better known ancient lake, called Lake Ballivian (Fig. 3). This lake, which was much smaller than its unnamed hypothetical precursor from the Miocene age, eventually shrank to become an even smaller second one called Lake Minchin. The latter lake was mainly responsible for the evolution of Lago Pequeño and the southern lakes and salares in present-day Bolivia. The northern part, which includes the more recent Lake Titicaca (Lago Grande) up to the plains of Junin (northern Peru), shrank again in the course of the following periods and most probably became separated from the southern part of the original lake as a result (Fig. 3).

The history of the glaciation periods is however somewhat controversial. It seems unlikely that the more shallow bodies of water – with the exception of Lake Titicaca with its large surface and its remarkable depth (>280 m) – remained suited for supporting higher plants and animals. On the other hand, it seems likely that the ancient ancestors of recent invertebrates, and vertebrates in particular, survived in this remnant of water – a lake or similar body.

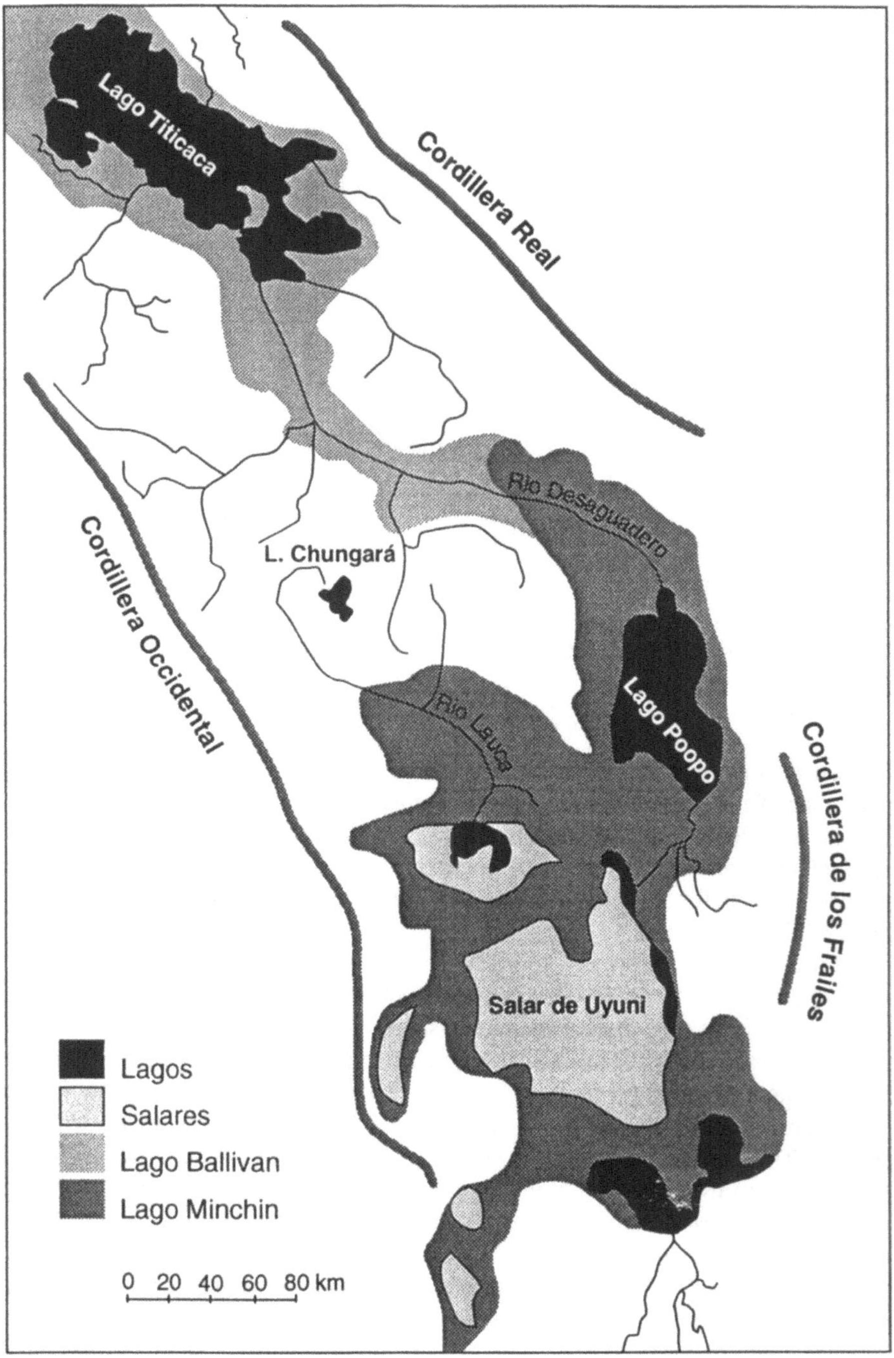

Fig. 3. Historical development of the Altiplano and its main bodies of water (according to Ahlfeld & Branisa 1960; computer illustration by Stiewe, Hamburg)

Ilustración 3. Desarrollo histórico del Altiplano y sus principales cuerpos hidráulicos (según Ahlfeld & Branisa 1960; illustración a ordenador de Stiewe, Hamburgo)

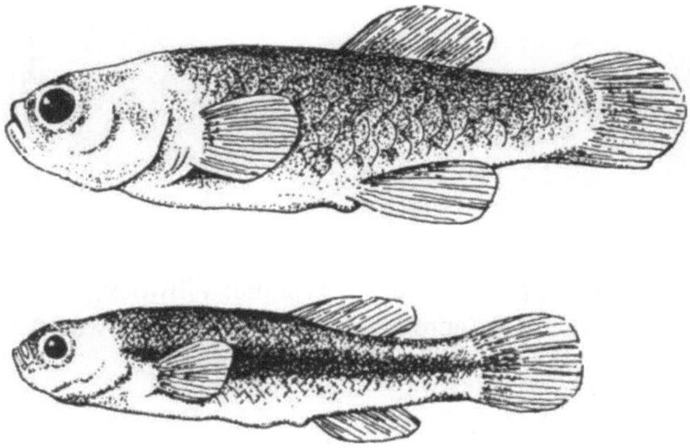

Fig. 4. *Orestias agassii agassii* CUV. et VAL. 1846, female above, male below, natural size

Ilustración 4. *Orestias agassii agassii* CUV. et VAL. 1846, hembra arriba, macho abajo, tamaño natural

2 Case study: Lake Titicaca and its fauna

Lake Titicaca is obviously much younger than was supposed in the past (Newell 1949). Accordingly, its evolutionary processes have not yet come to an end. This can be concluded, for instance, from the lack of progressive ecological differentiation in the autochthonous and endemic teleost genus *Orestias* spp. (Teleostei: Cyprinodontidae) that would lead to the formation of valid species. There may be a few exceptions. There are probably just three or four species that occupied or still occupy their own ecological niches.

Most of the representatives of *Orestias* did not even reach the stage of speciation arising from ecological adaptation, which requires much more time than speciation caused by geographic separation (c.f. Siberia's famous 30 million-year-old Lake Baikal). In case of Lake Titicaca the majority of representatives of the genus *Orestias* obviously did not develop beyond the stage of being ecological "generalists" (Villwock & Sienknecht 1993, 1995) (Fig. 4). Was this perhaps advantageous to their survival?

In contrast, the exotics had already developed special adaptations to their native habitats which enable them to compete successfully with endemics living in the same or similar habitats today. Therefore, it can be predicted that introduced exotics will ultimately dominate the autochthonous species completely in all areas due to their own ecological potency – and not least of all due to their much larger size. The latter situation probably caused the extinction of the largest of the endemics, the trout-like "umanto" (*O. cuvieri,* about 25 cm in total length; Fig. 5), whereas the second largest, the "boga" (*O. pentlandii,* Fig. 6), disappeared from the main lake and was reduced to smaller populations that are found mainly in Lake Arapa and Lago Pequeño (Kosswig & Villwock 1964, Villwock 1972, Courtenay & Stauffer 1984, Ferguson 1990).

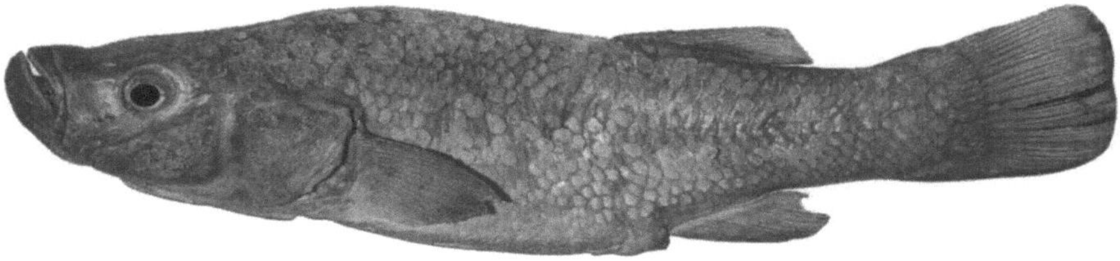

Fig. 5. *Orestias cuvieri* CUV. et VAL. 1846 (approx. ½ nat. size) (photo: Villwock)
Ilustración 5. *Orestias cuvieri* CUV. et VAL. 1846 (aprox. ½ tam. nat.) (foto: Villwock)

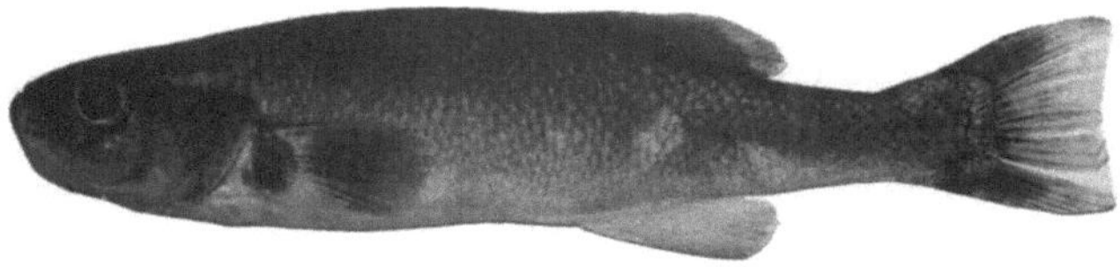

Fig. 6. *Orestias pentlandii* CUV. et VAL. 1846 (approx. ½ nat. size) (photo: Villwock)
Ilustración 6. *Orestias pentlandii* CUV. et VAL. 1846 (aprox. ½ tam. nat.) (foto: Villwock)

Disease is another negative consequence of the introduction of exotics: The endemics' lack of resistance to introduced parasites and diseases such as the ciliate *Ichthyophthirius multifiliis* caused severe damage among them. Wurtsbaugh & Tapia (1988) reported that 93% of the "carache" (*O. agassii*), one of the fish that is most frequently caught by native fishermen, had died. This is however just one well-known example.

3 Eco-social consequences

Although these aspects have far-reaching eco-social consequences and are among the most important, they have been neglected for many decades. This is because even today most of the native inhabitants along the coastline of Lake Titicaca and Lago Pequeño – especially the Urus, the inhabitants of the floating islands – refuse to consume various salmonids (Atlantic and Pacific salmon and certain trout). They are still much more accustomed to their "carache" (*O. agassii,* see Fig. 4) and "carache amarillo" (*O. luteus,* Fig. 7) than to the "truchas" which they sort out from their catches even today, or the "pejerreyes" which are also known as silversides (*Odontesthes bonariensis*), an atherinid that was introduced from the Rio Paraná system. "Pejerreyes," salmon and "truchas" are usually sold to tourist hotels or to tourists directly.

The reasons why native fishermen and their families refuse to eat either of them vary. According to an old fisherman, one of the reasons is the fact that larger salmonids now feed on local frogs (*Telmatobius* spp.) from the lake which is known as "pacha mama" or Mother Earth. And the people refuse to eat fish that

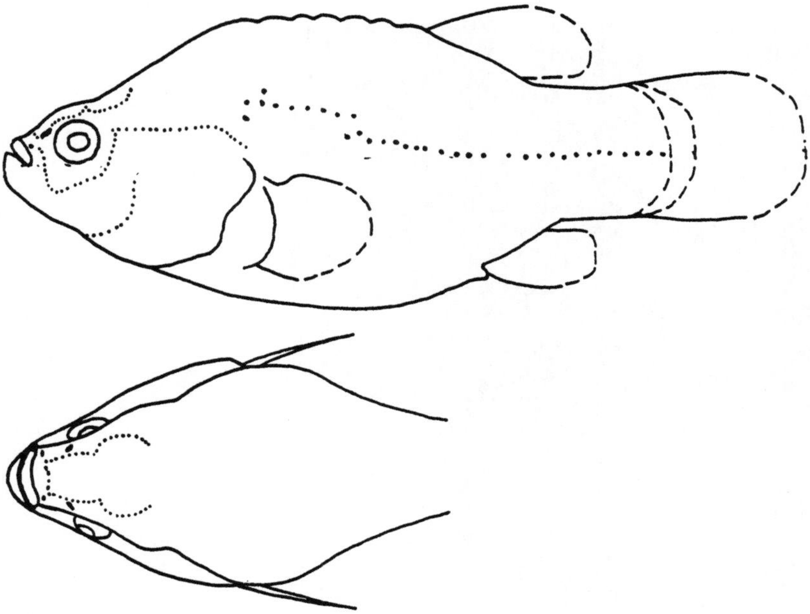

Fig. 7. *Orestias luteus* CUV. et VAL. 1846 (from Villwock 1986, changed)
Ilustración 7. *Orestias luteus* CUV. et VAL. 1846 (de Villwock 1986, modificado)

feed on "pacha mama" (Villwock 1994). This opinion is not very widespread, but is it justifiable to neglect or overlook it simply because it represents an "old-fashioned" point of view? No, it isn't!

4 Ecological consequences

Furthermore, there is another consequence worth mentioning: The inhabitants of the floating islands in Puno Bay once obtained their protein from different *Orestias* spp. but no longer have a sufficient protein supply. They naturally need other sources of protein. Besides the limited use of dwarf pig races (Fig. 8), they gather the eggs of birds such as herons living in the totora girdle. They either eat the eggs directly or hatch them to raise the young birds for their meat (Fig. 9). They also diminish other bird species by shooting and stuffing them for sale to tourists (Fig. 10). This allows them to earn the money they urgently need to buy goods that are not available on their floating islands. In earlier times, there were enough endemic fish to generate a surplus that could be sold or exchanged for the day-to-day goods they needed (Villwock 1994).

These are only some of the situations and consequences that followed the introduction of the exotics which included salmonids and "pejerreyes" (see Vaux *et al.* 1988, Loubens 1989). Since their introduction, the "pejerreyes" have formed self-maintaining populations in different locations and have destroyed the

Fig. 8. Dwarf pig on the floating islands of the Urus, Puno Bay, Lake Titicaca (photo: Villwock)
Ilustración 8. Cerdo enano de las islas flotantes de los Urus, la bahía de Puno y el Lago Titicaca (foto: Villwock)

Fig. 9. Juvenile red herons kept on the floating islands (photo: Villwock)
Ilustración 9. Jovenes de garzas rojas criadas en las islas flotantes (Foto: Villwock)

Fig. 10. Inhabitants of the floating islands, preparing and selling stuffed birds (photo: Villwock)
Ilustración 10. Habitantes de las islas flotantes preparando y vendiendo pájaros rellenos (foto: Villwock)

diversity of native fish even beyond these areas by using the littoral zone – which is the preferred habitat of the larvae and young of nearly all endemics – for reproduction.

5 Loss of evolutionary potency

Although policy-makers and people with only a commercial interest in the area might consider it to be of negligible importance, one other problem must be stressed here: Biologists (and evolutionary biologists in particular) fear that one of the last remaining examples of intralacustrine speciation may be lost as a result of the stepped up and uncontrolled breeding of exotics and their release into natural habitats in the region under discussion. Lake Baikal's unique fauna has been extensively destroyed by the tremendous changes that have occurred in its surrounding environments. These include logging in the forests and the manufacture of wood products. During the production of paper, waste water from the paper mills is released into the lake. As a result, the lake now contains a progressively contaminated, nearly sterile zone below a depth of 800 m. At one time, specialized organisms lived at depths of 1,400 m in Lake Baikal! Lake Aral has nearly disappeared thanks to the imprudent use of its water (e.g., for irrigating the huge cotton fields in its surroundings). Lake Lanano on the Philippine island of Mindanao lost its famous endemic species swarm of cyprinids (a relative of our common carp) following the introduction of a carnivorous exotic from the Pacific. Lake Victoria's brilliant ichthyofauna (more than 300 endemic percoid cichlids), known for their unique "preferential mating phenomena" which leads to rapid speciation, have been nearly destroyed by introduction of the "Nile perch" (*Lates niloticus*) (Oguto-Ohwayo 1989). Now Lake Titicaca and its endemic fauna are being threatened by a very similar fate – not to mention the eco-social consequences outlined above. And all this despite the Rio Convention from 1992 that was adopted to stop the unilateral exploitation of nature and to protect its biodiversity! The argument that these exotics were introduced to the lakes mentioned and raised solely to meet the urgent protein needs of the poor and hungry in underdeveloped countries will not hold: This was done merely to generate a profit by selling them to overdeveloped countries (USA, Japan, Europe) where consumers do not know and therefore do not care about the – lasting – environmental damage being done in the countries of origin. It has been proven over and over again: In most cases, the introduction of exotics – be they plants or animals – means only "fast money" but usually for only a limited time. Take Lake Victoria in East Africa as an example. The Nile perch destroyed 80% of the endemic cichlids there within less than 10 years. One consequence of this has been the virtually complete destruction of the original food chain, followed by dramatic changes in the lake's abiotic ecological factors. This in turn has led to the extinction of further endemics – and negative prospects for the "money-making" Nile perch itself (which is widely known as "Victoria Barsch" in Germany): Its own population is already declining in number because of the impact of its introduction. The Nile perch is an extremely carnivorous predator and has begun to feed on its own brood for lack of other, appropriate fish to feed on (a behaviour that has also been observed among salmonids and trout elsewhere). As a result, profits can be expected to decline within a few decades. Here and there the result will naturally follow the familiar pattern: Nature will become poorer and the people depending on it will have a similar fate. Therefore, those who are responsible for future development should make a stand for protecting what is left of our natural resources and biodiversity – which constitute the urgently needed genetic reservoir that future adaptive processes will draw on in a changing world.

6 Sustainable use of natural resources: Some suggestions

It would be naive to dream of a total renunciation of activities that destroy genetic and natural resources (which are actually the same thing!). However, we must all to come to a general agreement on the considerate (or, to use a modern term "sustainable") use of these resources and on putting an end to exploitation for short-term profit. This would be a major contribution toward allowing remaining endemics – plants, invertebrates and vertebrates as elements of an ecosystem – in Lake Titicaca, for instance to continue to exist in order to evolve.

Based on these considerations, some suggestions may be made for an alternative use of Lake Titicaca's natural resources (which would also be better than their present use):

1. The first step would to reduce the rate at which salmonids are being released or allowed to escape into the open lake itself (see Urquidi 1969).
2. The number of lake cages where salomonids are kept nowadays should be reduced because feeding them with pelleted food causes serious damage to the environment. The cages should be moved to different locations on a regular basis. In addition to this, it is extremely important that the cages be constructed in a way that prevents fish from escaping.
3. Hatching stations should be used to hatch and raise endemics, and protected areas should be established for released endemics (closed bodies of water that cannot be reached by exotics).
4. Scientific research should be conducted on the phenomenon of an odd "shifting" process that has been observed in a certain endemic of Lago Grande, the "ispi" (*O. ispi*, Lauzanne 1981) (Fig. 11). This species has not only withstood competition from various salmonids and the silversides but also seems to have gone from being a generalist to a specialist. It has developed a successful survival strategy by forming huge schools in deeper open waters where they no longer have to compete with exotics or their own relatives. The proposed research might even result in combining ecological and evolutionary considerations perhaps even with commercial use within acceptable limits – and thereby solve some of the previously discussed problems.

Although these four aspects have been touched upon only briefly, they may open up alternatives for the considerate use – the so-called "sustainable use" – of the remaining natural resources in Lake Titicaca (and elsewhere in the region) that combines the frequently stressed link between ecological *and* economic interests. Only when we make a serious attempt to arrive at an acceptable balance between what are normally regarded as contradictory and irreconcilable interests might coming generations of human beings have at least an inkling of the fascinating magnificence of a nature – that developed without man.

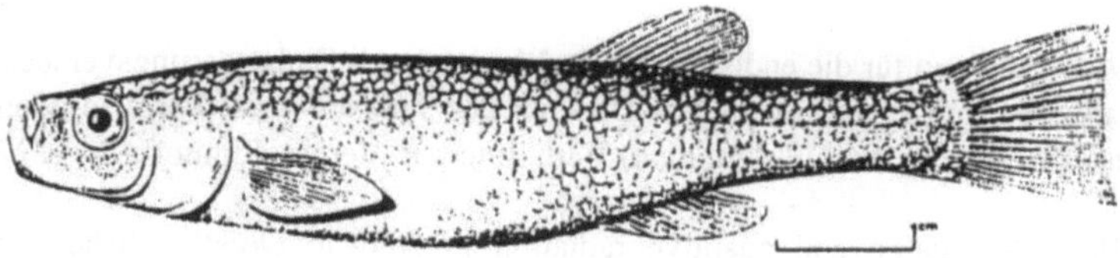

Fig. 11. *Orestias ispi* n.sp. Lauzanne 1991 (changed)
Fig. 11. *Orestias ispi* n.sp. Lauzanne 1991 (modificado)

References

Courtenay WR, Stauffer JR (eds) (1984) Distribution, biology and management of exotic fishes. John Hopkins Univ Press, Baltimore

Ferguson MM (1990) The genetic impact of introduced fishes on native species. Canadian J Zool 68: 1053-1057

Everett GV (1971) A Study of the Biology, Population Dynamics, and commercial Fishery of the Rainbow Trout (*Salmo gairdneri* RICHARDSON) in Lake Titicaca. FAO-Report 1-24 (c.f. 1973: J Fish Biol 5: 429-40)

Kött A, Gaupp R, Wörner G (1995) Miocene to recent history of the western Altiplano in northern Chile revealed by lacustrine sediments of the Lauca Basin (18°15'- 18°40' S / 69°30' - 69°05' W). Geol Rundsch 84 (4): 770-780

Kosswig C, Villwock W (1964) Das Problem der intralakustrischen Speziation im Titicaca- und im Lanao-See. Verh Dtsch Zool Ges Kiel (1964): 95-102

Lauzanne L (1981) Description de trois *Orestias* nouveaux du Lac Titicaca, *O. ispi n.sp.*, *O. forgeti n.sp.* et *O. tcherna-vini n.sp.* (Pisces, Cyprinodontidae). Cybium, 3e série, 5 (3): 71-91

Loubens G (1989) Observations sur les poissons de la partie bolivienne du lac Titicaca. IV.*Orestias ssp.*, *Salmo gairdneri* et problèms d'aménagement. Rev Hydrobiol trop 22 (2): 157-177

Monheim F (1956) Beiträge zur Klimatologie und Hydrobiologie des Titicaca-Beckens. Heidelb Geogr Arb H1: 1-152

Moon HP (1939) The geology and physiography of the Altiplano of Peru and Bolivia. Trans Linn Soc 1: 27-43

Newell N (1949) Geology of Lake Titicaca region, Peru and Bolivia. Geol Soc Am Memoir 36: 10

Ogutu-Ohwayo R (1989) The decline of the native fishes of lakes Victoria and Kyoga (East Africa) and the impact of introduced species, especially the Nile perch, *Lates niloticus*, and the Nile tilapia,*Oreochromis niloticus*. Environ Biol Fishes in press (cit. after Ferguson 1990)

Ochsenius C (1974) Relaciones paleobiogeograficas y paleoeco logicas entre los ambientes lenticos de la Puna y Altiplano Boliviano, tropico de Capricornio. Boletin de Prehistoria de Chile 6-7 (7-8): 101-137

Seyfried H, Wörner G, Uhlig D, Kohler I (1994) Eine kleine Landschaftsgeschichte der Anden Nordchiles. Jb Univ Stuttgart: 60-72

Urquidi WT (1969) Problemas de conversaciôn de los recursos pesqueros de Bolivia. Boletin Experimental, 40: 1-16

Vaux P, Wurtsbaugh W, Treviño H, Mariño L, Bustamante E, Torres J, Richerson P, Alfaro R (1988) Ecology of the Pelagic Fishes of Lake Titicaca, Peru-Bolivia. 20 (3): 220-229

Villwock W (1972) Gefahren für die endemische Fischfauna durch Einbürgerungsversuche und Akklimatisation von Fremdfischen am Beispiel des Titicaca-Sees (Peru/Bolivien) und des Lanao-Sees (Mindanao/Philippinen).Verh Internat Verein Limnol 18: 1227-1234

Villwock W (1986) Speciation and adaptive radiation in Andean *Orestias* fishes. In: Vuilleumier F, Monasterio M (eds) Oxford University Press, New York, Oxford, pp 387-403

Villwock W (1994) Die Titicaca-See-Region auf dem Altiplano von Peru und Bolivien und die Folgen eingeführter Fische für Wildarten und ihren Lebensraum. Naturwiss 80: 1-8

Villwock W, Sienknecht U (1993) Die Zahnkarpfen der Gattung *Orestias* (Teleostei: Cyprinodontidae) aus dem Altiplano von Bolivien, Chile und Peru. Ein Beitrag zur Entstehung ihrer Formenvielfalt und intragenerischen Verwandtschaftsbeziehungen. Mitt Hamb zool Museum Inst 90: 321-362

Villwock W, Sienknecht U (1995) Intraspezifische Variabilität im Genus *Orestias* VALENCIENNES, 1839 (Teleostei: Cyprinodontidae) und zum Problem der Artidentität. Mitt Hamb zool Mus Inst Suppl 92: 381-398

Wurtsbaugh WA, Tapia RA (1988) Mass Mortality of Fishes in Lake Titicaca (Peru-Bolivia) associated with the Protozoan Parasite *Ichthyophthirius multifiliis*. Trans Am Fish Soc 117: 213-217

Part 5.3

Policy and the conservation of biodiversity

Facing the challenges of biodiversity conservation in Bolivia

Alexandra Sánchez de Lozada
Ministry for Sustainable Development and Environment, La Paz, Bolivia

Key topics:

- Introduction
- The biodiversity of Bolivia
- Institutional reforms and the challenge of biodiversity conservation
- The institutional framework
- The National Directorate of Biodiversity Conservation
- The policies for biodiversity conservation.
- The National System of Protected Areas
- Popular participation in the conservation process
- Conservation and management of genetic resources
- Conservation and management of wildlife
- Main achievements of the NDCB
- Conclusions

Haciendo frente a los desafíos que supone la conservación de la biodiversidad en Bolivia

Temas clave:

- Introducción
- La biodiversidad en Bolivia
- Reformas de instituciones y el desafío que supone la conservación de la biodiversidad
- El marco institucional
- La Dirección Nacional para la Conservación de la Biodiversidad (DNCB)
- Las políticas para la conservación de la biodiversidad
- El Sistema Nacional de Áreas Protegidas
- Participación popular en el proceso de conservación
- Conservación y manejo de los recursos genéticos
- Conservación y manejo de la vida salvaje
- Principales logros de la DNCB
- Conclusiones

1 Introduction

Bolivia's public policies on the conservation of biodiversity have been directly influenced by the government's position on development. Since 1993, the government of Gonzalo Sánchez de Lozada has undertaken structural reforms to change the country's prevailing unsustainable development patterns. With this goal in mind, the Ministry of Sustainable Development and the Environment was created to plan and promote national development which would incorporate sustainable patterns of production as part of the management of Bolivia's natural resources. The Government's commitment to the conservation of biodiversity is exemplified by the work of the National Directorate of Biodiversity Conservation, whose main achievements are described in this paper.

2 The biodiversity of Bolivia

Although the real dimensions of Bolivia's biodiversity are not well known due to the short history of biological research here, recent scientific surveys have begun to document the country's extraordinary wealth of species and habitats. Among Latin American countries, it encompasses many natural ecosystems, with some 40 ecoregions described in its protected areas alone. Due to its low human population density and lack of economic development, Bolivia also has extensive areas of pristine ecosystems.

Approximately 22,000 species of native plants have been classified, among which orchid diversity is exceptional with more than 2,000 species known. There are also at least 90 species of palms in the country. In terms of fauna, some 1,380 species of birds have been documented, making Bolivia the seventh most diverse country worldwide for this taxonomic group. Among mammals, 360 species have been registered, representing about 40% of all Neotropic species, of which ten are endemic to Bolivia. It numbers fifteenth in primate diversity with 25 species identified to date. In addition, 260 species of reptiles and more than 100 species of amphibians roam the country's ecoregions.

Bolivia is also an important centre for domesticated and wild genetic resources. Many economically important species probably originated in this region, and/or were domesticated by its indigenous population. These include medicinal plants; crops, such as potatoes, maize, peanuts, quinua; and such animal species as llama, alpaca and guinea pigs. In addition, the cultural diversity of Bolivia is remarkable, and includes more than 40 ethnic groups, many of which conserve their traditional uses of natural resources.

The conservation of this biological wealth is one of the greatest challenges facing a country considered to be one of the poorest in Latin America.

3 Institutional reforms and the challenge of biodiversity conservation

In 1993 the president of Bolivia, Gonzalo Sánchez de Lozada, began an ambitious reform of the government's executive branch. The goal was to profoundly change the prevailing tendency toward environmental degradation and wasteful use of natural resources, and to improve the country's standard of living. The government adopted sustainable development as its creed and, to institutionalize this, fundamentally changed the government's structure by creating several new ministries including the Ministry of Sustainable Development and the Environment.

Complementing the institutional reform, important laws were enacted by Congress, such as the Law on Popular Participation which provides a venue for all citizens to participate in the establishment of democracy at all levels. The Law on Decentralization reduces the central government's power while giving local governments more decision-making authority, particularly in apportioning their budgets according to locally established priorities. These laws aim at redistributing the income of the country more equitably and, in particular, at improving living conditions in rural areas. Other important laws include the capitalization law which privatizes state enterprises and benefits Bolivian citizens in general and the educational reform law which was designed to improve the quality of the country's education system.

Two important bills were passed in the natural resources area. The Forestry Law regulates the sustainable use of forest resources, and the INRA Law created the National Institute of Agrarian Reform, addresses the issue of land tenure and ensures land holding rights for indigenous and peasant communities. This latter law also incorporates aspects relevant to biodiversity conservation by, in particular, recognizing and enhancing private land ownership with regard to conservation goals. Under previous law, land was seen only as a means of production.

These various laws also address the distortions that have accumulated over the decades and have had a detrimental effect on efficiency and social equity. Bolivia's future requires action on the part of an intelligent state with a long-term vision that will remove existing obstacles and facilitate the functioning of a market economy in a sustainable manner. The planning function of the state, coupled with a strategic vision in which the state is no longer the main producer of goods and services but the formulator of policies that will steer the national economy toward sustainability, is critical to resolving the acute structural problems that have long hindered national development.

The Bolivian government has, through institutional reforms and policies, incorporated sustainability and all its facets into the country's development strategy. This new concept of development provides guidelines for change based on the interaction of four variables: economic growth, social equity, governance, and the sustainable use of natural resources. The sustainable use of natural resources and biodiversity conservation are now critical components of public policy decisions. Further, the participation of citizens in this process at all levels assures democratic input in the development and implementation of policies and activities.

3.1 Institutional framework

The Ministry of Sustainable Development and Environment is charged with planning national development and fulfils its mission within the scope of the democratic and sustainable development reforms mentioned above, which include popular participation. This new system gives political power to representative government at all levels and provides for local autonomy in decision making.

The National Secretariat of Natural Resources and Environment, part of this Ministry, orients its policies, objectives and actions to conform to guidelines established by the Republic's General Economic and Social Development Plan. The National Directorate for the Conservation of Biodiversity (NDCB) forms part of this National Secretariat, and is also part of the Sub-secretariat for Natural Resources together with other directorates which include Forest Utilization, Watershed Conservation and Soil Conservation. In this context, the NDCB implements plans and the programmes required for the sustainable use and conservation of biodiversity.

4 The National Directorate for the Conservation of Biodiversity

The National Directorate for the Conservation of Biodiversity (NDCB) was created in 1992. Its principal objectives are to develop and propose policies and norms for the conservation and sustainable use of Bolivia's biodiversity. Other NDCB objectives are: to contribute to *in situ* and *ex situ* biodiversity conservation; to support scientific research and training of human resources in relevant fields; to develop planning processes for the conservation and sustainable use of biodiversity; to promote and arouse public awareness and education on these issues; and to promote and recover traditional practices and knowledge of indigenous peoples and peasant communities for the conservation and sustainable use of biotic resources. The Directorate is composed of the following departments: Protected Areas, Genetic Resources and Wildlife, thus providing for an integrated approach to biodiversity conservation.

5 Biodiversity conservation policy making

The Directorate's *institutional policies* are aimed at creating a normative framework based on national priorities and international agreements and commitments. Policies and norms include those relevant to the enforcement of international agreements on conservation such as: The Convention on

Biodiversity, CITES, the Ramsar Convention, the Natural Patrimony and Cultural Convention, the Decisions of the Cartagena Agreement, the Amazon Cooperation Treaty and others.

Management policies are oriented toward strengthening the administration and sustainable use of biodiversity, integrating the development of conservation concepts, and empowering citizens through participation in the management of protected areas. Policies and norms are developed for (1) the productive sectors so as to prevent unsustainable uses of biodiversity resources that exceed their capacity for regeneration; (2) the regulation, supervision and control of biodiversity resource use both outside and inside the National System of Protected Areas; (3) the establishment of agreements at local, national and international level for wildlife management, control, research and the development of proposals for the sustainable use of wildlife species; (4) the strengthening of genetic resource conservation in order to reduce the erosion of these resources, particularly those of major economic, medical and biological importance; and (5) the elaboration of norms and programmes for the development of sustainable tourism in Protected Areas.

Promotion and awareness policies are aimed at the development of activities which will allow the state and the social sectors to become better aware of the vital importance of the conservation of biodiversity. This provides for environmental education for local communities around protected areas as well as for national campaigns on these issues.

NDCB responsibilities: The NDCB has responsibilities at both national and departmental level. At national level, the Directorate develops norms for the management of biodiversity, defines policies and priorities for the conservation process, plans the management and sustainable use of biodiversity resources, negotiates financial resources and administers their distribution and use, and formulates coordination procedures with governmental bodies, operational organizations, territorial organizations, international organizations and others.

At departmental level, the NDCB enforces norms and policies generated at national level, implements policies and management planning processes in its departments, monitors the rational utilization of financial resources it has been allocated within the framework of national projects, and manages the resources available at departmental level. The Directorate also elaborates and implements departmental programmes for the conservation of biological diversity and participates in the management of wildlife resources.

6 Management of protected areas

The Protected Areas Department is responsible for the management, organizing, strengthening and expansion of the National System of Protected Areas. The objectives of this Department include insuring the conservation of representative samples of the country's ecosystems, as well as the *in situ* conservation of species and genetic resources. This Department also promotes citizen participation in conservation activities and it has, to this end, encouraged negotiations between local population groups and competent authorities at the governmental and citizen level. This unit is organized in four sections: Planning, Information and Monitoring; Protection; Training; and Promotion, Environmental Education and Tourism.

- The **planning** section's responsibilities, according to the established guidelines, are the monitoring and evaluation of the Protected Area Annual Operating Plan. In this connection, 15 protected areas are implementing their own annual operating plans during the course of 1997. This section is also charged with supervising the elaboration of management plans, which are typically prepared by specialized national and international NGOs. This year, management plans are either being implemented or prepared for eight protected areas. This offers a contrast to 1993 when only two of the nation's protected areas had management plans. Also, this section of the NDCB has gathered and systematized information about protected areas and biodiversity. A computerized information system has been developed, and a library of some 2,000 books and documents, and 160 maps has been established.
- The section focusing on **protection** concentrates on coordinating the local administration of the areas as well as the activities of the national park ranger staff. They control the activities of the areas, the application of norms and the sustainable use of natural resources.
- A systematic **training** programme for the park rangers, directors and personnel in the system has been developed since 1994.
- The **promotional** and **education** activities are centred on increasing public awareness through the mass media of the importance of conserving protected areas, as well as on providing environmental education for local populations. In addition, initial steps are being taken to develop policies and programmes for responsible and sustainable tourism within protected areas. The NDCB views ecotourism in the areas as a potential way to work towards the financial sustainability of some areas.

6.1 The National System of Protected Areas (NSPA)

The National System of Protected Areas (NSPA) contains 26 conservation units. In light of their ecological importance, these protected areas have been placed under special administration in order to conserve representative samples of Bolivia's main ecosystems. The primary purpose of the System is to implement rules, regulations and procedures, and to insure that representative samples of Bolivia's varied biogeographic regions are conserved. The following NSPA protected areas are currently under NDCB management:

- Madidi National Park and Integrated Management Natural Area
- KAA-IYA del Gran Chaco National Park and Integrated Management Natural Area
- Amboró National Park and Integrated Management Natural Area
- Noel Kempff Mercado National Park
- Carrasco National Park
- Beni Biosphere Reserve and Biological Station
- Ulla Ulla National Wildlife Reserve
- Eduardo Avaroa National Andean Wildlife Reserve
- Sajama National Park and Integrated Management Natural Area
- Cotapata National Park and Integrated Management Natural Area
- Isiboro Sécure National Park and Indigenous Territory
- Pilón Lajas Biosphere Reserve and Indigenous Territory
- Tariquía National Wildlife Reserve
- Cordillera de Sama Biological Reserve
- Torotoro National Park

7 Citizen participation in the conservation process

Popular participation is one of the most significant reforms implemented in Bolivia in recent years, not only due to its social and economic impact, but also because of the capacities that it is developing in Bolivian society in general. This has important ramifications for biodiversity conservation. Within the framework of the participation process, peasant communities and indigenous peoples are being incorporated into protected-area management and the decision-making process through the establishment of Management Committees and, in some cases, local co-management of the protected areas.

7.1 The Management Committees

The NDCB created Management Committees (MC) to ensure the active participation of local land holders in the management of protected areas. The MCs are comprised of representatives from the respective communities and other key local sectors including indigenous peoples, peasant communities, municipalities, public institutions, NGOs as well as from the Ministry for Sustainable Development and the Environment. Eight protected areas now have committees and three more are in the process of being established. The Committee takes part in the area's management by guiding, supervising and supporting the implementation of the Management and Annual Operation Plans. The conservation process gains a lasting and sustainable quality as a result of the involvement of vital interested sectors.

7.2 Management agreements

The protected areas are managed in two ways. One is directly through the NDCB, and the other is through agreements with other organizations. As a rule, direct management is carried out by the NDCB where local conditions do not permit, for any of a number of reasons, the administration of the area by a local group. Administrative agreements have to date been signed with NGOs, academic institutions and bodies, indigenous peoples and peasant communities. The agreements with private non-profit bodies and grass roots organizations helps make the management of the Protected Areas more efficient, democratic, decentralized and open. This model has been adopted for areas such as the Noel Kempff Mercado Park which is co-managed with the Friends of Nature Foundation (FAN); the Ulla Ulla National Wildlife Reserve, co-managed with the Canadian Center of International Research and Cooperation (CECI); the Pilon Lajas reserve which is co-managed by the international NGO, Veterinarians Without Borders; and the Kaa-Iya Park which is co-managed by the indigenous organization Capitanía of the Upper and Lower Izozog. Likewise, other agreements have been signed with public and private institutions for various purposes such as scientific research and ecotourism. Protected areas management in Bolivia is therefore being developed within the state's newly redefined roles where the government, while maintaining responsibility, develops collaborative agreements with relevant institutions from civil society for the administration of the country's protected areas.

8 Genetic resources conservation and management

Another important working area for NDCB is the field of genetic resources, both domesticated and wild, which could be considered Bolivia's greatest natural heritage. The Department deals with the conservation of genetic resources, particularly of species having economic, medical or biological importance. Its objective is to manage these resources so as to generate income for the Bolivian population. To date, the Department has concentrated on the elaboration of norms in conformity with international commitments toward the sustainable use and conservation of the country's genetic resources. It has been working under Decision 391 of the Cartagena Agreement which regulates access to genetic resources under a common regime for the Andean countries. This legal instrument was approved in 1996 and its principal objective is to ensure a fair and equitable distribution of the economic benefits arising from the use of these resources. It explicitly acknowledges the rights of indigenous peoples and peasant communities to their knowledge, innovations and traditional practices associated with the use of such genetic resources.

Among other things, the Genetic Resources Department has participated in the meetings held on the implementation of the Convention on Biological Diversity. It has been elaborating "biosafety" regulations to provide appropriate safeguards for the introduction of genetically modified organisms, recognizing their potential impacts on human health and biodiversity. Lastly, a National Conservation System for Genetic Resources has been established to provide inter-institutional coordination between the bodies dealing with these matters at governmental and non-governmental level.

9 Wildlife management and conservation

The NDCB's Wildlife Unit promotes the sustainable use of wildlife resources as well as endangered species protection, population restoration and species restocking. This unit has developed a National Program for Wildlife Conservation and Management. It supports and assists pilot management programmes for economically important species in order to improve living conditions among the indigenous and peasant population. It has been evaluating the potential for the sustainable culling of such species as vicuna *Vicugna vicugna*, alligators *Crocodilus yacare*, river turtles *Podocnemis expansa* y *Podocnemis unifilis*, capybaras *Hydrochaeris hydrochaeris* and quirquincho armadillos *Chaetophractus nationi.* A National Census of Vicunas was carried out in 1996 under the framework of the National Vicuña Program and with the assistance of several institutions. With regard to flora, a study was made of native palm trees *Palmae,* and another is being carried out on the status of mahogany *Swietenia macrophylla*, a species currently threatened with extinction due to commercial exploitation. This Unit has also developed regulations for the management of wildlife, and has worked towards the promotion of conservation activities through information workshops and public awareness campaigns.

10 Primary achievements

During the past four years, the government has made great progress in implementing its conservation policies. This is reflected by the provision of an infrastructure in many of the nation's protected areas and the implementation of a park ranger training programme. This progress is likewise reflected by the creation in 1995 of two very important Protected Areas: the Kaa-Iya Gran Chaco National Park and Integrated Management Area – a vast zone of dry tropical forest located in the southern part of the country – and the Madidi National Park and Integrated Management Area located north of La Paz. These two parks are the largest protected areas in Bolivia and contain vast expanses of pristine ecosystems. Due to its geographic location and variety of ecosystems, Madidi is considered by experts as perhaps the most biologically diverse park in the world. The Kaa-Iya is the largest and most significant dry forest protected area, a habitat type that is far more threatened than tropical moist forests. Together they represent an additional 5 million hectares under protection. The national system now has a total of 14 million hectares legally protected, covering approximately 10% of Bolivia's national territory. People living in these areas have traditionally used natural resources in a sustainable way and government policy emphasizes the involvement of local communities in the conservation process.

The NDCB has been working on the expansion of some areas such as the Noel Kempff Mercado National Park and the Beni Biosphere Reserve and Biological Station. Evaluations have been or are being undertaken for the creation of new Protected Areas such as Palmar Rodeo in Chuquisaca, which protects the dry inter-Andean range, and the Otuquis Tucavaca and San Matias reserves in the Pantanal in southeastern Bolivia. Two potential areas are also being evaluated in Pando, a region of Amazonian forest in the extreme north. The process of creating new areas implies *in situ* evaluation of biological diversity, evaluation of land ownership in the region, negotiations with local populations, institutional linkage at the governmental and public level, among other activities.

Another meaningful NDCB contribution has been the creation of a legal framework for the conservation of Bolivia's wild living resources. The Directorate now works with a body of regulations for the administration and management of protected areas, for the use and management of wildlife, for access to and conservation of genetic resources, and for biosafety. The NDCB has also been lobbying the National Parliament to approve the Biological Diversity Law. This Law will be its most important tool for the protection and conservation of the nation's great variety of ecosystems, animals and plants.

11 Conclusions

The process of conserving Bolivia's biodiversity has been launched with obviously a great deal of effort and enthusiasm at governmental level. During the past four years the National System of Protected Areas was established and consolidated and it now plays the key role in the *in situ* conservation of the nation's biological resources. First steps have also been taken to ensure the sustainable use and conservation of wildlife and genetic resources. Further, a political and legal framework for biodiversity conservation has been established as well. Action taken to empower the people and to promote popular participation, especially through the Management Committees in Protected Areas is producing satisfactory results. However, conserving biodiversity in a country with very limited economic resources poses an enormous challenge. It is to be hoped that the next presidential legislature, which will begin in August of this year, continues to pursue this goal.

The scientific and technical information for this paper was provided by Marco Octavio Rivera.

Table 1. National System of Protected Areas (April 1997)

Protected area and legal base **DS. (Decree of Creation and date)**	**Extension** (Hectares) **Location** (Department) **Altitude range** (Meters above sea level)	**Management**	**Ecological and cultural value**
1. Madidi National Park and Natural Integrated Management Area DS. 24123 - 9.21.1995	1,895,750 Northwest of La Paz, border with Peru 6,000 - 200 m	NDBC administration Management Committee	East of the Real Andean Cordillera, 10 ecoregions. Exceptionally diverse array of ecosystems, including snow peaks, Andean puna, Andean wet puna, Yungas paramo, subtropical montane forests, Andean cloud forest, Yungas subtropical forest, grasslands, Amazonian evergreen forests, deciduous forest, montane dry forests, wetlands and flooded savannas. Most biologically diverse park in Bolivia and one of the most biologically diverse in the world. Archaeological and historical colonial sites
2. KAA-IYA of Gran Chaco National Park and Integrated Management Natural Area DS. 24122 - 9.21.1995	3,441,115 South of Santa Cruz, border with Paraguay 200 m	Capitanía del Alto y Bajo Izozog (Indigenous organization) Management Committee	Two main ecosystems in excellent state of conservation. Lowland tropical dry and semideciduous forest. Thorn forest, low, dense, spiny forests. Chaco upland thorn forest. Species: Diversity of big mammals, Chaco Peccary and 22 other threatened species such as Guanaco (*Lama guanicoe*) almost extinct in Bolivia. Indigenous people: ethnic groups of Guarani origin, Izoceño and a nomadic tribe of Ayoreodes
3. Amboró National Park and Natural Integrated Management Area DS. 222934 - 10.11.1991 DS. 24137 - 10.3.1995	637,600 West of Santa Cruz border with Carrasco 3,330 - 300 m	NDBC administration Management Committee (in process)	Eastern part of the Andean Cordillera, 5 ecoregions. Montane dry forest, cloud rain forest, humid subtropical forest, transitions to the subhumid forest Tucumano-Boliviana. Rich in wildlife and endemism, particularly plants, birds (around 600 bird species) and amphibians. Archaeological sites around

4. Noel Kempff Mercado National Park DS. 21997 - 8.31.1988	1,540,000 Northeast of Santa Cruz border with Brazil 750 - 200 m	Agreement with Friends of Nature Foundation (FAN) Management Committee	Precambrian geological shell. Serranía de Huanchaca, spectacular beauty due to their waterfalls and cliffs, 4 ecoregions. Cerrado woodland, upland campo grassland, semideciduous forest, tropical dry forest, transition forest between Amazon and Cerrado, tropical rain forest, savannas of Cerrado
5. Carrasco National Park (includes Wildlife Sanctuary of Repechón Caves) DS. 22940 - 10.11.1991	622,600 West of Cochabamba border with Amboro 4,500 - 300 m	NDBC Management Committee	Sub-Andean, 5 ecoregions. Montane subtropical ecosystems, Yungas paramo, cloud mountain forest, rain forest, Yungas very humid sub-mountain forest (precipitation 5,000 mm). Archaeological sites
6. Beni Biosphere Reserve and Biological Station DS. 19191 - 9. 5. 1982.	135,000 Southwest of Beni 220 m	National Academy of Sciences Management Committee	Lowlands. Two main ecoregions. Variety of wetlands, Beni savannas seasonally flooded, rain forest, forest islands, swamps and grasslands. Archaeological sites of the Moxeno culture. Indigenous people: Chimane
7. Ulla Ulla National Andean Wildlife Reserve DS. 10070 - 1.7.1972	240,000 Midwest of La Paz border with Peru 6,200-2,800 m	Canadian Center of International Research and Cooperation CECI Management Committee	Andean Real Cordillera. Three ecoregions. High peaks, glacial lakes, Andean prairie, Andean dry and wet puna, Yungas Paramo, cloud montane forest. Centre of vicuna *Vicugna vicugna* conservation. Key species: condor *Vultur gryphus*, Andean bear *Tremarctos ornatus*, great diversity of aquatic birds. Traditional centre of Kallahuaya culture. Archaeological sites
8. Eduardo Avaroa National Andean Wildlife Reserve DS.11239 - 13.12.1973 DS. 18313 - 5.14.1981	714,745 Southwest of Potosí border with Chile and Argentina 6,000 - 4,200 m	NDCB Management Committee	South of the Western Andean Cordillera. Volcanic area with many salty lakes such as "Laguna Colorada" (Red Lake), first Ramsar site in Bolivia. One ecoregion. Cold deserts, Andean prairie, Andean dry puna. Three species of flamingos *Phoenicopterus chilensis* and *P.andinus*, and *Phoenicoparrus jamesi*. Archaeological sites

9. Sajama National Park DS. w/n - 8.2.1939 DS. w/n - 5.11.1945	120,000 (proposed) West of Oruro, border with Chile 6,540 - 4,200 m	NDBC Management Committee	Western Andean Cordillera, the Volcano Sajama (the highest peak in Bolivia 6,542 m.) One ecoregion. Andean dry puna, Andean grasslands. The world's highest forest of kenua *Polylepis tarapacana*. Centre of alpaca and llama wool production. Archaeological and historical colonial sites. Indigenous people: Aymara
10. Cotapata National Park and Natural Integrated Management Area DS. 23547 - 7.9.1993	58,620 Centre of La Paz 5,600 - 1,100 m	NDBC	Real Andean cordillera, 4 ecoregions. Variety of Andean ecosystems, Yungas paramo, cloud forest, Yungas montane subtropical humid forest. Great plant diversity. Archaeological sites. Indigenous people: Aymara
11. Isiboro Sécure National Park and Indigenous Territory DS. 7401 - 11.22.65 DS. 22610 - 9.24.1990	1,200,000 South of Beni and north of Cocha-bamba Chapare 3,000 - 200 m	Indigenous Central of Isiboro Sécure and NDBC	From Real Andean Cordillera to lowlands, 8 ecoregions. Great range of montane to lowland tropical ecosystems: cloud forest, very humid submontane, Yungas subtropical forest, flooded savannas, variety of swamps. Archaeological sites. Indigenous People: Moxeno, Yuracaré & Chimane
12. Pilón Lajas Biosphere Reserve and Indigenous Territory DS.23110 - 4.9.1992	400,000 Mideast of La Paz and Southwest of Beni; 2,500 - 250 m	Veterinarians Without Borders Management Committee	Sub-Andean humid tropics, 4 ecoregions. Cloud forest, very humid montane forest, tropical rain forest, palm swamp forest, savannas of high serranias. Archaeological sites. Indigenous people: Chimane and Mosetenes
13. Tariquía National Wildlife Reserve DS. 22277 - 8.1. 1989	246,870 South of Tarija 3,000 - 800 m	PROMETA For Tarija Environment	Three ecoregions. Tucumano-Boliviano Forest, cloud forest, cloud paramo and prairies. Dry montane forests and Chaco Serrano forest
14. Torotoro National Park DS. 22269 - 7.26.1989	16,570 North of Potosi 3,600 - 1,900 m	Torotoro Conser-vation Associa-tion (ACT)	One ecoregion. Montane dry deciduous forest. Paleontological sites: Huma Jalanta y Huaca Senca caves, Dinosaur footprints. Archaeological sites: Llama Chaqui with cave paintings

Table 2. Areas to be redefined or without management

Protected area and legal base (**DS. Decree of Creation and date**)	**Extension** (Hectares) **Location** (Department) **Altitude range** (Meters above sea level)	**Ecological value**
15. Blanco and Negro Rivers National Wildlife Reserve	1,400,000 Santa Cruz 400 - 200 m	Brazilian and Chiquitano Escudo, 3 ecoregions. Amazon forest, flooded forests palm forest swamps and lakes
16. Manuripi Heath National Amazonic Reserve DS. 11252 - 12.20.1973	1,884,375 Pando 200 - 180 m	Amazon forest
17. Incakasani-Altmachi National Andean Wildlife Reserve DS. 22938 - 10.11. 1991	23,300 Cochabamba 4,000 - 2,800 m	Three ecoregions. Andean ecosystems, Andean prairie, Yungas paramo, cloud forest
18. Yura National Wildlife Reserve DS. 11307 - 1.20.1991	96,856 Potosi 4,300 - 3,900 m	Dry puna
19. Llica National Park RM. 228/90 - 11.29.1991	997,500 Northwest of Potosi 5,000 - 3,500 m	Western Andean Cordillera, dry puna. Archaeological sites

20. Santa Cruz la Vieja Historic National Park DS. 22140 - 2.22.1989	17,080 South of Santa Cruz 400 - 300 m	Historical site: First capital of Santa Cruz, close to San Jose de Chiquitos
21. Palmares de Parajubea Protected Area	150,000 (proposed) Chuquisaca	Range and dry palm forest
22. Grande Masicuri River Forest Reserve DS. 17004 - 08.2 1979	242,000 Santa Cruz	Dry deciduous forest
23. Itenez Forest Reserve	1,500,000 Beni	Amazon forest
24. Otuquis Tucavaca National Reserve	1,000,000 Southeast of Santa Cruz, border with Brazil	Pantanal
25. Pantanal San Matias Biological Reserve	800,000 Santa Cruz	Wetlands
26. Cordillera de Sama Biological Reserve DS. 22721 - 30.1. 1991	108,500 Centre of Tarija 4,400 - 3,000 m	Cordillera of Tacsara, variety of Andean ecosystems, Puna and semihumid prairie, glacial lakes. Rich in Andean fauna, Guanaco (*Lama guanicoe*) Archaeological sites

NATIONAL SYSTEM OF PROTECTED AREAS (NSPA)

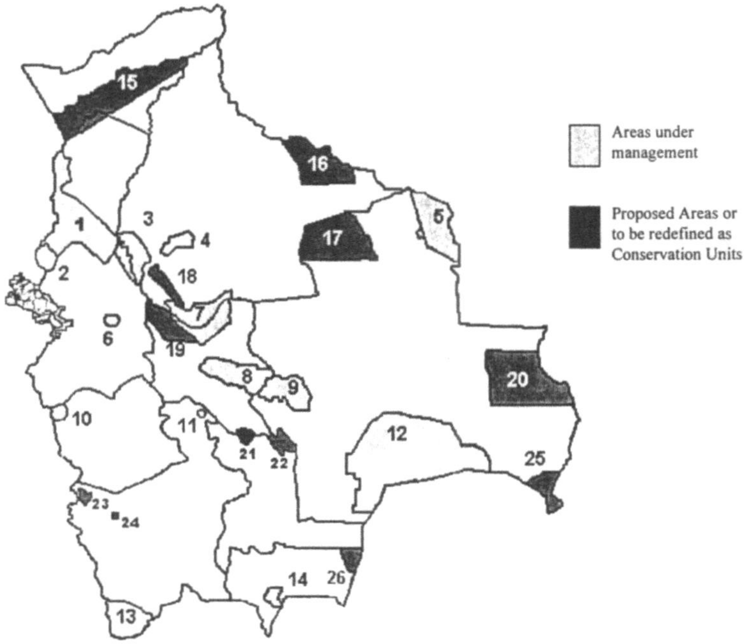

Fig. 1.

1 Madidi National Park
2 Ulla Ulla National Wildlife Reserve
3 Pilón Lajas Biosphere Reserve and Indigenous Territory
4 Beni Biological Station Biosphere Reserve
5 Noel Kempff Mercado National Park
6 Cotapata National Park
7 Isiboro Sécure National Park and Indigenous Territory
8 Carrasco National Park
9 Amboró National Park
10 Sajama National Park
11 Torotoro National Park
12 KAA-IYA del Gran Chaco National Park
13 Eduardo Avaroa Andean Wildlife Reserve
14 Tariquía National Wildlife Reserve
15 Manuripi Heath National Amazonic Reserve
16 Itenez Forest Reserve
17 Blanco and Negro Rivers National Wildlife Reserve
18 Eva Eva - Mosetenes Basin Protected Area
19 Incacasani - Altamachi National Andean Wildlife
20 Pantanal San Matías Reserve
21 Palmar Rodeo Protected Area
22 Grande Masicuri River Forest Reserve
23 Llica National Park
24 Potosi Area (Saltlake Island)
25 Otuquis Tucavaca National Reserve
26 Cabo San Juan National Wildlife Reserve

PROTECTED AREAS WITH MANAGEMENT COMMITTEES

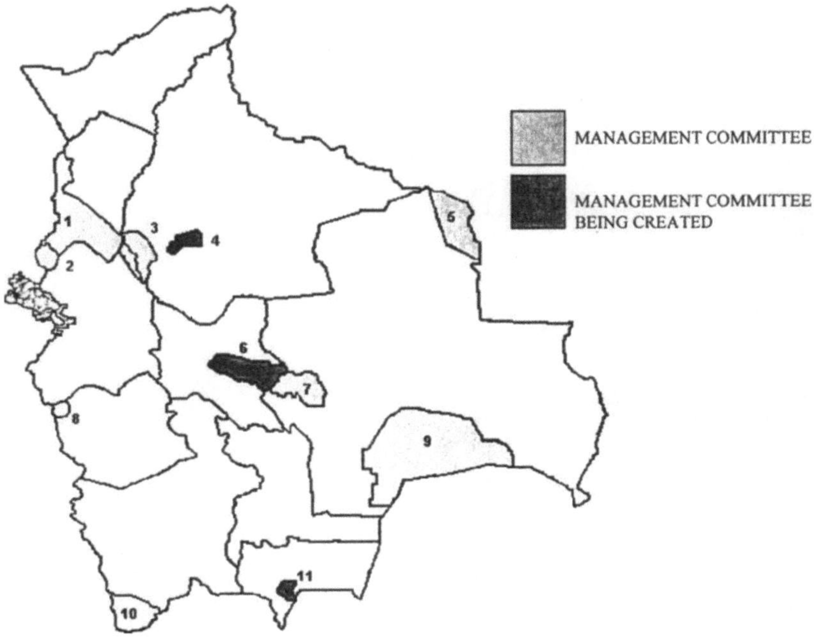

Fig. 2.

1 Madidi National Park
2 Ulla Ulla National Wildlife Reserve
3 Pilón Lajas Biosphere Reserve and Indigenous Territory
4 Beni Biological Station Biosphere Reserve
5 Noel Kempff Mercado National Park
6 Carrasco National Park
7 Amboró National Park
8 Sajama National Park
9 KAA-IYA del Gran Chaco National Park
10 Eduardo Avaroa Andean Wildlife Reserve
11 Tariquía National Wildlife Reserve

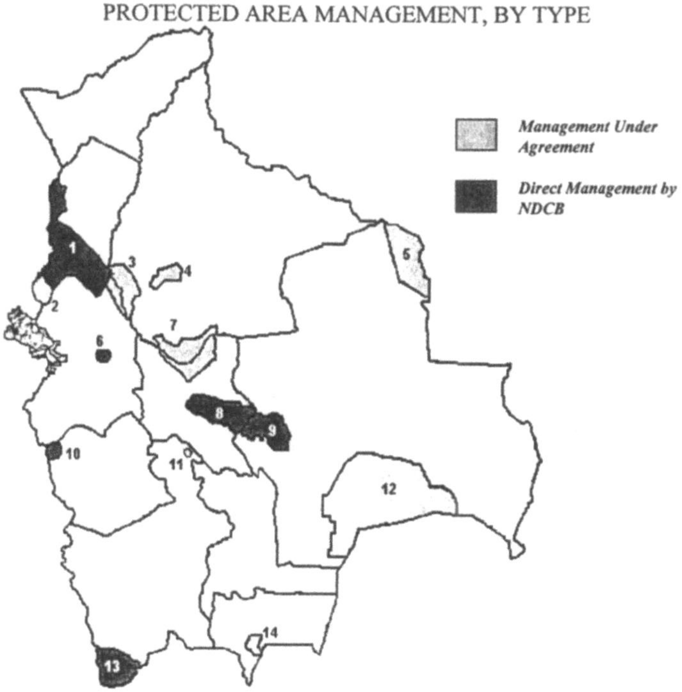

Fig. 3.

1 Madidi National Park
2 Ulla Ulla National Wildlife Reserve
3 Pilón Lajas Biosphere Reserve and Indigenous Territory
4 Beni Biological Station Biosphere Reserve
5 Noel Kempff Mercado National Park
6 Cotapata National Park
7 Isiboro Sécure National Park and Indigenous Territory
8 Carrasco National Park
9 Amboró National Park
10 Sajama National Park
11 Torotoro National Park
12 KAA-IYA del Gran Chaco National Park
13 Eduardo Avaroa Andean Wildlife Reserve
14 Tariquía National Wildlife Reserve

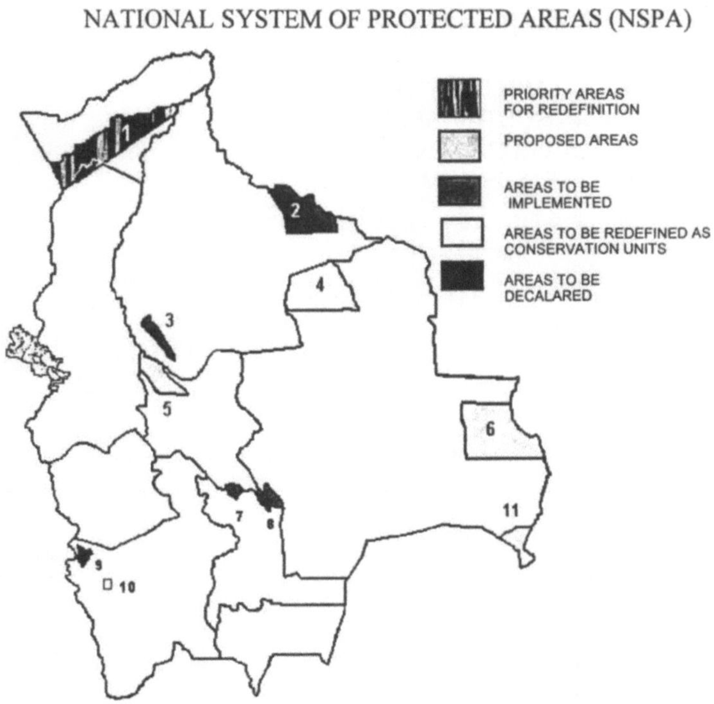

Fig. 4.

1 Manuripi Heath National Amazonic Reserve
2 Itenez Forest Reserve
3 Eva Eva - Mosetenes Basin Protected Area
4 Blanco and Negro Rivers National Wildlife Reserve
5 Cocapata - Altamachi Reserve
6 Pantanal San Matías Reserve
7 Palmar Natural Integrated Management Area
8 Grande Masicuri River Forest Reserve
10 Uyuni Saltlake Island Park
11 Otuquis Tucavaca National Reserve

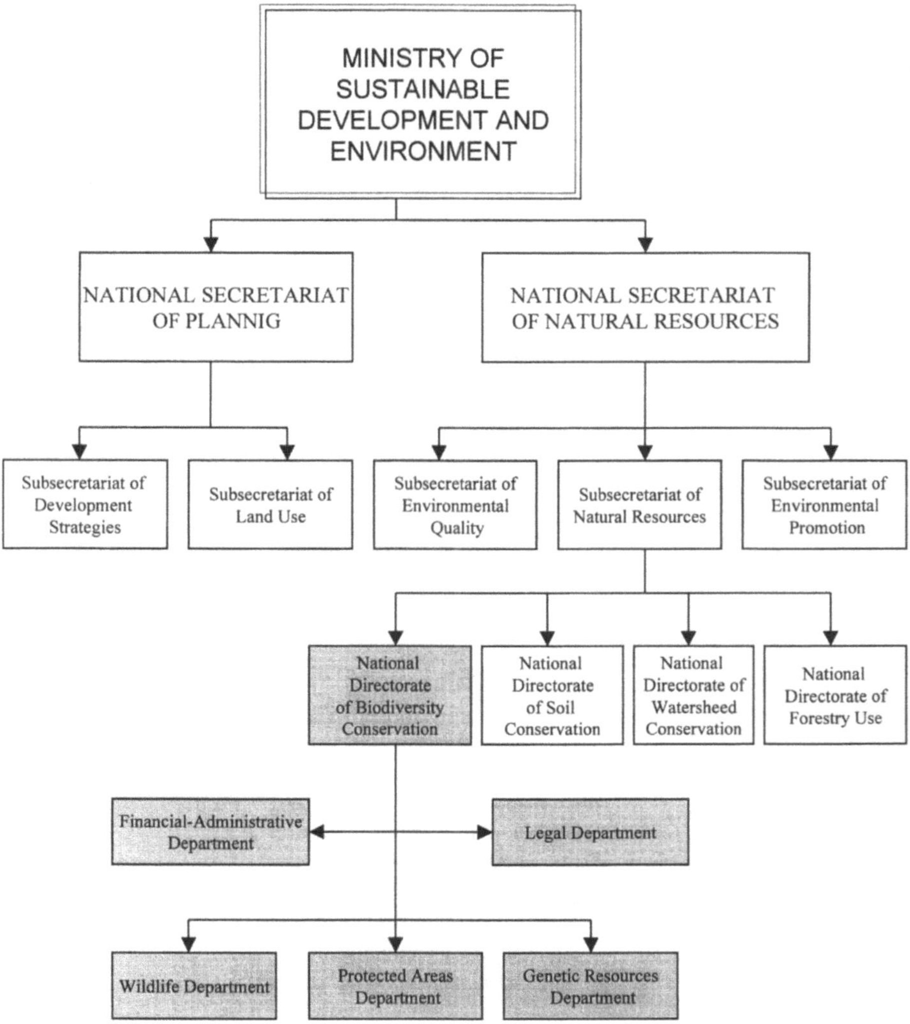

Fig. 5. Organization of the Ministry of Sustainable Development and Environment of Bolivia

Friends of Nature Foundation (FAN Bolivia) - The involvement of NGOs in conservation in Bolivia

Hermes Justiniano
Friends of Nature Foundation, Santa Cruz, Bolivia

Abstract. A private organization that was established to work on the conservation of nature in Bolivia has produced important results since its creation in 1988. Besides implementing pioneer projects in the country's national parks, new management standards have been established in for country's protected areas. The Friends of Nature Foundation (FAN) has acted as a catalyst in the government policy-generating process, having obtained as a result the drafting of environmental laws, the private administration of national parks and the lessening of human pressure on the protected areas. FAN's work and increasing experience has led to an important connection between the conservation of biodiversity and the improvement in the quality of life for people living next to protected areas.

La participación de las ONG en la conservación en Bolivia

Resumen. La creación de una organización privada destinada a trabajar en temas de conservación de la naturaleza en Bolivia ha resultado en importantes logros desde 1988. No solamente se han implementado proyectos en los parques nacionales, creando nuevos estándares de manejo en el campo. También se han catalizado exitosamente procesos para la generación de políticas gubernamentales, resultando en el diseño de leyes ambientales, la administración privada de los parques nacionales y la disminución de la presión humana sobre las áreas protegidas. La evolución institucional ha llegado a establecer una importante conexión entre la conservación de la biodiversidad y el mejoramiento de la calidad de vida de los vecinos de las áreas protegidas.

1 Historic background

FAN was established in 1988 in response to the need for grassroots involvement in conservation issues in Bolivia. The initiative to create a private, non-profit, conservation organization was driven by an awareness of the impact that development was having on landscape, flora and fauna in the eastern lowlands. At that time there was only one other environmental organization in the region and its main focus was on denouncing the illegal activities of timber loggers and colonists invading national parks and other protected areas, and the lack of appropriate reaction from the authorities. It was clear however that the government's capabilities were insufficient and that there was little hope for any real improvement. FAN's founding members were 17 young but very committed people, some of them professionals in the arts and biological sciences. The driving force behind the group was a clear appreciation of the value of biodiversity and the conviction that there had to be a balance between development and conservation. Right from the start, FAN found an array of contacts and sources of support to help it accomplish its mission, which it defined as the protection of Bolivia's biodiversity. During the early stages most of this support came from other non-governmental institutions abroad, such as The Nature Conservancy, Wildlife Conservation Society, Conservation International, World Wildlife Fund for Nature and a host of private foundations and individuals who faithfully committed themselves to a joint effort. In 1988, while providing aerial logistical support to Louisiana State University (LSU) to conduct a bird inventory in Noel Kempff Mercado National Park, several flights over mostly unknown areas of the park were made. A large illegal logging operation on the Brazilian side was discovered and denounced. Major efforts were made to follow up the park administration's subsequent actions which, to our dismay, were shamefully absent. This generated a strong determination to create a mechanism to ensure private participation in the management of that park. A strong and fully committed organization had to be created for this purpose. The LSU ornithology expedition was followed by numerous contacts with the scientific community, and soon other opportunities emerged to host and guide expeditions throughout previously unexplored areas of the country. Good pilots and the availability of aircraft were instrumental in obtaining access to many of these areas. Most of the first expeditions were made in collaboration with Conservation International's RAP Team and other national academic institutions. Trips were made to Madidi, the Rios Blanco y Negro area, Eastern Beni, Chaco, Pantanal, the Velasco Dry Forests, Pando, the Noel Kempff Mercado National Park, and other areas. Several reports were generated which later were used as scientific justification for the creation or expansion of national parks. In early 1989, FAN made good progress on its legal paperwork and was learning about the best way to structure its operational organization. The founding members had a number of highly productive meetings to define the organization's profile and the philosophy of its work. FAN began its conservation activities in July 1989 with financial support from The Nature Conservancy (TNC), a North American institution which was looking for a local partner in Bolivia at the time. The initial staff were an executive director and an administrative assistant who drafted the first one-year and three-year strategic plans with technical assistance from TNC. A small office with a telephone was rented, and it was possible to start working following the donation of a laptop computer and printer. It was decided to work hand in hand with the government, providing support wherever it was most needed. The Amboro and Noel Kempff National Parks were considered to be in the most urgent need of FAN's initial assistance.

2 Conservation projects that started to change the reality of Bolivian national parks

The first projects were started in 1990. The new Foundation was assigned responsibility to help Amboro and Noel Kempff Mercado National Parks with financial and technical support from USAID and TNC's Parks in Peril Programme. Opportunities were created to practice in-depth protection and management in certain areas of both parks. Efforts were made to establish new management standards, develop training seminars for rangers that continued over a period of several years, and to equip park personnel with uniforms, radios and vehicles. A special effort was made to understand more about the human population, the resources and the environment around the parks and to explore ways of improving the local communities' livelihood. An agronomist was hired for this purpose. The organization's first involvement in policy-making came in early 1990 with FAN's Development and Conservation Workshop – A Proposal for the Drafting of a Law on the Environment and Natural Resources. A group of senators, representatives and members of international cooperation institutions were invited to the five-day workshop. The results were very successful and produced recommendations that led almost immediately to the funding needed to draft the law. The Law on the Environment and Natural Resources was drafted with outstanding popular participation and adopted in less than two years. Almost simultaneously a first strategic move was made toward guaranteeing the protection of one of the country's most pristine forests which is also home to outstanding fauna. The Rios Blanco y Negro Wildlife Reserve was established by ministerial decree, providing an opportunity to prove that development (forest management) and conservation (wildlife protection) can coexist in a 1.4 million-hectare protected area. Funding was provided shortly thereafter for scientific studies for and the drafting of the Reserve's management plan which FAN and the Wildlife Conservation Society conducted in collaboration with the local Gabriel Rene Moreno University. By late 1990 FAN had been accepted as a member of the Noel Kempff Mercado National Park board of directors and had completed negotiations for acquire private land for conservation purposes for the first time ever in Bolivia. Flor de Oro was the only 10,000-hectare private ranch within Noel Kempff Mercado National Park. Up to then the area's naturally flooded savannas had been burnt yearly in order to maintain 500 head of cattle. In addition, the ranch was used for hunting and smuggling between Brazil and Bolivia. The ranch was bought with the financial assistance of several donors who supported the project through the TNC in the US. The objective was to transform the ranch into a centre for management, protection, scientific research and ecotourism. Since its acquisition, Flor de Oro has been transformed into a much sought after destination for national and international visitors. FAN was quite well organized by this time, and had a 12-member board of directors and a professionally staffed office with two departments (conservation and accounting). In 1992, two new departments were added: The Ecotourism Department was created to provide logistic and information services for visitors and scientists wishing to travel to Noel Kempff National Park. The Aviation Department was set up to ensure the quality and safety needed for the increasing number of flights for ecotourists and scientific research throughout the country.

A second strategic plan was drawn up in 1992 and contained a detailed review of FAN's mission and objectives. During the course of these activities it was determined that the human pressure around Amboro National Park had to be reduced, and the idea for a new project was born: the Community Participatory Planning Pilot Project. This project combined the ground-level implementation of sustainable development activities and the provision of environmental education for a handful of local communities that dedicated primarily to subsistence agriculture and cattle rising. The project focused mainly on encouraging a dialogue among the residents of the small communities on common goals for improving the quality of life in many areas. Special technical assistance was given to move away from the practice of farming annual crops that rely heavily on the use of pesticides, and to grow perennial fruit trees of improved quality instead. Guidance was also given for cattle raising, a special problem for Amboro National Park. Now at the end of its third year, the project – which was originally funded by the MacArthur Foundation and WWF – has proved itself very effective. It has even inspired a much larger project that is being funded by the British Government through CARE and will encompass many other communities around the perimeter of the park.

3 Initial efforts to enlarge and strengthen a very special park

A major conservation initiative was launched in 1992: The Expansion Project of Noel Kempff Mercado National Park which was aimed at establishing safer and more convenient natural boundaries, and at including critical areas of ecosystems, habitats and endangered species that were previously not represented. The park underwent a series of evaluations to justify its expansion from 0.7 to 1.6 million hectares. The initiative had to wait another three years due to financial constraints. In 1993, the Swiss government provided generous support for a pioneering initiative: the implementation of an Ecotourism Programme in Noel Kempff National Park. The project included the construction of infrastructure, acquisition of logistical equipment, funding of scientific studies on flora and fauna, promotional publications and studies to market the park as a desirable destination. The \$800,000 project is due to be completed in December 1996, and has greatly enhanced the park's already outstanding features for visiting and interpretative activities.

4 FAN grows to meet the challenges

In 1994 FAN decided to create a Science Department to support its decision-making process with solid scientific data. Further, most projects needed monitoring and evaluation with databases and GIS support. New professional capabilities within the Science Department led the way to exploring national genetic resources prospecting and planning for the sustainable use of these resources. The Foundation was actively involved in the drafting of the Law on Common Access to Genetic Resources for the Andean Countries. Meanwhile, the first agreement with the government on developing and commercializing genetic resources under the new law was signed. By the end of the year, internal planning brought up the idea of creating private for-profit enterprises – green companies – to support the ever-increasing array of FAN's conservation activities. It was becoming increasingly difficult to finance the Foundation's operating costs. FAN's initial staff of two had grown to 60 persons within a period of five years, and this rapid growth was having an impact on the organization's ability to find needed resources. Many projects that were funded by the government or through international cooperation did not consider the need for or approve enough funds for FAN's operating expenses. In the months that followed, several business plans were drafted in an effort to convert some of FAN's scientific, ecotouristic and aerial transportation capabilities into profit-making centres or businesses to support conservation. Partnerships with investors and other models were explored, some of which are being implemented at this time. The Foundation had been suffering from a lack of information within its organization and from an insufficient diffusion of its activities which led to the creation of a sixth department in 1995: the Communications Department. This department also provided an effective tool for stopping heavily biased attacks which the uninformed media launched with demoralizing and sometimes devastating local paper articles against FAN's quiet, low-profile work. The Communications Department is working on improving information about the institution's plans, projects and activities, and its internal and external dissemination. In 1994, FAN and TNC as a consortium were awarded a government contract in a public tender to develop management plans for the Amboro and Noel Kempff Mercado National Parks. Many areas of both parks were visited for the first time ever by scientists, who made outstanding discoveries. For instance, Amboro is the richest park in the hemisphere in terms of amphibians, Noel Kempff is one of the richest in terms of habitat and ecosystem diversity. Work on both management plans is to be finished within the next two months. The government publicly called for bids for the administration and management of several national parks in 1995. The proposals FAN submitted for Noel Kempff and Amboro won. In July 1995, a 10-year Long-Term Agreement for the Administration and Management of Noel Kempff Mercado National Park was signed, and activities in the area began shortly thereafter. However, the Amboro contract was never signed due to a shift in the government policies on park administration. It appears that this trend will continue well into the future. In 1995 FAN initiated the Medical Support Project for the Neighbouring Communities of Noel Kempff Mercado National Park to provide for the local population's most urgent health needs. Regular trips were organized to bring doctors and medicine to areas that were plagued by parasitism, malnutrition and malaria. On a typical flight, a team composed of a paediatrician, a gynaecologist, a general surgeon and a dentist travel by light plane with enough supplies and equipment to provide medical care for 1,200 people. These trips end in the park where the volunteer medical team treats the rangers and their families. The project is currently supported by the Wagner Foundation from the US and will be expanded in the future to encompass the entire watershed of the Paragua and Tarvo rivers under an agreement with the local government that includes co-financing.

5 Creative ideas for vital priorities

An effort to enlarge the NKMNP to its natural boundaries was launched in 1995 through an agreement with the government to compensate the five wood-logging companies in the expansion area and to work with the local communities in planning their future. When this was done, the government would sign a decree expanding the park. The actual compensation for the first 135,000 hectare concession was made in November 1995, followed by immediate cessation of all tree cutting and extraction from the forest. The compensation of another two concessions with a total of 385,000 hectares took place in April 1996. Compensation for two other concessions is pending for 1997 to complete the expansion area's 900,000 hectares. Meanwhile, FAN is in a middle of a fund-rising campaign to meet its financial needs. Because of the rather complex social and political arena in which the organization has to act to realize the expansion project, a second major effort to unite all the necessary elements was made in 1996 with the development of a large project: the Noel Kempff Climate Action Initiative. The project is about carbon sequestration from the atmosphere and has already gained support from the Bolivian government and international financial participants. The project contains components such as a trust fund for the park, a short-term protection programme for the expansion area, a community development support programme, funds for research and development of genetic resources, and a long-term monitoring programme. It is hoped that the project will start in early 1997. Presently, FAN continues to be a rather low-profile but effective organization that is committed to the conservation of Bolivia's biodiversity. Its members are aware that the conditions ruling conservation in the country are rapidly changing and that opportunities will be harder to find in the future. Although great efforts have been made and have met with relative success, much more needs to be done. In recent years, an increasing number of organizations like FAN have begun to play an important role in Bolivia's present development drama. These organizations will need much more support to balance the ever-increasing and devastating forces that impact the country's natural resources in the name of development.

In FAN's view, few countries in the world still have the opportunities to develop large integrated conservation and development models in ecologically healthy natural areas. Even fewer countries have the relatively low costs opportunities for such large scale conservation still found in Bolivia. Sustainable development must be supported by the conservation grassroots movement, for it to have a lasting and measurable success. Supporting conservation will lead the country to a judicious form of development. It means long-term, respectful planning and implementation that improves life for all.

Part 5.4

Call for research and action

Development and biodiversity conservation in Bolivia – a Call for Research and Action

Pierre L. Ibisch[1], Stefan G. Beck[2]
[1] Institute of Botany, University of Bonn, Germany
[2] National Herbarium of Bolivia, La Paz, Bolivia

Abstract. Various theses on development and biodiversity conservation in Bolivia were developed as a *Call for Research and Action* aimed at stimulating a critical and interdisciplinary discussion of the complex problems found in this field. The *Call* was distributed before and during the symposium documented in this book. In this article, the *Call* is juxtaposed with reactions and comments from the symposium's participants and other persons and institutions from a variety of countries and disciplines.

Desarrollo y conservación de la biodiversidad en Bolivia: Un llamamiento a la investigación y la acción

Resumen. Sobre desarrollo y conservación de la biodiversidad en Bolivia se desarrollaron varias tesis como un *Llamamiento a la investigación y la acción* destinado a estimular una discusión crítica e interdisciplinaria de los complejos problemas encontrados en este campo. El *Llamamiento* fue distribuido antes y durante el simposio documentado en este libro. En este artículo, el *Llamamiento* se yuxtapone a reacciones y comentarios de los participantes en el simposio y otras personas e instituciones de diversos países y disciplinas.

1 Introduction

With regard to research, Bolivia is one of the more neglected regions on earth. This is somewhat surprising given its biodiversity and its development problems. The organizers of the international symposium on biodiversity and development in Bonn chose only one 'case study country' to spotlight at their symposium: Bolivia. In light of the attention that this would draw to Bolivia, the authors prepared a short list of theses which revolve around Bolivia's biodiversity-development problems and together constitute a *Call for Research and Action*. Of course, it was not possible to draw up a comprehensive, complete and definitive list of all relevant facts and necessary claims and theses. The result would have been a document exceeding the biodiversity convention's dimensions. But comprehensiveness was not the authors' aim: Their intention was to provoke reflection on and discussion of some of the facets of these complex problems.

Prior to the symposium the authors sent the *Call* to several institutions and persons and asked them to provide comments on it. The symposium's participants were also invited to contribute their ideas regarding the paper. Now, it turns out that the idea achieved its desired effect: The thesis has provoked controversial statements. And these statements may help to attract research and action to this special country!

The text of the original *Call* as it was distributed is shown in the grey boxes below. The authors cite parts of the statements received and juxtapose them with the *Call's* text.

The authors wish to thank everyone who prepared statements in response to the *Call* and especially Brian Boom, New York Botanical Garden, and Laura Siklossy, Bonn, for their valuable comments on first drafts.

2 The Call for Research and Action and reactions to it

A. BOLIVIA IS A MEGADIVERSITY COUNTRY

"Bolivia has been called a 'megadiversity country' and if any country deserves this title, it is certainly Bolivia."

Brian M. Boom, Vice President for Botanical Science and Pfizer Curator for Botany, The New York Botanical Garden, New York, USA

1. The territory of Bolivia includes Amazonian, Andean, Chaco and Cerrado ecosystems as well as several tropical biomes and ecotones. Bolivia's **ecosystem diversity** is one of the highest in the world.
2. Bolivia is one country that encompasses a major portion of global **species diversity** within its territory.
3. Bolivia is one of the few centers of **crop-genetic diversity** and source of a multitude of cultivated species and wild relatives.

"Bolivia, as a tropical-subtropical Andean country, represents one of the richest regions in the whole world with regard to the types of ecosystems, the variety of sites, and the number of plant, and animal species. This became evident, for example, by the activities of the Instituto de Ecología at the university (UMSA) of La Paz as well as by the collections joint with it, namely the 'herbario Nacional de Bolivia' and the 'Museo Zoológico'. However, further decades (or even centuries)

of continued and broadened research are needed in order to know most of the species really present, and to organize an effective conservation of this natural richness."

Prof. em. Dr. Dr. h.c. mult. Heinz Ellenberg, University of Göttingen

"In Kenntnis der ungeheuren abiotischen, biotischen wie kulturellen Vielfalt Boliviens begrüßt der Verein zur Förderung von Landwirtschaft und Umweltschutz in der Dritten Welt (VFLU) die Durchführung des Symposiums zur Biodiversität in Bonn, insbesondere die Einbeziehung und Hervorhebung des Andenstaats [Bolivien] als 'megadiversity country'."

Verein zur Förderung von Landwirtschaft und Umweltschutz in der Dritten Welt (VFLU), (Geschäftsstelle Wiesbaden: Axel Goldau, Andengruppe: PD Dr. Rainer Buchwald, Freiburg)

B. BOLIVIA IS A MODEL COUNTRY DESERVING PRIORITY CONSIDERATION FOR DEVELOPMENT AND BIODIVERSITY CONSERVATION RESEARCH

"Model country für was? Es könnte unter bestimmten Bedingungen eine Art Modell-Land für den nachhaltigen Umgang mit Biodiversität werden. Diese Bedingungen müssen allerdings erst geschaffen werden"

Dr. Elmar Römpcyk, Friedrich-Ebert-Stiftung, Bonn

1. The mega-biodiversity of Bolivia contributes to the **high cultural diversity** of the country. This implies a broad knowledge of the **traditional use of biodiversity**
2. In comparison with the majority of the tropical countries, Bolivia until now **has preserved a considerable percentage of its natural ecosystems**
3. Bolivia is the **least developed country in South America** and one of the poorest in Latin America. This is at the same time a cause of and a risk for the ecosystem's intactness
4. **The country's development recently has started to affect its biodiversity on a large scale. Biodiversity conservation is fighting on many fronts**, e.g. colonization of virgin forests by "ecorefugees" from regions of traditional settlement which are losing their carrying capacity due to anthropogenic degradation, drug production, timber extraction, booming oil production, dynamic agroindustry, rapidly expanding cities with rising demands for resources like food, land and water. Conservation concepts and actions including a powerful conflict management are urgently needed

5. The following **topics of development and biodiversity research** must be given priority:
a) **Inventories and mapping of biodiversity**
b) **Inventories of indigenous use and conservation of biodiversity** (ethnoecology)
c) **Quantity and quality of acceptable use in different ecoregions to reach sustainability**
d) **Policies and conflict management between land use and biodiversity conservation, especially in biodiversity-rich regions**
e) **Conservation, use and commercial value of genetic resources and biodiversity to guarantee the sustainable development of rural communities** (bioprospecting, "eco"-products, ecotourism)

"Die von den Autoren des Thesenpapiers für prioritär erachteten Forschungsthemen können wichtige Ergebnisse und Datengrundlagen für die weitere Implementierung [der] Politikvorgaben [des bolivianischen Ministeriums für Nachhaltige Entwicklung und Umwelt; s.u.] liefern, und sind daher absolut förderungswürdig."

Dr. Hans Schoeneberger, Programmkoordinator "Ländliche Regionalentwicklung", GTZ, Bolivien

"Fortunately, there is a growing body of scientific knowledge about how to effectively make this combination [of development and conservation concerns]. Bolivian and foreign researchers are poised to make the combination work now, but there is a great lack of funding to support even the simplest and most effective sorts of projects. There is a shortage of funds in Bolivian universities to pay faculty, provide materials and field trip opportunities for students, and to purchase equipment and supplies for research. What is needed is scarcely the problem; the [authors of the CALL] have identified the relevant areas for priorization: inventories and mapping of biodiversity, inventories of indigenous use and conservation of biodiversity, quantification and qualification of acceptable use in different ecoregions to achieve sustainability, land use policies that balance development and conservation, bioprospecting, green products derived from biological resources, and ecotourism."

Brian M. Boom, Vice President for Botanical Science and Pfizer Curator for Botany, The New York Botanical Garden, New York, USA

"... not only research, but to give good education for all Bolivians, especially indigenous people! - It is very dangerous for protection of biodiversity to announce to investors the natural resources, exotic timber, iron for mining to be used in industries and medicine industry."

Angelika Meents, Bonn

"[B.5.d)] ist ein sehr zentraler Aspekt und bedarf sehr hoher Aufmerksamkeit. [B.5.e)] ist ein weiterer zentraler entwicklungspolitischer Aspekt und bedarf ebenfalls höchste Aufmerksamkeit."

Dr. Elmar Römpcyk, Friedrich-Ebert-Stiftung, Bonn

Nach Ansicht des Vereins [zur Förderung von Landwirtschaft und Umweltschutz in der Dritten Welt] müssen schnelle und spezifische Anstrengungen unternommen werden, um die natürlichen wie auch die kulturbetonten Lebensräume in ihrer Vielfalt zu erforschen und zu verstehen. Eine solche Aufgabe ist i.d. Regel nur in Zusammenarbeit in- und ausländischer Wissenschaftler zu bewältigen, die eine Vielzahl verschiedenster Methoden, Arbeitsweisen und Erfahrungen beitragen. Ohne eine Inventarisierung, Analyse und bewertung der natürlichen sowie der land- und forstwirtschaftlichen Ressourcen und Lebensräume ist deren Erhaltung nicht möglich: nur wer die 'Schätze' eines Landes kennt und zu würdigen weiß, kann effektiv zu ihrem dauerhaften Schutz beitragen!"

Verein zur Förderung von Landwirtschaft und Umweltschutz in der Dritten Welt (VFLU), (Geschäftsstelle Wiesbaden: Axel Goldau, Andengruppe: PD Dr. Rainer Buchwald, Freiburg)

"The causes and effects of the loss of biodiversity are often global in nature: poverty and high population densities lead to over-exploitation of flora and fauna and the degradation of habitats. Protecting biological diversity has therefore become an absolutely vital task to which development policy of the German government is also committed."

Prof. Dr. Bohnet, Ministerialdirigent, Federal Ministry for Economic Cooperation and Development (BMZ), Bonn, Germany

C. BOLIVIA: BIODIVERSITY CONSERVATION AND DEVELOPMENT HAVE THE SAME GOALS

"[Dies] ist so ohne weiteres nicht akzeptabel. Allenfalls aus negativer Sicht, indem es bisher keine wirkliche Politik zum Biodiversitätserhalt gab und auch keine überzeugende Entwicklungspolitik, sondern nur eine Wachstums- und Teilindustrialisierungspolitik."

Dr. Elmar Römpcyk, Friedrich-Ebert-Stiftung, Bonn

"I do agree with the CALL in general but a primary goal of biodiversity conservation and development should / must be the satisfaction of the basic needs of the local population. We should be conscious of the fact that those who define the goals of development also define those of conservation - and those who define

the goals are still - and once more - the developed countries. If we don't change our attitude and our point of view we'll just improve once more "goals" shaped by our own needs."

Eva Koenig, Hamburgisches Museum für Völkerkunde, Hamburg

"We would suggest a change (...) in the sense that biodiversity conservation and development should have the same goals. In reality this still remains to be proven, but we understand the proposed CALL as an attempt to reach exactly that. Thus, the thesis shouldn't figure as a premise, but an objective instead."

Imke Oetting, Carl-von-Ossietzky-Universität Hannover, Hannover
Heike Knothe, Nationalpark "Sächsische Schweiz", Freiberg

"(...) Development must ensure biodiversity conservation in order to needs."

Pablo Canevari, Bonn

1. Loss of biodiversity is irreversible - **all development actions should contribute to biodiversity conservation**
2. **Local, regional and national land use planning** should be orientated to biodiversity conservation and sustainable development
3. In order to preserve intact ecosystems, major efforts should be concentrated on **intensifying the utilization of anthropogenic land currently used and where the natural ecosystems have already been disturbed**. It is vital to stop degradation through the implementation of sustainable land use systems (e.g. agroforestry)

"Within the framework of German development co-operation, measures to protect and care for nature reserves in order to safeguard biodiversity and an ecological balance have been growing in importance since the mid-80s. The goals of development policy in the field of nature conservation are to support partner countries in their efforts to preserve their natural habitats in accordance with their ecological, socio-cultural and economic significance and to practice sustainable land use, whilst giving due regard to methods which are traditional and close to nature.

The BMZ is currently promoting more than 200 projects in the field of nature conservation and biodiversity (including some in Bolivia). Within the framework of bilateral financial and technical co-operation the BMZ gives support not just to the demarcation of nature reserves and the establishment of nature conservation authorities, but also to measures for sustainable, ecologically sound use of natural resources. Having learnt by experience that traditional conservation-based protection of nature often fails because of socio-economic frame conditions in the developing countries, these development projects link the protection of nature with the goal of improving living conditions for the population."

Prof. Dr. Bohnet, Ministerialdirigent, Federal Ministry for Economic Cooperation and Development (BMZ), Bonn, Germany

"[C.3] scheint gegenüber [B.5.e)] unverständlich und gegenüber dem zuvor Gesagten sogar widersprüchlich."

Dr. Elmar Römpcyk, Friedrich-Ebert-Stiftung, Bonn

"Es sollte (...) nicht verkannt werden, daß für die Entwicklungspolitik, um deren praktische Umsetzung sich die deutsche Gesellschaft für technische Zusammenarbeit (GTZ) mit ihren Partnern in den Entwicklungsländern bemüht, der Mnsch und die Verbesserung seiner Lebenschancen im Mittelpunkt stehen. Daß diese angestrebte Verbesserung nachhaltig nur möglich ist, wenn die natürlichen Lebensgrundlagen, und damit auch die Biodiversität, bewahrt werden, bedarf wohl keiner weiteren Erläuterung.

Aus dieser Sichtweise ist Biodiversitätserhaltung kein Selbstzweck oder Ziel für sich, sondern Voraussetzung für eine wirklich nachhaltige Entwicklung der menschlichen Gesellschaft, die ihre natürlichen Lebensgrundlagen schützt und bewahrt. Den menschen nur als Risiko oder Störfaktor zu betrachten, der die Biodiversität gefährdet, und den es von der intakten, jungfräulichen Natur fernzuhalten gilt, widerspricht dieser entwicklungsspolitischen Grundauffassung.

Die derzeitige bolivianische Regierung ist in diesem Sinne mit der Einrichtung eines Ministeriums für nachhaltige Entwicklung und Umwelt (in dieser Reihenfolge!) und dem beginn der Umsetzung einer entsprechenden sektorübergreifenden Politik einen wichtigen Schritt in die richtige Richtung gegangen und wird dabei von einer großen Zahl von geberländern technisch und finanziell unterstützt"

Dr. Hans Schoeneberger, Programmkoordinator "Ländliche Regionalentwicklung", GTZ, Bolivien

"When I first visited Bolivia in 1983, efforts to develop the country's biological diversity were in their infancy. Since then, however, the situation has changed dramatically. These changes give a new urgency to biodiversity conservation studies, as development is proceeding rapidly, unchecked, due to oil production, agroindustry, non-sustainable extractive economies (e.g. harvest of palm heart from Euterpe precatoria), timber cutting, and so on. A greatly expanding human population compounds the problems; for example, each woman in the Amazonian town of Riberalta gives birth to 8.5 children, a fact I learned during my most recent trip to the region in August of this year! The difficulty of all this is that development is drastically needed to improve the lives of Bolivians, who are among the poorest people of Latin America. Reduction in population growth is a high priority, but even if the rate were to be lowered to 2 per family and eventually stabilized, that would not be enough. The solution is not to oppose development, per se, but rather to make sure that development schemes are

conducted on a solid scientific basis that ensures the sustainability of the activity, with ample guarantees of maintaining the biological and cultural integrity of the systems being manipulated. This combination of development and conservation concerns is a priority of highest order for the future of Bolivia, its people, and its natural resources."

Brian M. Boom, Vice President for Botanical Science and Pfizer Curator for Botany, The New York Botanical Garden, New York, USA

D. BIODIVERSITY RESEARCH AND ACTION – BOLIVIA REQUIRES INTERNATIONAL COOPERATION

"Dem ist dann voll zuzustimmen, wenn auch hier die Kooperationsbedingungen und -zielsetzungen im Sinne von nachhaltiger Entwicklung geklärt sind."

Dr. Elmar Römpcyk, Friedrich-Ebert-Stiftung, Bonn

1. **Sustainable development and biodiversity conservation should be a main and universal goal of international development programs for Bolivia**
2. It is not enough to simply identify and avoid all possible negative development project impacts on the environment. Each project should be assessed by a multidisciplinary and independent team to evaluate its potential towards making an active contribution to biodiversity conservation. **A constructive biodiversity assessment is needed to complement the traditional environmental impact assessment**
3. The data base for the evaluation of development impact on biodiversity and the planning of a sustainable use of biodiversity in Bolivia is totally inadequate. **Management and planning of sustainable land and biodiversity use require a long term investment in biodiversity and development research**
4. The current trend to **integrate specific development projects into programs** is very positive. This integration at different counterpart levels (e.g. government, *prefecturas*, *muncipios*) and of distinct types (e.g. rural development, food security, buffer zone management, strategy counseling, education) offers the opportunity to increase synergistic effects and the overall significance of development action
5. **Promoting sustainability is a long-term task which requires long-term perspectives, projects and commitment**
6. Bolivian scientists and scientific centers have been supported during the last years. However, there is still a need to strengthen the capacity for the active involvement of Bolivian professionals and technicians in environmental sciences

7. **Bolivia**, by conserving its biological diversity, **offers a variety of environmental services to mankind** which, up to now, have been taken for granted and enjoyed free of charge. This must change. The **industrialized countries** which derive benefits from these services and have a direct interest in propagating them **should be called upon to make financial compensation**
8. **We appeal to the international community** to demonstrate its commitment and responsibility towards the **promotion of sustainable development and biodiversity conservation** by providing the funds required to establish a "Center of Excellence" for biodiversity in Bolivia

"The Convention on Biodiversity signed in 1992 at the Conference for the Environment and Development in Rio de Janeiro has created a new legal frame of reference for development co-operation in the field of natural resources. The obligation upon the industrialised countries to give the developing countries financial support for the implementation of the obligations under the convention derives from their particular reponsibility for the alarming state of global biodiversity. To this ende the Global Environment Facility (GEF) was set up. The BMZ is an active contributor to the management of the GEF. Germany's financial participation in the current phase amounts to US$ 240 million or 12%. GEF biodiversity measures in Bolivia are therefore also partially financed with German funds. When the Convention was debated it was agreed that a number of immediate measures should be undertaken; these include priority measures such as described in the paper received, for example mapping biodiversity and making inventories of its possible uses. For this purpose a sectoral project - 'Prompt Start Measures for implementing the Biodiversity Convention' - was set up in 1993. Under this project institutions in selected partner countries have been helped with the development of national strategies and sustainable exploitation of biodiversity. There is also a need to intensify efforts with regard to the development of concepts for the implementation of the central theme of the Convention, namely the equitable distribution of any gain from sustainable exploitation of biological resources. The possibility of promoting promising approaches in Bolivia could be investigated to the extent that funding is available within the framework of the project.

The conservation of biodiversity is one of the key areas of German-Bolivian development co-operation. Bilateral co-operation takes into account both the environmental and the resource protection aspects, with evironmental impact assesments being carried out regularly for all projects. Many of the German-Bolivian projects make a contribution to the protection of natural resources. Examples of such projects are the promotion of the Ecology nstitute of the University of La Paz, the support for the Cochabamba School of Forestry, the Bufferzone Management Project and the Natural Resources Management Project in Santa Cruz.

Of these, the Ecology Institute in La Paz, which has been receiving support under German technical co-operation for a number of years and which seeks to develop teaching and research capacity in the area of biodiversity, is a project working in an area closely related to the proposed Centre of Excellence. If required, intensive co-operation, for example in connection with the procurement of experts, could be sought."

Prof. Dr. Bohnet, Ministerialdirigent, Federal Ministry for Economic Cooperation and Development (BMZ), Bonn, Germany

"Bolivia, of course, is still a poor country, and not able to finance sufficient activities in research, collections, and intensified conservation. International help is urgently needed, especially a German support. This is important also in view of another call of action, the 'Agenda Systematics 2000', which aims at intensifying biosystematic research in all parts of the globe concerning biodiversity."

Prof. em. Dr. Dr. h.c. mult. Heinz Ellenberg, University of Göttingen

"I do agree with the CALL in general but No. D 7 should be specified by some examples! Which are the 'variety of environmental services' offered to mankind?"

Dr. Horst Korn, Federal Agency for Nature Conservation, Vilm, Germany

"I do agree with the CALL in general but I want to stress "D.6" as of outmost importance for Bolivia, and you should have given an short- and long-term suggestion for the "economy-ecology-problem".(...)

Dr. Tjitte de Vries, Universidad Católica, Quito, Ecuador

"Uneingeschränkt und unterstützenswert erscheint die Forderung nach synergetischen Effekten bei der Konzeption integrierter Entwicklungsprojekte, an denen sowohl die lokalen und regionalen staatlichen Instanzen beteiligt sind (Entwicklungsplanung) als auch Vertreter zivilgesellschaftlicher Organisationen mit unmittelbaren Interessen für ländliche Entwicklung, Ernährungsssicherung und Bildung. Uneingeschränkt und unterstützenswert erscheint genauso die Forderung, daß die forscherischen Kapazitäten in Bolivien qualitativ verbessert werden sollen."

Dr. Elmar Römpcyk, Friedrich-Ebert-Stiftung, Bonn

"Den Forderungen der Autoren nach einer langfristigen internationalen Förderung der Strategien zur nachhaltigen Entwicklung und zur Unterstützung beim Aufbau entsprechender Forschungs- und Umsetzungskapazitäten in Bolivien kann nur voll und ganz zugestimmt werden."

Dr. Hans Schoeneberger, Programmkoordinator "Ländliche Regionalentwicklung", GTZ, Bolivien

"[D.8:] 'We appeal' is not enough. Why not raise funds to buy special land and preserve it, as e.g. it is done in the South of Chile, as José Lutzenberger has mentioned!?"

Josef-Thomas Goeller, freelance journalist, Bonn

"Angesichts verschiedenster Beeinträchtigungen, Degradierungen und Zerstörungen (Erosion, abnehmende Bodenfruchtbarkeit, Landflucht in den andinen lebensräumen; Abholzung und intensive Besiedlung der berg- und Tieflandsregenwälder; expandierende Kokainwirtschaft; starlke Ausdehnung der Städte; usw.) ist ein schnelles Handeln unverzichtbar. Obwohl die notwendigen politischen, sozialen und ökologischen Veränderungen von Bolivien ausgehen und in diesem Staat geschehen müssen, ist die Kooperation des Auslandes durch verstärkung der gesllschaftlichen, finanziellen, entwicklungspolitischen und wissenschaftlichen Aktivitäten im Sinne einer nachhaltigen, d.h. nicht gegen die natürlichen ressourcen des Landes gerichteten Entwicklung gefragt."

Verein zur Förderung von Landwirtschaft und Umweltschutz in der Dritten Welt (VFLU), (Geschäftsstelle Wiesbaden: Axel Goldau, Andengruppe: PD Dr. Rainer Buchwald, Freiburg)

(...) These sorts of [relevant] activities are not pursued effectively in an ad hoc manner. There is a serious need for long-term commitment. Clearly, Bolivia's economy does not have the capacity to do this alone. The international community must join Bolivian researchers in this effort. I strongly support the symposiums organizers' idea to establish a Center of Excellence for Biodiversity in Bolivia."

Brian M. Boom, Vice President for Botanical Science and Pfizer Curator for Botany, The New York Botanical Garden, New York, USA

3 General statements and comments

I do agree with the CALL and I support it:

Dr. Klaus Bosbach (Botanical Garden, University of Osnabrück, Osnabrück, Germany), Thomas Claßen (Gladbeck, Germany), Prof. em. Dr. Dr. h.c. mult. Heinz Ellenberg (University of Göttingen), Antonista Noli H. / PROBONA (Programa de Bosques Nativos Andinos, La Paz, Bolivia), Hans-Christian Offer (Institute of Biology, Plant Physiology, Humboldt-University, Berlin, Germany), Friedhelm Keil (Department of Plant Ecology, University of Bayreuth, Bayreuth, Germany), Dr. Michael Kessler (Systematisch-Geobotanisches Institut, Universität Göttingen, Göttingen, Germany), Michael Römer (Diedorf, Germany), Alejandra Sánchez de Lozada (Ministerio de Desarrollo Sostenible y Medio Ambiente, La Paz, Bolivia), Prof. em. Dr. Paul Seibert (Univ. of Munich), Evy Thies (freelance, GFA-Agrar, Hamburg, Germany), Prof. Dr. Anne Valle Zarate (Tierzuchtwissenschaften, Universität Bonn, Bonn, Germany), Prof. Dr. Villwock (Zoologisches Institut und Zoologisches Museum Hamburg, Hamburg, Germany).

"Most of the scientists and politicians [who are] *actually important studied in the time from 1925 to 1965 when plant and animal systemtics were looked upon as 'antiquated' and 'less scientific' thn molecular biology and physiology. With the growing estimation of ecology, biodiversity and nature conservancy during the*

last decades, this underestimation turned out to be a serious handicap. Mainly biodiversity needs to be studied and promoted or at least maintained, chiefly in tropical-subtropical areas, where it is maximal."

Prof. em. Dr. Dr. h.c. mult. Heinz Ellenberg, University of Göttingen

"Grundsätzlich ist den Autoren des Thesenpapieres zuzustimmen, wenn sie auf die herausragende Bedeutung der Biodiversitätserhaltung hinweisen, ohne die eine nachhaltige Entwicklung für die Menschen in Bolivien nicht denkbar ist."

Dr. Hans Schoeneberger, Programmkoordinator "Ländliche Regionalentwicklung", GTZ, Bolivien

"Protection of biodiversity has to be linked to the population of the areas affected, i.e. especially the indigenous people, and their social situation. Your approach is too scientific, i.e. too abstract."

Dr. Feeke Meents, Presse- und Informationsamt der Bundesregierung, Bonn

"Selbstverständlich unterstützt die gtö [Deutsche Gesellschaft für Tropenökologie] jede Initiative, die die Situation der Tropenforschung verbessert und die Erhaltung und ökologisch verträgliche Nutzung der Biodiversität fördert. Wir sind aber ein Verein, der viele Mitglieder umfaßt, die in den verschiedensten Ländern ihre Forschungs-, Interessens- und persönliche Zuneigungsschwer-punkte haben und deren Meinungen sehr weit auseinandergehen, wenn es darum geht, hier eine Wahl zu treffen. Als Gesellschaft können wir mit Sicherheit keine Prioritäten festlegen (bzw., wenn überhaupt, dann nur durch einen demokratisch gefaßten Beschluß in einer Mitgliederversammlung) und eines von mehreren Megadiversitätsländern bevorzugt zur Forschung empfehlen. Sie kann sich in einem solchen Fall nur ideell hinter Personen stellen, die - wie bei Ihrem Vorschlag zweifellos der Fall - gute Programme entwerfen und die Initiative zu ihrer Umsetzung ergreifen.(...)"

Prof. Dr. K.-E. Eduard Linsenmair, Präsident der Deutschen Gesellschaft für Tropenökologie, Lehrstuhl für Tierökologie und Tropenbiologie der Universität Würzburg, Würzburg, Germany

"El Programa de Bosques Nativos Andinos (PROBONA), realiza sus actividades sobre los 2,000 msnm, en cuyas áreas la degradación y destrucción de la biodiversidad es significactiva; en los callejones interandinos quedan pocas manchas de bosques relictos, degradados por el sobrepastoreo y la extracción de leña, así como por las necesidades crecientes de las poblaciones humanas en cuanto a beneficios y productos de los bosques. Hemos realizado algunos estudios de evaluación de la biodiversidad, pero aún son insuficientes, quedando mucho por hacer en este área. Habiendo recepcionado la carta "A call for research and action", concerniente al futuro del desarrollo y la conservación de la biodiversidad en Bolivia, me permito comunicarle que este Programa está de acuerdo con la propuesta (...)".

Antonista Noli H., Coordinadora PROBONA, La Paz, Bolivia

"I have been interested in and conducted research projects in Bolivia for the past thirteen years. I was drawn to the rich biological and cultural diversity for which that country is renown. I was not disappointed with what I found. My particular interest and expertise is in the Amazonian portion of northern Bolivia, where I am currently engaged in a project on the plant diversity of the Chácobo Indian reserve in the vicinity of Alto Ivón in the Beni, following-up on a study I did earlier. Conducting research in Bolivia can often be challenging, but the most diffficult thing for me was to settle on one particular ecosystem and one site within this for long-tem research given, especially the tremendous diversity of biomes (Amazonian, Chaco, Cerrado, and Andean). (...)"

Brian M. Boom, Vice President for Botanical Science and Pfizer Curator for Botany, The New York Botanical Garden, New York, USA

"Wir begrüßen die Durchführung des internationalen Symposiums 'Biodiversity - a challenge for development research and policy' sehr, insbesondere die Einrichtung eines Schwerpunkttages, der sich mit der Erhaltung der Biodiversität in Bolivien beschäftigt. Als eine, in Bolivien tätige politische Stiftung, die dem christdemokratischen Gedankengut verpflichtet ist und damit der Erhaltung der Schöpfung besondere Bedeutung beimißt, fördern wir u.a. auch in Bolivien Projekte zur Förderung des Umweltschutzes. Hervorgegangen aus einem Projekt zur Förderung des Umweltschutzes in den Anden unterstützt die Konrad-Adenauer-Stiftung seit 1993 die Arbeit des Umweltdachverbandes von LIDEMA, dem derzeit 22 Umweltinstitutionen angehören. Ziel ist es, durch die Unterstützung von LIDEMA ein Umweltbewußtseins in der bolivianischen Gesellschaft, aber auch in Politik und Wirtschaft zu erreichen, Vorschläge für die Umweltpolitik zu erarbeiten und die Rahmenbedingungen für eine nachhaltige und umfassende Umweltpolitik zu schaffen. Daneben fördert die Konrad-Adenauer-Stiftung die Arbeit einer NGO, die zugleich Gründungs-mitglied von LIDEMA ist. Im Rahmen dieses Projektes wird auf der Ebene von Dorfgemeinden ökologisch und ökonomisch sinnvoller Boden- und Ressourcenschutz propagiert, wodurch die Lebensbedingungen der dort lebenden Menschen verbessert werden sollen. Wir versuchen, durch die oben skizzierten Maßnahmen einen Beitrag zur Erhaltung der Biodiversität zu leisten, soweit uns dies als politische Stiftung möglich ist. Wir wünschen Ihnen für das Symposium und für die weiteren Maßnahmen zu diesem Thema viel Erfolg."

Carolin Strobel, Konrad-Adenauer-Stiftung, Sankt Augustin

"Der Verein zur Förderung von Landwirtschaft und Umweltschutz in der Dritten Welt (VFLU) hat sich zum Ziel gesetzt, Informationen über Staaten der sog. Dritten Welt in den Bereichen Landwirtschaft, Flora/Vegetation, Fauna sowie Natur- und Umweltschutz zu sammeln und in Form von Vorträgen, Seminaren und Publikationen für eine interessierte Öffentlichkeit zu verbreiten. Wesentliches Organ für solche Publikation ist die vereinseigene Zeitschrift 'Kritische Ökologie'

(ehemals: 'Umweltzeitung'), welche im gesamten deutschssprachigen Raum vertrieben wird. Ein deutlicher Schwerpunkt der Informationsarbeit des VFLU stellt der Andenraum, speziell Bolivien dar. Als Themen wurden dabei u.a. der Einfluß menschlicher Tätigkeit auf Landwirtschaft und Vegetation (P. Seibert), die Degradierung und Regenerationsmöglichkeiten von Agrar-Ökosystemen (A. Schulte, P. Ibisch), historische Entwicklung und aktuelle Situation indigener Dorfgemeinschaften (R. Buchwald), Anbau und Erhaltung autochthoner Kulturpflanzen, v.a. Quinoa (F. Trauzettel/B. Altrieth) und die Coca/Kokain-Problematik (G. Janzing, R. Buchwald) behandelt. (...) Im Rahmen seiner Möglichkeiten wird der VFLU weiterhin zur Information über Bolivien und die angrenzenden Andenstaaten sowie zur Vernetzung der an Landwirtschaft und Natur-/Umweltschutz des Landes interessierten Institutionen und Personen des In- und Auslandes beitragen. Der Verein drückt mit seiner Stellungnahme die Hoffnung aus, daß mit einem erfolgreichen Sympsoium 1996 in Bonn ein großer Schritt zur umfassenden Erforschung und Erhaltung der Biodiversität Boliviens hin getan sein wird."

Verein zur Förderung von Landwirtschaft und Umweltschutz in der Dritten Welt (VFLU), (Geschäftsstelle Wiesbaden: Axel Goldau, Andengruppe: PD Dr. Rainer Buchwald, Freiburg)

"Stichwortartiges Resümee für nachhaltige Entwicklungspolitik unter Einschluß einer Biodiversitäts-Sicherungspolitik:

1. Auf nationaler Ebene

a) Eigner des geistigen Eigentums an biodiversitären Informationen sind die indigenen Völker
b) Die Unterstützung für naturwissenschaftliche Forschung muß in Koordination und Kooperation mit den indigenen Völkern geplant werden
c) Der Staat muß sich um die Regelung der gesetzlichen und institutionellen Rahmenbedingungen für eine nachhaltige Entwicklungspolitik bemühen

2. Internationale Ebene

a) Alle bisherigen Erfahrungen zeigen, daß vor allem internationales Privatkapital an der Inventarisierung der nationalen Biodiversität interessiert ist
b) Bolivien als Nationalstaat muß für sich selbst und für die international Interessierten an der Biodiversität sicherstellen, daß zentrale Vereinbarungen der UN-Konvention zur Biodiversität eingehalten werden (vor allem Artikel 15 und 16)
c) Wenn Bolivien Modell-Land werden will, muß es darauf drängen, daß die Frage der sogenannten intellectual property rights im Interesse der bolivianischen Bevölkerung (das schließt die indigene Bevölkerung ausdrücklich ein) geklärt wird

d) *Eine zentrale Frage, für die es bislang international keine zufriedenstellende Beispiele gibt, ist das Thema Controlling und Monitoring bei der Umsetzung von zu vereinbarenden Nutzungsverträgen*

e) *An der Ausarbeitung von Nutzungsverträgen müßte die nationale Regierung, die authentischen Sprecher der betroffenen Bevölkerungsgruppen und die international Interessierten gleichberechtigt beteiligt sein*

(...)

Friedrich-Ebert-Stiftung, Internationale Entwicklungszusammenarbeit, Referat Lateinamerika und Karibik:

Interesse der Friedrich-Ebert-Stiftung an Kooperation

Die Vertretung der Friedrich-Ebert-Stiftung in Bolivien arbeitet auch am Thema Biodiversität, u.a. mit der Liga de defensa del medio ambiente (LIDEMA) zusammen und unterstützt nationale Vorbereitungsseminare für den Cumbre de las Americas sobre desarrollo sostenible, der für Ende 1996 in Santa Cruz vorgesehen ist."

Dr. Elmar Römpcyk, Friedrich-Ebert-Stiftung, Bonn

List of Contributors

Marc Auer
Bundesministerium für Umwelt, Naturschutz und Reaktorsicherheit
Godesberger Allee 90, 53175 Bonn, Germany
present address: Secretariat of the Convention on Biological Diversity
World Trade Centre, 393 St. Jaques Street, Office 300
Montreal, Quebec, Canada

Prof. Dr. Wilhelm Barthlott
Botanisches Institut der Universität Bonn
Meckenheimer Allee 170, 53115 Bonn, Germany

Dr. Stephan G. Beck
Herbario Nacional de Bolivia, Instituto de Ecología
La Paz, Bolivia

Dr. Nadja Biedinger
Botanisches Institut der Universität Bonn
Meckenheimer Allee 170, 53115 Bonn, Germany

Prof. Dr. Thomas S. Fiddaman
Insitute for Policy and Social Sciences Research, University of New Hampshire, Hood House
Durham, NH 03824, USA

Dr. Jon Fjeldså
Centre of Tropical Diversity, Museum of Zoology, University of Copenhagen
Universitetsparken 15, 2100 Copenhagen, Denmark

Dr. Werner Hanagarth
Balthasar-Neumann-Str. 6, 76646 Bruchsaal, Germany

Prof. Helen Mayer Harrison
University of California at San Diego
La Jolla, California 92093, USA

Prof. Newton Harrison
University of California at San Diego
La Jolla, California 92093, USA

Prof. Dr. Gerhard Haszprunar
Zoologische Staatssammlung
Münchhausenstr. 21, 81247 München, Germany

Prof. Dr. Uwe Holtz
Institut für politische Wissenschaft, Universität Bonn
Am Hofgarten 15, 53111 Bonn, Germany

Dr. Rainer Hutterer
Zoologisches Forschungsinstitut und Museum Alexander Koenig
Adenauerallee 160, 53113 Bonn, Germany

Dr. Pierre L. Ibisch
Botanisches Institut, Universität Bonn
Meckenheimer Allee 170, 53115 Bonn, Germany
present address: Fundación Amigos de la Naturaleza
Casilla Postal 2241, Santa Cruz, Bolivia

Prof. Dr. Norbert Jürgens
Botanisches Institut der Universität Köln
Gyrhofstr. 15, 50931 Köln, Germany

Hermes Justiniano
Fundación Amigos de la Naturaleza
Casilla Postal 2241, Santa Cruz, Bolivia

Martina Keller
Journalistenbüro Keller & Paulus
Kasselerstr. 1 a, 60486 Frankfurt, Germany

Dr. Michael Kessler
Systematisch-Geobotanisches Institut
Untere Karspüle 2, 37073 Göttingen, Germany

Prof. Dr. Frank Klötzli
Geobotanisches Institut der Eidgenössischen Technischen Hochschule Zürich
Zürichergergstr. 38, 8044 Zürich, Switzerland

Jörn Köhler
Zoologisches Forschungsinstitut und Museum Alexander Koenig
Adenauerallee 160, D-53113 Bonn, Germany.

Dr. habil. Horst Korn
Bundesamt für Naturschutz, Außenstelle Vilm
185081 Lauterbach, Germany

Prof. Dr. Walter Kühbauch
Landwirtschaftliche Fakultät der Universität Bonn
Meckenheimer Allee 174, 53115 Bonn, Germany

Prof. Dr. Hartmut Leser
Geographisches Institut der Universität Basel
Klingenbergstr. 16, 4066 Basel, Switzerland

Stefan Lötters
Geographisches Institut, Universität Bonn
Meckenheimer Allee 166, D-53115 Bonn, Germany.

Prof. Dr. José A. Lutzenberger
Fundaçao Gaia
Jacinto Gomez 39, 90040 Porto Alegre, Rio Grande do Sul, Brazil

Nicolas Mateo
Instituto Nacional de Biodiversidad INBio
Apto. Postal 22-3100, Santo Domingo, Heredia, Costa Rica

Prof. Dr. Dennis Meadows
Insitute for Policy and Social Sciences Research, University of New Hampshire, Hood House
Durham, NH 03824, USA

Dr. Mónica Moraes R.
Herbario Nacional de Bolivia, Instituto de Ecología
La Paz, Bolivia

Prof. Dr. Wilfried Morawetz
Institut für Spezielle Botanik, Universität Leipzig
Johannisallee 21-23, 04103 Leipzig, Germany

Dr. Werner Nader
Instituto Nacional de Biodiversidad INBio
Apdo. 22-3100, Santo Domingo, Heredia, Costa Rica

Prof. Dr. Peter Nagel
Institut für Natur-, Landschafts- und Umweltschutz (NLU) / Biogeographie, Universität Basel
St.-Johanns-Vorstadt 10, 4056 Basel, Switzerland

Prof. Dr. Clas M. Naumann
Zoologisches Forschungsinstitut und Museum Alexander Koenig
Adenauerallee 160, 53113 Bonn, Germany

Carsten Rahbek
Centre of Tropical Diversity, Museum of Zoology, University of Copenhagen
Universitetsparken 15, 2100 Copenhagen, Denmark

Steffen Reichle
Zoologisches Forschungsinstitut und Museum Alexander Koenig
Adenauerallee 160, D-53113 Bonn, Germany.

Prof. Dr. Michael Richter
Institut für Geographie, Universität Erlangen
Kochstr. 4, 91054 Erlangen, Germany

Alexandra Sánchez de Lozada
Secretaria Nacional de Biodiversidad
Ministerio de Desarollo Sostenible y Medio Ambiente
Casilla Postal No. 31116, La Paz, Bolivia

Prof. Dr. Wulf Schiefenhövel
Forschungstelle für Humanethologie, Max-Planck-Gesellschaft
Von-der-Tann-Str. 3, 82346 Andechs, Germany

Andrzej Szwagrzak
Cruz Verde de Bolivia
Casilla Postal 3515, La Paz, Bolivia

Prof. Dr. Wolfgang Villwock
Zoologisches Institut und Zoologisches Museum, Universität Hamburg
Martin-Luther-King-Platz 3, 20146 Hamburg, Germany

Dr. Udo Vollmer
Bundesministerium für wirtschaftliche Zusammenarbeit und Entwicklung
Friedrich-Ebert-Allee 40, 53113 Bonn, Germany

Dr. Michael von Websky
Bundesministerium für Umwelt, Naturschutz und Reaktorsicherheit
Godesberger Allee 90, 53175 Bonn, Germany

Prof. Dr. Matthias Winiger
Geographisches Institut, Universität Bonn
Meckenheimer Allee 166, 53115 Bonn, Germany

Subject Index

A

MIX
Papier aus verantwortungsvollen Quellen
Paper from responsible sources
FSC® C105338
FSC
www.fsc.org

If you have any concerns about our products,
you can contact us on
ProductSafety@springernature.com

In case Publisher is established outside the EU,
the EU authorized representative is:
Springer Nature Customer Service Center GmbH
Europaplatz 3, 69115 Heidelberg, Germany

Printed by Libri Plureos GmbH
in Hamburg, Germany